REVOLUTION

Ice Age Re-Entry

CARLTON B. BROWN

Revolution[(RE)]
Ice Age Re-Entry

By Carlton B. Brown

Print Edition: Published October 28th 2018, 1st Edition.

ISBN: 9780992775063

Editor: Jon Harrison

Cover Designer: Jeanine Henning https://jhillustration.wordpress.com/

Connect with the Author: http://grandsolarminimum.com.

Free eBook Download: Amazon (https://amzn.to/2PyQsxV), Google Play (http://bit.ly/2JFHz08), Kobo (http://bit.ly/2F3DdRQ).

Table of Contents

Introduction

The IPCC's Version of Climate Science is Fundamentally Flawed and Its Articles 1 and 2 Hide Global-Scale Catastrophic Risks

We are led to believe by the Intergovernmental Panel on Climate Change (IPCC), an intergovernmental body of the United Nations, through its advice to governments, that greenhouse gas emissions caused by human activity constitute the dominant cause of recent global warming.[1] This global warming message is being used to direct a risk mitigation strategy designed to keep the 21st century global mean temperature increase below a certain threshold, based on specific emissions reduction scenarios.[2,3] The IPCC's laudable intent is to motivate a needed switch of the world's energy system to renewable energy, and to move humans toward living sustainably.

The IPCC's scientific paradigm entrenching Articles 1 and 2 has dictated the IPCC's strategic and politicized version of climate change since its founding in 1988. The IPCC's version of climate change is not a scientific consensus, but existed from the outset in its founding articles.[4,5] The climate science field outside of anthropogenic global warming has identified several *alternative climate risk factors* that the IPCC has chosen not to bring to the public's attention, most likely to prevent the undermining of its political agenda.

Surprisingly, the IPCC's climate risk assessment dismissed, ignored, or failed to review the natural climate risks associated with climate-forcing volcanism,[6] the current grand solar minimum,[7] and the prospect of rapid climate change.[8,9] Solar scientists expert in climate change have warned in a consensus-like manner that the current grand solar minimum, or low solar activity phase, will lead to Little Ice Age-like conditions in the decades ahead.[10,11,12,13,14,15,16,17]

The IPCC's risk assessment also dismissed or ignored the plethora of climate-related human catastrophes that occurred throughout the

Little Ice Age (13th–19th centuries).[18,19,2021,22] The Little Ice Age climate catastrophes included large magnitude volcanic eruptions like Rinjani (1257) and Tambora (1815), which devastated global agriculture and caused widespread human catastrophes.[23,24,25,26,27] On two occasions during the Little Ice Age, China lost nearly half its population,[28] while Europe lost about one-third of its population in the 14th century. These climate-related catastrophes included famines, epidemics, and wars.

This book promotes the urgent need to prepare the world, not for global warming, but for a 21st century switch to a global cooling phase. This climate switch will be associated with catastrophic risks linked to the impact of colder climates and precipitation extremes (drought, rainfall, snow) on agriculture. We now live in an era of enhanced risk for climate-forcing volcanic eruptions that, if they occur, will rapidly cool the planet and trigger glacier ice expansion. Increased earthquake risks are also likely. Linked to this cold climate switch and grand solar minimum is the enhanced prospect of pandemic flu, something to which humans are highly vulnerable.

To provide context for the above, some of these historic cooling phase switches since the Holocene Climate Optimum 8,000 years ago were termed rapid climate change events.[29,30,31,32,33,34,35,36,37,38] These events were associated with the collapse of several ancient civilizations. Centennial-scale (100 plus years) Arctic ice accumulation and its associated Little Ice Age and rapid climate change cannot simply be dismissed, given the evidence for it.[39,40,41,42,43]

Polar ice cores retrieved by scientists reveal the peak in the Holocene interglacial temperature occurred 8 and 10.5 millennia ago in the Arctic and Antarctic respectively. This means earth actually entered the current Ice Age 8 and 10.5 millennia ago in the Arctic and Antarctic respectively.[44,45,46] Since the Holocene Climate Optimum 8,000 years ago, the Greenland ice core declined in an oscillating manner to its lowest temperature trough by 1700 (a 4.9°C decline), before increasing to 2016's global warming sub-peak.[47]

By 1880, or the start of the modern instrument era for recording global temperatures and the period used by the IPCC to begin its anthropogenic global warming story, the Greenland ice core temperature was still 3.6°C lower than at the Holocene Climate Optimum.[48] Despite being told recent temperatures are the highest on record,[49] the temperature was actually 2°C to 4°C higher in the Arctic during the Holocene Climate Optimum than it is today.[50,51,52]

It is important to keep in mind that current global warming trends peaked in 2016, but this temperature peak was still about 2°C lower than at the Holocene Climate optimum 8,000 years ago.[53] To give you some perspective, the Greenland GISP2 ice core registered a temperature rise of 24.5°C from the last glacial maximum (coldest trough) 24,000 years ago to the Holocene Climate Optimum (warmest peak) 8,000 years ago.[54] This means that by 1700 CE, the ice core had actually dropped about 20 percent of its full decline for one glacial cycle, and we were at about 8 percent of the full temperature decline of one glacial cycle in 2016.

We are completely disoriented as regards what stage of the glacial cycle we are living in. Science tells us the last ice age ended 11,700 years ago,[55,] whereas in reality that was the end of a rapid climate change event called the Younger Dryas.[56] The last ice age ended 19,000 and 24,000 years ago in the Antarctic[57] and Arctic[58] respectively, after the last glacial cycle's deepest temperature trough. There was also less ice at both poles during the Holocene Climate Optimum than exists today.[59,60,61,62,63]

Beginning about 5,000 years ago, ice began to accumulate at the poles, and northeast Greenland was ice-locked by 3,000 years ago.[64,65,66,67] Glacier ice rapidly accumulated during the Little Ice Age, reaching its peak buildup by the mid-19th century.[68,69] Much of this glacier ice melted after the mid-19th century[70,71,72] as the sun passed through its 20th century grand solar maximum phase.

Human Activity's Resource-Depletive Impact Increases Our Vulnerability to Climate Switching Risks

Don't get me wrong, human activity does interfere with the climate system, and it has exacerbated the 20th century's grand solar maximum to help melt millennia of ice buildup in just a matter of decades. Human activity has also depleted two-thirds of the world's largest aquifers beyond sustainable limits, and increased the amount of time when there is insufficient water available in water basins and major transboundary river systems to meet our needs.[73,74]

Human activity has deforested much of the planet and destroyed huge swathes of biodiversity. We have polluted the atmosphere and the oceans. The demands of profligate growth for corporate profit have left little under and on earth's surface for tomorrow's generations. So yes—*human activity interferes with the climate, and its environmentally destructive and depletive impact is real and unsustainable.*

World population and economic growth accelerated the depletion of world energy reserves, such that we only have 50 years of proven oil and gas reserves left.[75,76,77] Making matters worse is that 50 percent and 70 percent of the world's oil and gas reserves are (respectively) in fact *unproven guesstimates.* This includes shale resources.[78,79] The prospects for new oil and gas discoveries are not as bright as the promises made by the oil and gas industry might lead one to believe.[80,81],[82,83,84] The profitable extraction of fossil fuels in the future will require higher oil and gas prices, rendering their extraction less attractive compared to lower risk, renewable energy investments.

How do you think world agriculture will cope in extreme drought, after the water basins it relies upon have lost much of their drought-buffering capability? How will the world's nations provide for their energy needs in a colder world, while also facing severe pressures on their agricultural systems? What are you going to do for your children's sake when you realize that peak oil and gas discoveries and production are things of the past?

What can we do about this collective, profligate insanity? How can the world rapidly prepare for a switch to a global cooling phase, and the risks this portends?

The second half of this book ambitiously attempts to answer these questions.

Rapidly Switching to Renewable Energy and Decentralized Sustainable Living Will Reduce Our Vulnerability to Risks

The second part (Section 3) of this book reviews the global supply and demand for energy, water, and food resources, the reserves available, and the key factors impacting their sustainable future supply. Section 3 focuses on how we can live sustainably while preparing for a switch to a global cooling and drought regime, as well as climate-forcing volcanism. The material in Section 3 is highly relevant for those living in the Northern Hemisphere and in arid, semi-arid, and monsoon regions.

Section 3 reviews how major sectors of the world economy use energy and the possibilities for switching to renewable energy systems. The strategies available for implementing decentralized sustainable development are *pitched at three different levels*: for central governments, local governments, and for you at home. This book provides best practice solutions utilized by some governments, municipalities, and homes around the world.

Rapidly *switching the world's energy system* to renewable energy and implementing *decentralized sustainable development* are pivotal strategies to mitigating 21st century climate and resource supply risks. Such a dual strategy will permit a greater degree of self-sufficiency in the home and in urban areas, aimed at reducing our vulnerability during a climate switch and its associated risks. A focus on decentralization means that individuals, communities, and cities embrace partial self-sufficiency in energy, water, and food. Decentralization anticipates that government, corporate, and commodity market actions will restrict resource supply in a crisis event, and so aims to reduce vulnerability to that risk.

In order to remove impediments to change, there is an urgent need for a scientific revolution, to amend the IPCC's Articles 1 and 2 to emphasize both natural climate change *and* human activity. A switch to a global cooling phase will happen in the 21st century, and the grand solar minimum we have entered is prime time for that climate switch. As such, we need to urgently accelerate this fledgling renewable energy revolution, and switch the world's energy system away from its current dependence on fossil fuels.

The combination of the right climate message, economic incentives, and the full realization that our oil and gas reserves are shrinking will be more effective in rapidly switching the world's energy system than the current anthropogenic global warming message.

Economic incentivization can be achieved by implementing a carbon tax priced at an appropriate level, so we can incentivize the energy system switch and drive energy efficiency innovations. A relatively low price for oil and gas means the world currently is squandering precious energy resources, instead of leaving something for today's youth to use in tomorrow's colder world. By letting the world know that peak oil and gas discovery are behind us, a higher non-manipulated market price for oil and gas can prevail—a price that reflects future scarcity in an ice age world.

The big challenge, beyond switching the three-quarters of electricity use not currently supplied by renewable energy, is in switching the non-electricity uses of fossil fuels by industry and transportation to renewable energy systems. Massive quantities of renewable energy capacity will be required for this transition to renewables. To cope with an accelerated switch to renewable energy, we will need regional super-grids and local smart grids, and plenty of high voltage direct current transmission interconnections. Gigawatt-scale battery storage capacity will also be required to ensure energy system resilience. All of these systems are reviewed in Section 3.

The way we fuel our transportation system requires an overhaul. The technology and fuel options that would allow us to move transportation to renewable energy are reviewed in Section 3. This includes transitioning internal combustion engine transportation to renewable energy fuels, and migrating people to electric road, rail, and public transport systems.

During the Little Ice Age's four grand solar minima, extreme drought was commonplace in Asia, Africa, North America, and the Middle East. If the current grand solar minimum replicates the Little Ice Age experience, then we urgently need to anticipate and solve looming water supply problems. Growing demand for water by agriculture, industry, and urban centers will also further stress water basins and depleted groundwater stores, so we have a double reason to act with a sense of urgency now.

Integrated water resources management must come to the fore, particularly with respect to the need to both naturally and artificially recharge groundwater and aquifer supplies (in order to improve their capacity to withstand drought). With one-third of the world's population living within 100 kilometers of the coast, renewable energy-powered desalination systems offer a means of providing sustainable water supplies to coastal communities. Building pipelines for transporting bulk water from water-rich sources to cities, industries, and agricultural areas must be made a priority.

Global food supply will need to increase substantially to feed our growing urbanized, affluent, and populous world as we re-enter a cold age. Unsustainable growth, the fact that the majority of nations depend on food imports from a relatively few nations, and our reliance on fewer food producers per capita, means that we face major vulnerabilities to our food supply should a crisis occur. Options for increasing global food supply, mitigating food supply risks, and building resilience into our fragile global food system are all reviewed.

Decentralizing food supply for cities and communities, and promoting food production at home, must become key priorities for central and municipal governments. Urban agriculture, high tech indoor farming with renewable energy, soil-less food production, and home food production systems are all reviewed as means for improving food supply in cities and reducing urban vulnerability in a food crisis.

Sustainable agriculture and reducing yield gaps, supporting smallholder farmers who feed the world's poor, reducing food waste, and sustainable aquaculture all have roles to play in improving food supply and security. Cold- and drought-adapted crops and farming practices exist for the main crop staples, but our preparation for their deployment in a climate switch scenario or climate catastrophe is not publicly apparent.

The increased prospect of sun-blocking, large magnitude or climate-forcing volcanism means we need new food supply solutions independent of the prevailing climate and the sun. Industrialized greenhouses surrounding cities, high-tech indoor farming, and single cell protein manufacture all offer climate- and sunlight-independent food production capabilities to complement existing food stockpiles. Our national and municipal emergency food stockpiles will need to be reassessed for their ability to supply food in a climate switch or climate catastrophe event.

Rapidly switching the energy system and moving to living sustainably has a cost, but it is a fraction of the future cost of doing nothing. As a benchmark, it's been estimated that about 1 percent of global GDP annually will be required to pay for climate change mitigation and preparation.[85,86,87]

We have financing options available, such as implementing a global carbon tax, using more renewable energy feed-in-tariffs (i.e., higher priced renewable energy supply contracts), and partially redirecting fossil fuel subsidies to allow the poor to gain access to affordable, renewable energy systems. Global pension funds also hold substantial investment firepower, if the rules governing their investments (i.e., the upper and lower limits of asset class investment holdings) are amended to facilitate de-risked public-private partnerships (i.e., insurance-backed infrastructure bonds). The Paris agreement's $100 billion per year pledge is more important than ever if we are to help the developing nations prepare for climate change and ensure both their supply of energy and global food security.

A chapter is dedicated to reviewing best practice methods for how to live sustainably at home and prepare for global cooling and climate extremes. This review focuses on the sustainable supply and efficient use of energy, water, and food at home. Best practice methods are reviewed for growing climate-adapted food and ensuring that you have food stockpiled, a seed bank, and food growing systems at the ready. In this manner, you and your family will have a head start on developing ready-to-go emergency solutions. The impact of extremely cold winters and summers, of climate-forcing volcanism, and of rapid climate change will affect communities very quickly. Being prepared will make the difference between survival (at best) and being left vulnerable.

Grand solar minima and climate switches represent high-risk times for pandemic flu outbreaks. As a species, we lack immunity against highly pathogenic avian H7N9 and H5N1 flu strains, which are already causing small-scale epidemics of animal-to-human infections that kill 25 to 50 percent of those infected.[88,89,90,91,92] Progress in vaccine technology since 2009's swine flu pandemic offers bona fide solutions to protect people against emerging pandemic flu strains before a pandemic arises.[93,94,95,96,97] However, the underlying pandemic flu vaccine technology and industry's paradigm for vaccine supply has not changed since 2009. The overriding reason for this lack of change is reviewed in Chapter 14.

This book is intended to be an *information bomb* that will shatter the illusory climate change consensus before the climate switch occurs. It calls for four peaceful and cooperative revolutions: (1) a scientific revolution that allows us to see nature's climate change risks; (2) a renewable energy Revolution[RE] designed to rapidly switch the world's energy system to renewable energy; (3) a pandemic flu vaccine revolution and pre-pandemic immunization to prevent unnecessary and horrific deaths in a pandemic outbreak; (4) a voting revolution, to vote in leaders who can quickly help us switch the world energy system and move humanity to living sustainably.

Read on in the *Spirit of Revolution[RE]*. (RE = Renewable Energy).

Key Themes

The Entrenchment of a Scientific Paradigm

The United Nations Framework Convention on Climate Change (UNFCCC) and Intergovernmental Panel on Climate Change (IPCC) and their Articles 1 and 2

The Mechanics of Scientific Revolution

CHAPTER 1

Without a Scientific Revolution Catastrophic Risks Will Remain Hidden

The big question I have for the following two chapters is this: Do the normal rules of science and its transformation through paradigm shifts[98] apply to the IPCC's version of climate science that is provided to our governments?

On the one hand, I admire what the United Nations and its intergovernment affiliates such as the United Nations Environment Programme, the International Renewable Energy Agency (IRENA), and the Intergovernmental Panel on Climate Change (IPCC), and conventions such as the United Nations Framework Convention on Climate Change (UNFCCC) have collectively achieved from a strategic business perspective.

The strategic business perspective I am referring to is the United Nations championing the switch of the world's primary energy system from fossil fuels to renewable energy. This switch offers the ability to mitigate anthropogenic greenhouse gas emissions, which the IPCC claims is largely responsible for global warming.[99,100,101] In reality, this energy system switch is required before the world runs out of proven oil and gas reserves in the decades ahead. However, running out of oil and gas is never explicitly mentioned in the IPCC's publications as the primary motivation for switching energy systems.[102,103,104]

Yet, on the other hand, the IPCC's founding Articles 1 and 2, published in 1988, entrenched its anthropogenic global warming scientific paradigm. Articles 1 and 2 were provided to the IPCC by the UNFCCC, under the direction of the World Meteorological Organization (WMO) and United Nations Environment Programme (UNEP).[105,106,107]

Articles 1 and 2 *obstruct* our ability to perceive global-scale natural climate change risks, such as low solar activity and climate-forcing volcanism, and their impact on global cooling, centennial-scale glacier ice expansion, and extremes of precipitation.[108],[109],[110,111,112,113] This natural climate change and its associated risks have been ignored, downplayed, or dismissed by the IPCC in their assessment reports provided to governments.[114,115,116,117,118] It is this obstruction of natural climate change risks and the grand solar minimum potential for climate-related catastrophes that *creates the need for a scientific revolution.*

The version of climate science promoted by the IPCC to governments, and portrayed as a consensus of the international scientific community, *has nothing to do with normal academic science.* It is a narrowly focused, highly constrained, and politicized opinion dictated by Articles 1 and 2, and by specially selected government scientists.

The above paragraph's last sentence summarizes the harsh criticism provided by the InterAcademy Council, which is composed of many of the world's national science academy leaders,[119] and by the Nongovernmental International Panel on Climate Change (NIPCC), comprised of leading climate science experts whose scientific disciplines have been marginalized by the IPCC process.[120]

Do Normal Rules for Scientific Revolutions Apply to the IPCC's Politicized Version of Climate Change?

Science is the discovery of knowledge, and a better understanding of both the natural and man-made world involving the application of knowledge and systematic research methods. A scientific theory explains some aspect of the natural world, based on reproducible observations that are established by measurement and analysis.

A theory has predictive value and can be assessed downstream by analyzing the accuracy of its forecasts against measured data, given a sufficient time interval.

Karl Popper is regarded as one of the 20th century's greatest scientific philosophers, and he tells us a scientific theory can never be proven, but it can be *falsified*. All it takes to falsify a theory is a single reproducible and quality experiment.[121] As such, a scientific theory is temporary and only has validity until it is successfully disproven. A good scientist will try to disprove his or her own theory, and the independent peer review process has the same function.

On the other hand, Thomas Kuhn, also one of the most influential 20th century philosophers of science, tells us a scientific theory must be *rejected* when the outcome predicted by that theory is different from nature.[122]

On the basis of Kuhn's ideas, the IPCC climate science theory (anthropogenic global warming-induced climate change) should be rejected, given its highly inaccurate forecasts. Alarmingly, more than 97 percent of the IPCC's promoted climate forecasts between 1998 and 2012 overstated the global mean air temperature. Wind the clock back, and we find that 82 percent of their forecasts between 1985 and 1998 understated the temperature, while also missing the fifteen-year climate hiatus between 1998 and 2012.[123]

Despite this abject failure in forecasting the global mean surface temperature, the IPCC made no attempt to modify or reject its climate change theory. Instead the IPCC reiterated its "very high confidence" that the climate forecasts showed long-term trends consistent with real data. The IPCC then suggested its climate forecasts should be scaled down by 10 percent to fit better with this real world data. However, even with that rescaling, their forecasts still missed the climate hiatus (see previous citation). *That is not normal science.*

The development of normal science and its paradigms of scientific understanding do not happen in a uniform manner. This evolutionary development is comprised of alternating "normal" and revolutionary phases. Revolutionary phases are always brought about by the occurrence of scientific anomalies.[124]

A scientific paradigm in its discovery phase does not normally stand still for three decades, as the IPCC's anthropogenic global warming theory has. Normal science evolves to accommodate new scientific discoveries, but this has not happened with the science promoted by the IPCC to our governments.

For example, climate science discoveries known since the mid-1990s indicate that an increase in carbon dioxide concentration follows a global temperature increase. This conclusion is based on scientists analyzing hundreds of thousands of years of ice core data, as well as annual data obtained over multi-decade periods from the lower and upper atmosphere. These scientific discoveries fundamentally challenge our common understanding of carbon dioxide's role in climate change. These studies demonstrate that the rise in air temperature precedes the rise in atmospheric carbon dioxide concentrations by 9-12 months on decadal timescales,[125] and by centuries on glacial cycle timescales.[126,127,128,129]

Based on independent scientific studies, superior climate correlations exist over the IPCC's climate reference period (since 1880), and over much longer periods. These superior climate correlations involve other mechanisms of climate control and regulation (Chapter 6, Figures 4.4 and 5.4).[130,131,132,133,134,135,136,137,138] Yet these alternative climate mechanisms do not feature in the climate science, forecasts, and theory promoted by the IPCC to governments.

This highlights a very important point, viz., that once a paradigm becomes entrenched, then theoretical alternatives are strongly resisted by *vested interests*. You can read about the reasons for this entrenchment of paradigms in *The Structure of Scientific Revolutions*.[139]

Why do scientific revolutions occur? Because the incumbent theory and its predictions no longer fit the facts of nature.

Under the rules of normal science, a scientific revolution will occur when a scientific anomaly is discovered, resulting in a recognized scientific crisis. A new, plausible theory explaining the scientific anomaly must then be proposed. The old paradigm and scientific ways of working (i.e., assumptions, framing of scientific problems, techniques, and methodologies) must be overthrown, and the scientific field concerned must be fundamentally reconstructed (in terms of common assumptions, techniques and methods). The history and textbooks associated with that scientific discipline must also be rewritten. The hearts and minds of scientific leadership, scientists, and the vested interests must also be won over. Being able to reconcile a new theory with the commonly cited historical data is fundamentally important to enabling a scientific revolution.[140] Falsification of a theory alone is insufficient to cause a scientific revolution.

While the planet has been warming, it has been very difficult for the climate science field to change the anthropogenic global warming scientific paradigm mandated by Articles 1 and 2. The *only way a scientific revolution will take place* in the climate science field is with a climate switch to a cooling phase, a climate-forcing volcanic eruption that induces global cooling and glacier ice expansion, or the onset of a rapid climate change event. Hopefully, change can take place before such a climate catastrophe occurs, when it is realized that the well-meaning, strategically motivated, and visionary Articles 1 and 2 are inadvertently hiding catastrophic risks.

IPCC Articles 1 and 2 Entrenched a Scientific Paradigm Three Decades Ago

The IPCC is an intergovernmental agency established in 1988 by the World Meteorological Organization (WMO) and the United Nations Environment Programme (UNEP), and works under their direction and in support of the United Nations Framework Convention on Climate Change (UNFCCC). The UNFCCC is the main international treaty on climate change.[141] That the WMO and UNEP established the IPCC helps explain the politics associated with climate change, and how Articles 1 and 2 came to entrench the global warming paradigm in 1988.[142,143]

Maurice Strong, who was the Secretary General of the United Nations Conference on the Human Environment in 1972, and of the Rio Environmental Summit in 1992, as well as the first Executive Director of the United Nations Environment Program,[144] helped found the IPCC. The Nongovernmental International Panel on Climate Change (NIPCC)[145] was formed to provide a counter-balancing second opinion on the IPCC's science relating to global warming. The NIPCC's website informs us that *Maurice Strong played a key role*, together with the WMO and UNEP, in providing the IPCC with its "narrow" Article 1 definition of climate change. According to the NIPCC website, the IPCC was given responsibility for finding the evidence of human influence on the climate system, as well as to build the case for its policy-led mitigation.[146]

The IPCC provides periodic scientific assessment reports on climate change to the governments of all member states (1990, 1995, 2001, 2007, and 2014). Of *critical importance* is that these assessment reports provide governments a scientific basis on which to develop their climate-related policies and adaptation and risk mitigation plans.

The problem for humanity is that our *governments' publicly stated plans are only as good as the scientific and risk mitigation advice they receive from the IPCC.*

The IPCC claimed in its first scientific assessment (1990) that its method of scientific review, which was reportedly conducted by the international scientific community, resulted in the most authoritative statement about climate change.[147] In reality, however, it focused its attention solely on anthropogenic global warming, something that is clearly stated in the preface of its first scientific assessment report in 1990,[148] and all other scientific assessment reports since.

In reality, all of *these reports are replete with confirmation bias*[149] *dressed up as an authoritative review.* The IPCC process makes it very easy for us to see this bias by using a language that science does not use, viz., "with high or low certainty or confidence" depending upon what happens to support its theory, or, conversely, what it wishes to dismiss.

We are told in the prefaces to the IPCC's 2007[150] and 2014[151] assessment reports that its focus of scientific review *caters to policymakers, and does not review all of the climate science literature.* It nonetheless reiterates its strong commitment to thoroughly assessing the scientific literature without bias. Yet the conflict inherent in this attitude is readily apparent.

This narrow focus of scientific review, catering to policy makers, was dictated by the IPCC's founding Article 1 definition for climate change, and by its Article 2 defined objective (see citation endnote for details of Articles 1 and 2).[152] Importantly, you can see *how this paradigm was entrenched in 1988, and how it effectively subordinates or trivializes natural climate change* relative to human activity. But you can't dismiss nature when it comes to climate-related risks.

On a positive note, and under the above assumption, Articles 1 and 2 were visionary way back in 1988. We should all be incredibly grateful for the United Nation's decision to control science's message on climate change, designed to promote a switch of the world's energy system to renewables *before we run out of oil and gas.*

Nevertheless, these two articles explain why the IPCC's version of climate science largely ignores, downplays, or dismisses the other climate science sub-disciplines (see Figure 1.1). The consequence of this is *the IPCC inadvertently misses the broader climate-related risks* that are on the horizon.

The IPCC's Policies and Procedures that Enable Its Scientific Bias

The version of climate science promoted by the IPCC to our governments as a consensus of the international scientific community has nothing to do with normal academic science. It is a biased scientific opinion mandated by Articles 1 and 2 since 1988. This scientific bias manifests itself in the IPCC's use of key words associated with anthropogenic global warming, as opposed to the use of key words for other mechanisms of climate control (Figure 1.1).

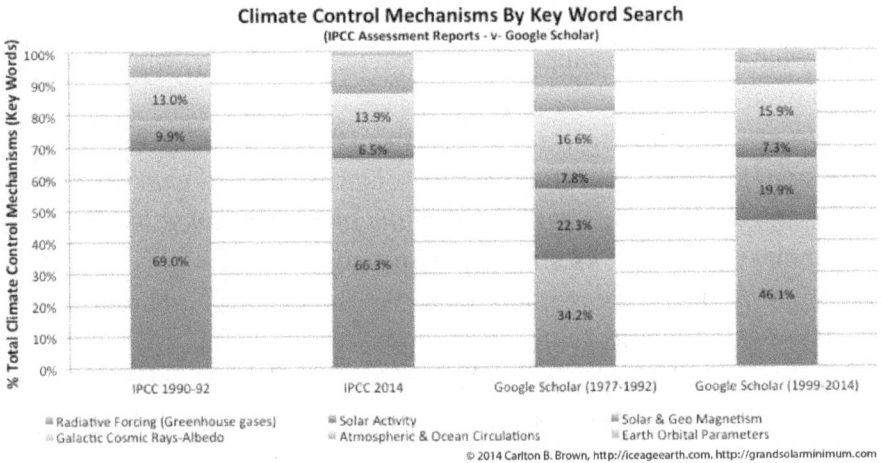

Figure 1.1. The key words associated with all the main mechanisms of climate control and climate science sub-disciplines were used to search IPCC assessment reports and Google Scholar (1977-1992, 1999-2014). The radiative forcing climate mechanism accounts for 66 percent of the total 22,449 key words used by the IPCC (2014), while solar and magnetism combined account for only 6.7 percent of these key words. In contrast, Google Scholar shows that solar and magnetism mechanisms account for 27 percent of the 8,431,419 key word hits.[153] (IPCC Assessment Reports: 1990-1992,[154,155,156,157,158] 2014[159,160,161,162]) **Conclusion**: The IPCC's review of the climate science is biased, and this bias has been maintained for a quarter of a century. The fact that multiple mechanisms of climate control and regulation exist, supported by a significant body of peer-reviewed scientific literature, implies that there is no scientific consensus supporting anthropogenic global warming. *A consensus exists only in the media.* If you wish to understand more about the *IPCC's procedures which enable its bias*, as well as a critique of its science by IPCC-opposed climate experts, then the following cited documents will give you a good perspective.[163,164,165]

8

Two of these organizations cited above have issued strong critiques of the IPCC. First is the InterAcademy Council, which is an Amsterdam-based organization whose scientific leadership is made up of presidents of many of the world's national science academies.[166] The InterAcademy Council was commissioned to conduct an independent review of the processes and procedures of the IPCC, and it provided a *damning, though diplomatic, criticism* of the IPCC. Second is the Nongovernmental International Panel on Climate Change (NIPCC), which is comprised of leading experts from the different climate science fields, and which provided a machine gun-like volley of scientific rebuke to the IPCC's promoted science.[167]

Scientific bias is facilitated by the IPCC's process for selecting volunteer scientists to do its work. The IPCC author-scientists are selected from *lists of national experts provided by governments.*[168] According to the InterAcademy Council's review, political considerations are given more weight than scientific expertise in the IPCC's selection process. Additionally, governments did not always nominate the best scientists from among its list of national volunteers.[169]

In the IPCC procedures document there is no mention of selecting scientists to represent all the different mechanisms of climate control, or climate sub-fields.[170] This helps explain how the reviewed science in the assessment reports is replete with confirmation bias[171] toward anthropogenic global warming, while at the same time dismissing, downplaying, or ignoring the other mechanisms of climate control and their associated risks.

The co-chairs of the IPCC's working group are very influential people in the IPCC hierarchy with respect to developing the assessment and synthesis reports. These co-chairs select lead and coordinating authors from a list of nominees provided by governments.[172] The co-chairs also have significant influence and control over the scientific assessments, and in leading the preparation, review, and finalization of their working group's reports.[173] A truly independent review of the assessment reports has never been done, because the working group co-chairs also select the review editors.[174]

In a similar manner, final synthesis reports are not written by independent expert scientists, but are the product of negotiations among government representatives and the IPCC chair and working group co-chairs.[175,176] The synthesis reports are the ones that really matter for busy policy and business decision makers. After all, with the most recent IPCC assessment reports being between 1,000 and 1,500 pages long, who is going to be able to read them and sift out data and facts from the projections, and know what constitutes confirmation bias?

What's more, government representatives negotiate and agree to the final synthesis report wording line by line[177] over marathon sessions that last several days and nights. This line-by-line negotiation results in differences between the assessment reports and the final politicized synthesis report provided to governments.[178]

This means that the UNFCCC's Articles 1 and 2, combined with the IPCC's procedures, have held the IPCC's scientific paradigm rigidly in place for 30 years. Therefore, this is *not a bonafide scientific paradigm operating by normal rules of science* (i.e., in Kuhn- or Popper-like fashion), but rather by intergovernmental mandate and procedures.

Normal Science Does Not Operate by Consensus

Normal science does not operate by consensus. It operates by scientists proposing testable hypotheses, developing reproducible data to prove or falsify them, and then publishing their results in peer-reviewed journals, so that other scientists can attempt to reproduce the results.

All it takes is a single reproducible and quality experiment to falsify a theory.[179] Consensus is irrelevant in the face of one well-conducted and reproducible experiment.

Of course, the above is only true if there are no top-down obstructions to getting those experiments conducted and the results published in the first place. Obstructions could result from government funding priorities, such as what constitutes fundable or non-fundable areas for research, or how much funding is directed to specific sub-disciplines (i.e., anthropogenic global warming versus solar activity or cosmic rays), or institutional funding strategies (i.e., which academic institutes will receive the funding for building research capacity, infrastructure and tools). Scientific dissenters can also be weeded out through the grant-funding process (i.e., scientific advisory committee decisions). Dissenters or scientific sub-disciplines that could undermine the popular paradigm or vested interests can be deselected, deprioritized, or scientifically discredited.

Whoever controls the funding of climate science controls its direction. The dominance of publications associated with the current paradigm (i.e., anthropogenic global warming), assure that the paradigm is maintained.

Another form of obstruction would be biased vetting of a manuscript submitted to a scientific journal for publication. This biased vetting could be masked as scientific criticism of the methodologies, of the data analysis, or of any conclusions reached. A decision could then be taken, with or without a delay, not to publish the manuscript. Public and private undermining of a scientist's reputation could also be used to raise doubt about their work (i.e., word of mouth within the academic community).

The fact that such a term as scientific consensus is discussed in the media should raise alarm bells among discerning scientists. Reflect on the arguments presented in this chapter and in Chapter 2, and you will see that the foundation upon which the IPCC's theories and forecasts are based is actually nothing more than quicksand.

Illuminating the IPCC's Highly Inaccurate Climate Forecasts

Every scientific theory is based on causal relationships, determined through correlation and statistical analysis of empirical data. If a climate theory is correct then it should accurately predict nature, as evidenced through its climate forecasts.

A critical omission in the IPCC scientific review has been present since 1990. The IPCC has failed to provide a theory-justifying correlation analysis between the global mean surface temperature and atmospheric carbon dioxide concentrations. To support the basis of any scientific theory, scientists would normally plot both parameters on the *same* graphic and provide scatter plots with a trend line and its regression equation and R-squared value (at the least). This correlation analysis should also provide correlation values, their levels of statistical significance, and any attempt made to optimize the correlation through the rephasing of parameters or rendering of the data. This correlation would need to be maintained over numerous relevant timescales for it to have any scientific merit. There is a good reason for the omission of this carbon dioxide correlation, as you will see below and in Figure 4.3.B.

Based on the IPCC's failure to accurately forecast the climate from the years 1985 to 2018, or justify the scientific foundation of its theory in 1990 and since, normal science should have falsified (via the peer review process) and replaced this theory a long time ago.[180] However, this hasn't happened.

To make matters worse, in order that the IPCC could justify a full 21st century global warming, they had to defer the next ice age and ignore or dismiss natural climate change phenomena such as centennial-scale climate oscillations, climate-forcing volcanism, and secular changes in solar activity (i.e., grand solar minima and maxima, solar magnetism).[181] At the foundation of IPCC climate forecasts lies a fundamental assumption that another ice age will not occur for 30,000 years.[182] This chapter tears that dangerous assumption apart.

Don't get me wrong, climate change is real, and global warming was real between 1700 and 2016. Likewise, human activity via fossil fuel combustion (water vapor, carbon dioxide), irrigation (water vapor), pollution (aerosols, chemicals), deforestation, and land use changes, etc., is disruptive to the climate system. Human activity has helped the sun melt in just a few decades the Arctic ice that had built up over two millennia. But that's not the point.

The IPCC Climate Change Theory and Forecasts Are Fundamentally Flawed

The radiative forcing theory attributes earth's climate to the difference between the amount of sunlight energy absorbed by the earth (i.e., measured as watts per square meter at the top of the atmosphere), and that radiated back into space. Under this theory, changes in insolation (solar irradiance) and the concentrations of an array of anthropogenic greenhouse gases and aerosols are the primary drivers of earth's climate.

The problem with the IPCC's theory and forecast models, however, is that they ignore natural climate change factors that also impact earth's radiation balance—such as volcanic aerosols (see Chapter 5) and secular changes in solar activity (see Chapter 6). These longer-term changes in solar activity not only affect the level of solar irradiance reaching earth's upper atmosphere, but also the strength of the magnetized solar wind reaching into space. This magnetized solar wind modulates earth's magnetic shield and therefore the level of cosmic rays entering earth's atmosphere from space. Cosmic rays play an important role in modulating cloud cover over the planet. Cloud cover increases the level of incoming solar radiation that is reflected back into space, thus cooling the planet (see Chapter 6).

The IPCC's theory also ignores the impact of increased atmospheric water vapor resulting from global warming, as well as that resulting from daily human activity (e.g., irrigation, fossil fuel combustion). Importantly, water vapor is more plentiful and a more potent greenhouse than carbon dioxide. Similarly, the IPCC's theory ignores the increased atmospheric concentration of carbon dioxide resulting from ocean degassing, due to global warming.[183] The bottom line is the climate system is far more complex than can be explained by simply blaming human activity while ignoring nature, or re-branding natural climate change as human-induced climate change.

Under the IPCC's radiative forcing theory, the impact of anthropogenic greenhouse gases on the global mean surface temperature is dependent on the sensitivity of the climate system to the concentration of those greenhouse gases.[184] Ignoring the natural climate system obviously renders the IPCC's assumptions on climate sensitivity erroneous.[185] The erroneous nature of the IPCC's climate sensitivity assumptions is obvious, because the IPCC suggested it should scale back its climate forecasts by 10 percent to account for this sensitivity when attempting to explain the inaccurate forecasts.[186]

Carbon Dioxide Levels Lag the Global Temperature

What is generally not realized is that from the late 1990s higher resolution long-term climate data have demonstrated that historical global temperature changes preceded changes in carbon dioxide concentrations by many centuries,[187,188,189,190] and by many months over the multi-decade time scale.[191] This means that global mean land surface, ocean surface, and stratospheric temperature changes precede changes in atmospheric carbon dioxide concentration. This emphatically tells us that increased *carbon dioxide levels are consequential, not causal, of global warming.*

According to a more recent study covering the period 1981-2011, broadly coinciding with the start of the IPCC's climate forecasting period, a clear phasing relationship exists between atmospheric carbon dioxide and various temperature compartments (ocean, atmosphere, and stratosphere). Changes in the atmospheric carbon dioxide concentration always lagged behind changes in temperature by between 9 and 12 months. This study also indicated that the bulk of the atmospheric carbon dioxide increase during the period 1981-2011 was due to ocean degassing consequent to solar activity (i.e., a grand solar maximum).[192]

Nearly All IPCC Climate Forecasts Overstated or Understated the Global Average Temperature (1985-2018)

The acid test for any scientific theory is its ability to predict the future of the natural system that is being investigated. A failure to predict nature means the theory is either incomplete or wrong, and should be revised or rejected.

In general, the global climate models promoted by the IPCC for forecasting the global average surface temperature perform poorly when their projections are assessed against empirical data.

This forecasting inaccuracy is exemplified in 2014's fifth assessment report. In this report, the IPCC shared a graphic detailing one hundred and thirty-eight of their promoted forecasts developed under the Representative Concentration Pathways from their CMIP5 models (Coupled Model Intercomparison Project Phase 5).[193] The IPCC also shared data from 42 of their historical models since 1985, overlaid with four sets of real climate data, while providing a weak self-critique of these inaccurate forecasts.

Between 1998 and 2012 there was a hiatus in the rise of global mean surface temperature, which nearly all (111 of 114 forecasts, or 97 percent) of the IPCC's promoted forecasts had failed to predict. This 15-year hiatus occurred despite the fact that the concentration of atmospheric carbon dioxide continued its unabated increase during this time (i.e., by 7.4 percent).[194] Moreover, the IPCC then informs us that during the 15-year period prior to 1998, the actual global mean surface temperature was higher than 82 percent (93 of 114) of these CMIP5 model forecasts.[195]

This means that for a *30-year period the IPCC promoted forecasts either underestimated or overestimated the global mean surface temperature.* In short, these inaccurate forecasts demonstrate a lack of causative correlation between carbon dioxide and the global mean surface temperature.

Recall the above-cited study covering the 1981-2011 period[196] in which carbon dioxide lagged the sea, air, and stratospheric temperatures by 9–12 months. This cited study utilized the same 30-year period covered by the IPCC's forecast. This study refutes the IPCC climate change theory by showing that carbon dioxide lags temperature by 9–12 months in multiple different climate compartments (i.e., ocean, atmosphere, and stratosphere).

It gets worse. In the 2014 Physical Science Basis assessment report, the IPCC predicted with high confidence that global warming will be pervasive throughout the 21st century. It forecasts the global mean surface temperature will increase between 0.3°C and 0.7°C from 2016 to 2035 relative to the 1986–2005 average.[197] Let me bring to your attention the fact that between 2016, or the start of the IPCC new forecast period, and October 2018 the global mean surface temperature declined by 0.21°C, and in the Northern Hemisphere by 0.27°C.[198]

In their attempt to alleviate the concerns of readers (i.e., government, business, global finance etc.), the IPCC suggested this forecasting inaccuracy could be caused by internal climate variability, or a reduced trend in external forcing (i.e., declining solar activity). Climate variability in the real world includes important naturally occurring climate change like volcanic activity (cooling aerosols), atmospheric water vapor and aerosols, cloud feedbacks at different altitudes and latitudes, atmospheric and ocean circulations, etc. In other words the *inaccuracy could simply be due to the influence of natural climate change*, which the IPCC have ignored or dismissed because of Articles 1 and 2. Most of these natural factors impact radiative forcing, yet are not accounted for in the IPCC's version of radiative forcing.[199]

The IPCC's response to this high forecasting inaccuracy was merely to suggest that their near-term climate forecasts should be reduced by 10 percent to better fit the real world data.[200] *In fact, the data* refuted *their own theory.*

IPCC Climate Forecasting Models have not been Validated by Forecasting Experts

Forecasting is a scientific discipline spanning more than seventy years of empirical research. This discipline is professionally overseen by the International Institute of Forecasters,[201] and has its own peer-reviewed journals.[202,203] This body of forecasting research has been summarized in 140 scientific principles,[204] which must be observed in order to make valid forecasts.

Scientific forecasting experts found the IPCC forecasting procedures violated more than half of the 140 established forecasting principles, and no justification was provided for those violations. These global warming forecasts were also not subject to peer review by experts in scientific forecasting.[205]

Expert forecasters inform us that it is not possible to improve upon a naïve (i.e., no change) climate forecast, because many conditions involved in controlling earth's climate are still unknown, leading to high forecasting uncertainty.[206,207] These experts tell us that forecasting models should only be used if they can be shown to provide forecasts that are more accurate than those from a naive model.[208]

Forecasting accuracy is normally determined using backtesting, or the ability of models to accurately hindcast historical climate data. The accuracy of IPCC forecasts was assessed by expert forecasters, who found the errors were 7–12.6 times larger than a no change forecast.[209,210] This would help explain the high level of forecasting inaccuracy demonstrated in the IPCC forecasts between 1985 and 2018.

The Unstated Theories and Assumptions at the Foundation of IPCC Climate Forecasts

At the foundation of the IPCC's global climate forecasts are a series of fundamental assumptions made, and oversights (of existing climate data) required, in order to support the IPCC's climate change theory of anthropogenic global warming. We must be aware of these forecasters' unstated assumptions and oversights to understand the tenuous foundation upon which the IPCC's promoted climate forecasts are based.

The only way to justify global warming across the entire 21st century—and obtain four Representative Concentration Pathway global warming forecasts[211]—would be to assume that the start of an ice age still lies ahead of us today (reviewed below).

It would be imperative that climate forecasters also ignore the Arctic[212,213] and Antarctic[214,215] climate data depicting sea-saw glacial cycles (see Figures 3.2 and 4.2), while disregarding where we are today relative to the Holocene Climate Optimum. Importantly, forecasters would need to ignore the existing climate data depicting centennial-scale climate oscillations since the Holocene Climate Optimum (see Figures 4.2 and 4.3). In Chapter 4 you will see that today's global temperature sits atop a trough-to-peak warming phase that started in 1700, and is the biggest global warming phase in 8,000 years (i.e., a statistical outlier).

By ignoring the climate data before 1880 and back to the beginning of this current global warming phase in the early 1700s,[216,217] forecasters would not need to explain why the temperature started to rise two-plus centuries before human activity became a factor for consideration in climate change (see Figure 4.3).

In order to project a full 21st century global warming, it would be essential that global warming forecasters and their IPCC chaperone dismiss, ignore, or even discredit any dissenting science opposed to 21st century global warming. For example, these forecasters would need to ignore the solar activity and other scientists expert in climate change who predict a global cooling phase will accompany the early to mid-21st century grand solar minimum.[218,219,220,221,222,223,224,225]

To assist the IPCC forecasters' base assumptions, the IPCC arbitrarily delayed the next ice age by 30,000 years. The IPCC posited this 30,000-year ice age delay without subjecting that theory to scientific peer review. Meanwhile, the IPCC told governments that this delay was a "robust finding."[226] The Milankovitch orbital pacemaker theory upon which the IPCC's ice age delay is based is scientifically contentious, with other areas of science contesting the validity of the theory.[227,228,229,230,231,232]

This claimed robust finding, that earth would not enter another ice age for at least 30,000 years, is the opposite of what climate scientists expert in the Artic climate and its glacier dynamics are saying. These Arctic climate specialists tell us that since the Holocene Climate Optimum and thermal maximum, the Northern Hemisphere's summer temperature has declined in line with the decline in Northern Hemisphere summer solar irradiance (insolation) from its peak 8,000–10,000 years ago. This decline in summer insolation, on the order of 40–50 watts per meter squared, is a result of changes in earth's orbit.[233,234,235,236,237] This decline in summer insolation is about 15 times greater than the equivalent climate-forcing impact of carbon dioxide at today's atmospheric concentration.[238]

The IPCC's 30,000-Year Ice Age Deferral Is Statistically Falsifiable

There are crucially important statistical implications relating to this "new IPCC theory" for delaying the next ice age by 30,000 years.[239] The statistical implications have been ignored (I am assuming), and would automatically lead to the dismissal of this theoretical delay. The statistical implications relate to how this delay adjusts the interval back to the previous ice age, the interglacial period duration, and the relative phasing gap between the Antarctic and global climate optima, when compared with all previous glacial cycles in the last 800,000 (Antarctica) and 2,000,000 (global) years.[240],[241]

By extending the start of the ice age another 30,000 years from its "real start" 2,100 years ago (global data), the interglacial period duration is extended. The last interglacial period started 19,600 years ago and ended 2,100 years ago at the Holocene Climate Optimum. Consequently, the interglacial period would increase from its existing 17,500 years to 49,600 years, rendering the revised 49,600-year interglacial period duration a statistically significant outlier (see the data table summary embedded in the citation).[242]

Similarly, by ignoring the global climate optimum 2,100 years ago and extending the interglacial period by another 30,000 years, one would be positing a new Holocene Climate Optimum 30,000 years in the future. Creating a new Holocene Climate Optimum would extend the already longest interval, going back to the previous climate optimum, from 122,700 years to 154,800 years.

Compared with all thirty-two preceding climate optima intervals in the last 2,026,800 years, this revised 154,800-year interval would become a statistically significant outlier. The original non-delayed climate optimum interval, while an outlier, was not a statistically significant one (See the data in the following citation).[243]

In the third statistical analysis, the timings for the last nine glacial cycle climate optima in Antarctica were compared with their corresponding global climate optima timings, to determine the phasing gap for each glacial cycle. In all but two glacial cycles, the climate optimum was reached first in Antarctica, and on average 2,100 years before it was reached globally. The global climate optimum 2,100 years ago already had the longest phasing gap with its corresponding Antarctic climate optimum, compared with all previous Antarctic and global phasing gaps in the past 800,000 years.

By delaying the Holocene Climate Optima 30,000 years, the phasing gap between Antarctica EPICA Dome-C (10,500 years ago) and the global (2,100 years ago) data changes from 8,400 years to 40,500 years, rendering this revised phasing gap a statistically significant outlier (See the data in the following citation).[244]

Delaying an ice age by 30,000 years cannot be statistically justified on at least three counts. By delaying the ice age 30,000 years a statistical outlier is created for interglacial durations, inter-climate optima intervals, and Antarctic-to-global climate optima phasing gaps (P-value <0.05 in all three of the above cited cases). In addition, by delaying the ice age another 30,000 years, the distribution of the data is changed from a normal distribution to a non-normal distribution profile. If normal science were operating in an unimpinged manner, it would emphatically reject the IPCC's 30,000-year delay to the start of the next ice age.

Key Themes

"Paradigm Shift—Scientific Revolution"

The Ice Age Started 8-10.5 Millennia Ago
The 1700-2016 CE Global Warming is the Biggest Trough-to-Peak
Outlier in 8,000 Years and Is Already Two Centuries Late for a
Switch to a Cooling Phase
The Higher the Trough-to-Peak Phase, the Bigger and More Abrupt
the Drop in Temperature
Large Magnitude Volcanism and this Grand Solar Minimum Pose
Catastrophic Risks (Dismissed by the IPCC)

CHAPTER 3

The Ice Age Started Millennia Ago, after the Holocene Climate Optimum

It is my contention that we are disoriented as regards the stage of the glacial cycle that we occupy today. The biggest impediment to gaining a correct orientation on today's climate is failing to look back beyond 1880, and indeed back to the Holocene Climate Optimum, when interpreting climate data. Additionally, the polar ice core temperatures are more relevant to climate forecasting than the global mean surface temperature. This is because the global climate optima lag the polar climate optima. Thus, the global climate data is consequential to what is happening in the polar regions (Chapter 2).

We do not have a correct bearing on today's climate relative to the Holocene Climate Optimum, and are in error as to the correct stage of the glacial cycle that we are currently in. To correct or readjust our glacial cycle orientation we must bind key points from Chapters 2 and 4 with key points in this chapter. And the sooner we do that the better.

First, the ice age did not end 11,700 years ago. It was the Younger Dryas that ended then.[245] Second, the Holocene Climate Optima for the Antarctic, Arctic, and global climates are millennia behind us, and these were comparatively late relative to all other glacial cycles over the past 800,000 (Antarctica data) and 1,000,000 (global data) years. Implicit in the IPCC's statistically flawed 30,000-year ice age deferral is the assumption that the Holocene Climate Optimum lies ahead of us, an assertion that is contradicted by the ice core data.

Third, today's Greenland ice core temperature is between 2^0C and 4^0C lower than at the Holocene Climate Optimum 8,000 years ago,[246,247,248] which means current temperatures are *not* the highest on record.[249] Today's temperatures are only the highest since 1880, when the modern instrument era for recording surface air temperatures began. Additionally, three of the last five glacial cycles recorded climate optimum temperatures greater than at the Holocene Climate Optimum.[250]

Fourth, today's global warming is part of a centennial-scale climate oscillation that is currently in its trough-to-peak or warming phase, and possibly passed this peak in 2016. This trough-to-peak phase is superimposed over a millennial-scale decline in Northern Hemisphere summer temperatures.[251,252,253,254,255] The warming phase of this centennial-scale climate oscillation will be followed by a peak-to-trough phase, when the climate switches back to a cooling phase. This is what happened for all 38 preceding climate oscillations since the Holocene Climate Optimum, whose warming exceeded 0.99^0C from trough-to-peak (Figure 4.2).

Fifth, there is more ice today than at the Holocene Climate Optimum, in both Greenland[256,257,258] and Antarctica,[259,260] even with the recent acceleration of ice melt. Ice began accumulating in earnest by 5,000 years ago at both poles, and by 3,000 years ago part of the Arctic (northeast Greenland) was ice-locked. Ice accumulation gained momentum during the second millennium CE, and even more so during the Little Ice Age, when glaciers reached their peak size.[261,262 ,263]

Sixth, starting in the mid-19th century, and accelerating over the last five-plus decades, human activity has helped a grand solar maximum melt two millennia worth of accumulated ice. Until we understand the relevant historical climate data, this recent ice melt simply disorients us as to the stage of the glacial cycle that we are in today.

Glacial Cycles Are Long-Term Climate Oscillations During Ice Age Epochs

Climate change is real and happens on many different timescales—tens of millions of years, glacial cycle length, or millennial, centennial, and even smaller scales. *Denying climate change would be futile, but blaming it only on human activity while ignoring natural climate change would be wrong, and dangerous.*

We need to understand climate change on glacial cycle, millennial, and centennial timescales. This understanding needs to be pegged to this glacial cycle's climate optimum as the common point of reference, as opposed to the climate since 1880. Each of these trough-to-peak then peak-to-trough oscillations in the temperature, across all of these different timescales, has its own set of causes.

Only by understanding these period-specific climate oscillations will we have the necessary insight to produce a climate forecast of meaning and utility.

The journey to understanding what will happen with the 21ˢᵗ century climate starts with understanding centennial-scale climate oscillations since the climate optimum. One thing I promise to do in the following chapters is *undo* all of your preconceptions about climate change, based on scientifically validated and verified data.

Earth displays grand climate cycles lasting approximately 135 million years. The big climate troughs are referred to as *ice age epochs*, each lasting tens of millions of years.[264] During the last half billion years (Phanerozoic Eon) polar ice caps have not been a regular feature of the earth's topography. Global temperatures during most of this time have been significantly higher than they are today. In fact, Antarctica's ice sheet only began forming about 40 million years ago,[265] while Greenland's ice sheet only began forming in earnest about 7 million years ago.[266] Polar ice sheets are consistent with being in an ice age epoch.

The current ice age epoch is said to have begun 2.6 million years ago at the start of the Pleistocene Era, with the Arctic and Antarctic ice sheets waxing and waning in cycles. In Antarctica, these cycles have averaged 97,000 years during the last 800,000 years.[267] According to conventional timelines the Pleistocene Era ended 11,700 years ago, whereupon the Holocene interglacial is said to have started.[268]

During the Pleistocene Era earth's climate passed through a series of temperature cycles, commonly referred to as glacial cycles. Glacial cycles are composed of periods of freezing (glacial periods) when snow and ice accumulate in the polar regions and at high altitudes, and sea levels decline. Glacial periods are followed by periods of warming (interglacial periods) when the polar ice melts and sea levels rise. For the purposes of this book I refer to the glacial period as an "ice age," which typically lasts for about 75,000 years.

This Ice Age Started at the Poles after the Holocene Climate Optimum 8–10.5 Millennia Ago

Each glacial cycle displays recurring phases and common points of reference such as a climate optimum, a glacial maximum, an interglacial period, and a first main trough after the climate optimum (Figure 3.1).

Interglacial periods are warming phases that extend from the glacial maximum (i.e., peak of the ice age) to the climate optimum. The glacial maximum represents the lowest temperature of the glacial cycle, during which the maximum ice mass and lowest sea levels exist. A *climate optimum* represents the highest temperature period of the glacial cycle at the end of the interglacial period, during which ice mass is at its lowest and sea level is at its highest. Ice ages start after the climate optimum. Glacial cycles last on average for 93,000 years, and are comprised of 75,000 years of ice age and 18,000 years for the interglacial period.[269]

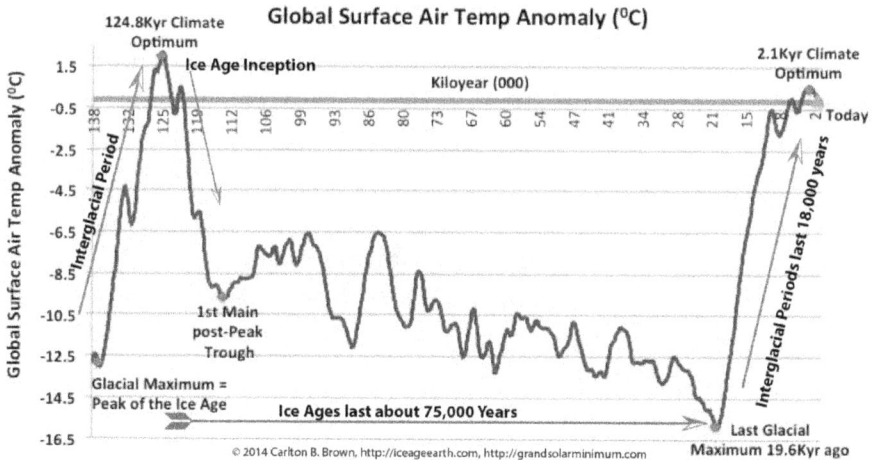

Figure 3.1. During the last 1 million years the average glacial cycle has lasted for 92,900 years, the average interglacial period was 18,200 years long, and the interglacial temperature rose on average 13.5°C (global data). The average glacial period, or period of freezing, lasted about 74,700 years, from the climate optimum to the next glacial maximum.[270]

A number of datasets were used to understand glacial cycle temperature changes, their relative phasings, and what stage of the glacial cycle we occupy today. The reconstructed temperature data analyzed spans of 3,000,000 years (global),[271] 800,000 years (Antarctica Dome EPICA C),[272] 360,000 years (Antarctica Dome Fuji),[273] 248,000 years (Greenland Oxygen-18 isotope),[274] and 11,700 years[275] and 49,000 years[276] (Greenland). These provide an informative panorama of the earth's changing climate and its historical glacial cycles.

Based on analyzing the above climate reconstructions, it becomes clear that climate optima are specifically phased (Figure 3.2). In general, a climate optimum is reached in Antarctica before it is reached globally. Based on two glacial cycles, the Arctic's climate optima[277] have occured after Antarctica's, but prior to the global climate optima. The Holocene Climate Optimum was reached 10,500 years ago in Antarctica (Dome-C data), 8,000 years ago in the Arctic (Greenland data),[278] and 2,100 years ago globally, yielding a phasing gap of 8,400 years. This 8,400-year phasing gap is the longest in the last 800,000 years of glacial cycles.

Likewise, the interval between the Holocene Climate Optimum and its preceding climate optimum was the longest recorded in the above-cited global and Antarctic climate data.

Based on the above-cited climate phasing gaps and climate optimum intervals, there is no justification for proposing the ice age lies ahead of us. There is even less justification for thinking the ice age lies ahead of us today when you realize there is more polar ice present now than there was at the Holocene Climate Optimum (see below).

This means that the climate optima temperature peaks for Antarctica 10,500 years ago, Greenland 8,000 years ago, and globally 2,100 years ago, represented the end of the previous interglacial period. As stated above, ice ages start after the climate optimum. The conclusion therefore is that *this Ice Age started* 8,000 and 10,500 years ago in the north and south poles respectively, and 2,100 years ago globally.

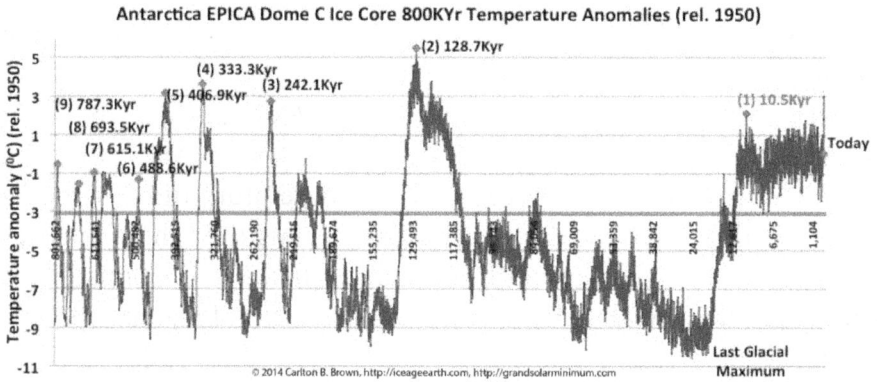

Antarctica EPICA Dome C Ice Core 800KYr Temperature Anomalies (rel. 1950)

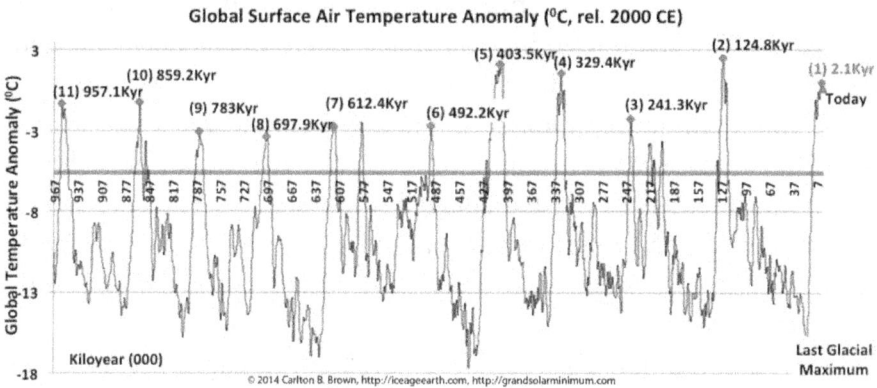

Global Surface Air Temperature Anomaly (°C, rel. 2000 CE)

Figure 3.2. Nine of Antarctica EPICA Dome C's glacial cycle temperature peaks (climate optima) were compared with their corresponding global climate optima over the last 800,000 years. Red dots and numbered dates are used to highlight the corresponding Antarctic and global climate optima. In all but two glacial cycles the climate optimum was reached in Antarctica first, on average 2,100 years before it was reached globally. The Holocene Climate Optimum was reached 10,500 years ago in Antarctica (Dome-C data) and 2,100 years ago globally, yielding a phasing gap of 8,400 years. This 8,400-year phasing gap is the longest in 800,000 years of glacial cycles. See the data table summary in the citation for all climate optima timings, phasings, and intervals.[279]

Our Glacial Cycle Disorientation

The Last Ice Age Ended 20,000–24,000 Years Ago, Not 11,700 Years Ago

We are told by science and in our schools that the last ice age (glacial period) ended 11,700 years ago, after which the Holocene interglacial began. However, the reality is this timing actually coincided with the end of the Younger Dryas, or a period that represented the worst rapid climate change event since the last glacial maximum (global data) 19,600 years ago.

Prior to the end of the Younger Dryas 11,700 years ago, and within the space of a few decades, the temperature in the Arctic dropped by about nine degrees Celsius.[280] The temperature did not recover for another three hundred years. During this time the Arctic ice sheets advanced and the most pronounced fauna extinctions of the Holocene interglacial took place, including dozens of mammalian and avian species.[281],[282]

Climate data reconstructions show that the lowest temperature at the last glacial maximum in Greenland (GISP2 ice core)[283] occurred 24,098 years ago, in Antarctica at Dome Fuji 19,300 years ago,[284] and globally the lowest temperature occurred 19,600 years ago.[285] The Antarctica Dome-C[286] and Greenland Ice Core Project (GRIP)[287] climate data reveal similar glacial maximum and climate optimum timelines to those displayed in Figure 3.3. The correct date for the last glacial maximum (i.e., end of the last ice age) can be seen in Figure 3.3 relative to the 11,700-year date for the end of the Younger Dryas.

By the end of the Younger Dryas 11,700 years ago, when the current Holocene interglacial had "officially" started (as we are told), nearly two-thirds of the Holocene's total sea level rise, and three-quarters of the Holocene's total temperature rise had already taken place (see table summary in the citation).[288] Therefore, equating the end of the Younger Dryas with the end of the last ice age means we are in error as to the correct stage of the glacial cycle that we are in now.

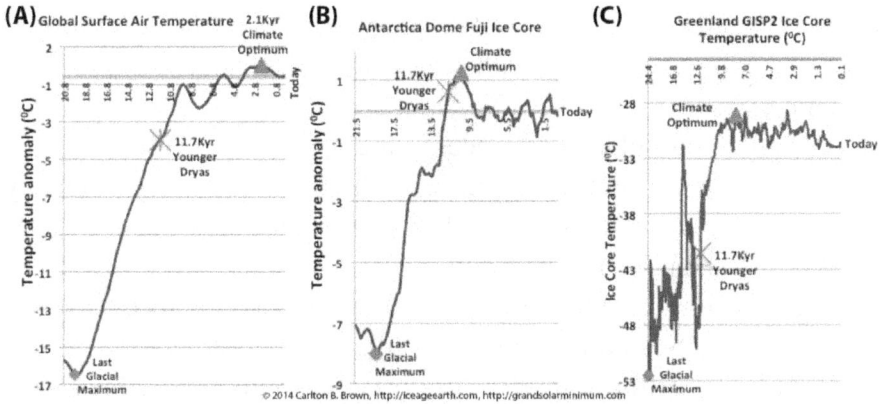

Figure 3.3. Reconstructed global,[289] Antarctic,[290] and Arctic[291] glacial cycle temperatures from the glacial maximum (lowest temperature, red diamond shape) to just past the climate optimum (highest temperature) delineate three points of reference i.e., the last glacial maximum, the end of the Younger Dryas 11,700 years ago, and the Holocene Climate Optimum (red triangle shape). The glacial maxima and the climate optima are at either end of the interglacial period. The supposed end of the last ice age 11,700 years ago (11.7 kiloyear) is marked above. If the ice age ended 11,700 years ago the 11.7 kiloyear marker should be close to the glacial maximum marker at the bottom of the graphics, but this is not the case. **Conclusion**: The ice age ended between 19,300 (Antarctica) and 24,000 years ago (Arctic). The Younger Dryas ended 11,700 years ago. The 8,000–12,000 intervening years between the glacial maximum and the 11.7-kiloyear timeline hides most of the real Holocene interglacial (sea level and temperature rise) and a number of rapid climate change events that caused major species extinctions.

Millennia of Ice That Accumulated after the Climate Optimum Has Melted Since the Mid-19th Century

More generally, the Arctic's Holocene Climate Optimum occurred between 8,000 and 5,000 years ago, varying regionally in onset. Temperatures were in general two to four degrees Celsius higher than today.[292,293,294]

Greenland's ice sheet margins retreated to less than their extent today between seven and four thousand years ago,[295] reaching their minimum extent between five and three thousand years ago.[296] The zone of coastal ice melt in the Arctic also retreated five hundred kilometers farther north, and there were summers free of sea ice.[297]

After the climate optimum, ice began to accumulate once again. This is evidenced by an abrupt ice accumulation along Greenland's north coast starting 5,500 years ago; northeast Greenland was ice-locked by about 3,000 years ago.[298] In Greenland's southeast, today's Kulusuk glacier region had been ice-free during the climate optimum.

Then, between 4,100 and 1,300 years ago, there were six major glacial advances, which coincided with major cooling episodes in the North Atlantic Ocean.[299]

The number of glacial advances in the second millennium CE was greater than in the first millennium, with most of the geographically widespread and extensive advances taking place during the Little Ice Age between the 13th and mid-19th centuries (see Figure 3.4B).[300] During this time winter sea ice closed off previously accessible sea routes between Scandinavia and Greenland.[301]

Beginning in the mid-19th century, as temperatures increased again, glacier ice began to melt, with this accelerating over the past five decades.[302,303,304] Based on Figure 3.4.B, we can appreciate that a significant part of the last two millennia of ice accumulation has melted since the mid-19th century.

The *key take-home message* is that there was less glacier ice in the Arctic at the Holocene Climate Optimum than exists today. Ice then began to accumulate starting about five millennia ago, accelerating through the Little Ice Age in an oscillatory manner until the mid-19th century. From the mid-19th century, the newly accumulated ice began to melt, with this accelerating in recent decades.

The ice melt since the mid-19th century happened during the current global warming ph ase, or the trough-to-peak phase of what is a centennial-scale climate oscillation that started in 1700 CE (Chapter 4). With most of the last two millennia of ice accumulation having melted recently, it *is difficult for us to perceive the correct stage of the glacial cycle that we are living in.*

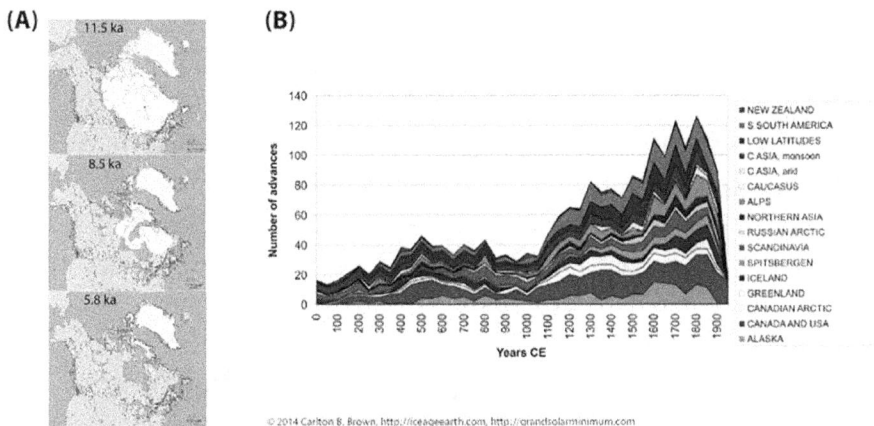

Figure 3.4. This image presents three smaller images (from 15) extracted from a time series of the Laurentide and Greenland ice sheets during the Holocene. The complete time series shows a stepwise reduction in ice extent from 11,500 years ago to its minimum extent 5,800 years ago.[305] **B)** A stacked time series of glacier advances and retractions during the last two millennia. A small number of glacier advances occurred in many regions between the first and twelfth centuries CE. There was a sharp increase in the number of glacier advances from the 13th century to the mid-19th century, after which glaciers started to recede.[306] **Conclusion**: We entered the current ice age after the Holocene Climate Optimum. There were significant glacial advances during the current era, especially during the Little Ice Age, but much of this ice has melted since the mid-19th century.

Antarctica Has More Ice Today than at the Climate Optimum

During the last glacial maximum, about 20,000 years ago, some parts of the Antarctic ice sheet reached the continental shelf edge.[307,308] Initial ice retreat from the last glacial maximum was under way by between 17,000 and 14,000 years ago, and between 10,000 and 8,000 years ago melting extended into Antarctica's interior, with deglaciation continuing until about 5,000 years ago.[309]

A widespread early Holocene Climate Optimum took place between 11,500 and 9,000 years ago, with a secondary optimum between 8,000 and 5,000 years ago. *By 5,000 years ago most of the Antarctic glaciers had retreated to, or behind, their current positions.*[310] During Antarctica's climate optimum *the central interior domes of the ice sheet were actually about one hundred meters lower than today,* telling us there was less ice than exists today.[311]

During the *last eight centuries Antarctica's ice mass has waxed and waned.* Periods of high ice accumulation occurred during the last millennia, most notably between the 14th and early 17th centuries, coinciding with the Little Ice Age. Since the 1960s ice accumulation has increased in the high coastal regions and over the highest part of east Antarctica.[312]

In conclusion, the maximum ice retreat was reached by the end of Antarctica's secondary climate optimum 5,000 years ago, and its inner ice domes were 100m lower than today. Since then Antarctica's ice mass has increased, *which is consistent with having entered a new ice age after the climate optimum.*

The Slowest Descent into an Ice Age in 2 Million Years

Since the Holocene Climate Optimum the Northern Hemisphere's summer temperature has slowly declined in line with the decline in Northern Hemisphere summer insolation, or the amount of solar irradiance reaching earth's upper atmosphere. This millennial-scale decline in summer insolation from its peak between 8,000 and 12,000 years ago[313,314,315,316,317] creates an underlying "ice age context" for the Northern Hemisphere, one from which there is no escape (and no 30,000-year delay). The Northern Hemisphere's millennial decline in summer temperature is arrived at in a centennial-scale oscillatory manner (Figures 3.6 and 4.1).

The above-cited decline in summer insolation resulted from changes in earth's elliptical orbit, primarily resulting from the precession of the summer solstice. This precession effect places the midsummer position of the Northern Hemisphere at an ever-increasing distance from the sun in its elliptical orbit, thus steadily reducing the amount of solar radiation reaching the earth's upper atmosphere.

The temperature decline since the Holocene Climate Optimum 2,100 years ago (global data)[318] is the smallest decline compared to all 33 preceding glacial cycles during the last 2,026,800 years.[319] Likewise, the temperature decline 10,500 years after the Antarctic's Holocene Climate Optimum is the smallest decline compared with all glacial cycles in the past 800,000 years.[320] While this current ice age inception temperature decline is the biggest outlier in both cases, it is not a statistically significant outlier (i.e., P-value >0.05). See Figures 3.5 and 3.6.

On the basis of the above data, probability indicates we're late for a date with a colder climate. The average glacial cycle temperature was 1.26°C lower 2,100 years ago globally, and 3.1°C lower 10,500 years ago in Antarctica, than it is at present. This suggests the odds are greater for a *change to a global cooling phase* than remaining is a warming phase during the 21st century.

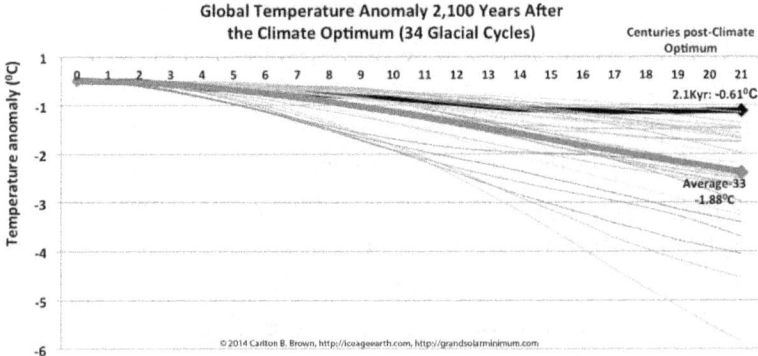

Figure 3.5. The first 2,100 years of temperature data after a climate optimum was extracted for the last 34 glacial cycles, and was rebased to zero degrees and zero time. The temperature declined by 0.61°C after the Holocene Climate Optimum, which was 1.26°C above the average of all other glacial cycles in 2,026,800 years.[321]

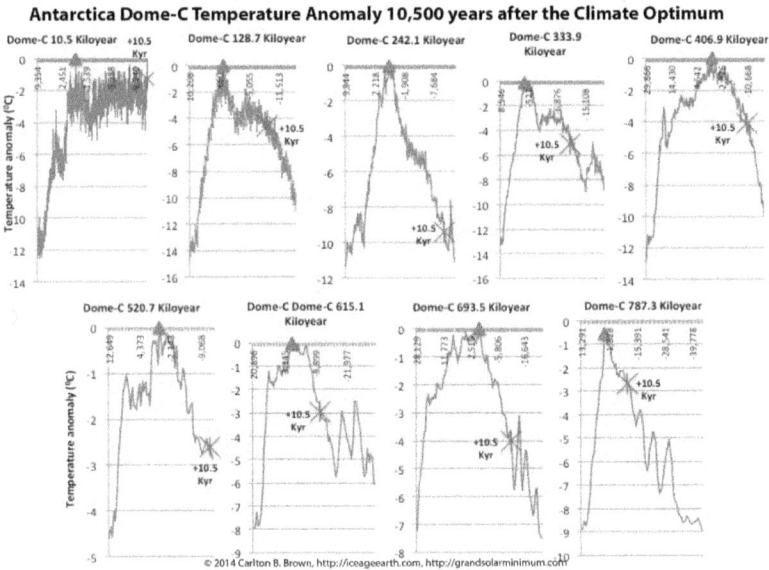

Figure 3.6. Ice core temperature data from Antarctica's EPICA Dome C reveals that the temperature has only declined by 1.2°C (top left figure) since the Holocene Climate Optimum 10,527 years ago. The comparator group of eight glacial cycles declined on average by 4.3°C, 10,500 years after their respective climate optima (red triangle shapes). This difference was not statistically significant.[322]

Conclusion (A and B): This current ice age inception shows the smallest decline in temperature compared with all other glacial cycles in 800,000 (Antarctica, 3.1°C higher) and 2,000,000 (global, 1.26°C higher) years.

How Our Ice Age Will Unfold

How does the typical ice age inception unfold after a climate optimum? If this current ice age follows the trend of the past 1,000,000 years of glacial cycles, then between the climate optimum 2,100 years ago and the next glacial maximum the global temperature will fall about 13.5°C over the course of 75,000 years.[323]

The temperature will plunge three-quarters of this maximum amount (-10.6°C) in the first 12,700 years after the climate optimum. There will be little respite in the ever-colder climate until we reach the first big temperature trough after the climate optimum. How volatile or extreme the climate oscillations will be during this plummet is indeterminable from this dataset, because each data point represents one century. The Greenland ice core data suggests the temperature will oscillate on centennial timescales.

During this voyage from the (average) climate optimum peak to the average first main trough (or inflexion point) 12,700 years later, the planet will accumulate one-fifth of the next glacial maximum's total ice volume, with most of this ice forming in the Arctic.

CHAPTER 4

Extreme Global Warming Temperatures Fall Abruptly when the Climate Switches to a Global Cooling Phase

There are three important points to convey in this chapter, which together will help you understand why this current global warming phase will end sooner rather than later.

First, to fully comprehend the 21st century's climate, we must understand centennial-scale climate oscillations, especially those taking place in the Arctic. Embedded in the largest trough-to-peak then peak-to-trough temperature oscillations in the Greenland ice core data are important clues to our *climate switching fate and its timing.*

Second, leading solar scientists expert in climate change predict the sun will reach the depths of this current grand solar minimum in the decades ahead, and during this time there will be a *return to a Little Ice Age-like climate.*[324,325,326,327,328,329,330]

Third, based on my analysis of the Northern Hemisphere climate and solar activity data, and in support of what the leading solar scientists expert in climate change have advised, a cold climate *always* follows a grand solar minimum.

The climate data revealed to scientists by the Greenland ice core is comprised of a series of 39 centennial-scale temperature oscillations, with trough-to-peak and then peak-to-trough phases exceeding first plus and then minus 0.99°C. These sequential rises and falls brought the ice to a lower temperature as the earth has moved further into this ice age. Since the Holocene Climate Optimum and up to 1700, when the lowest ice core temperature trough was reached, the ice core temperature declined by 4.86°C. Since 1700, at the start of this last trough-to-peak or warming phase, the temperature rose 2.87°C by 1940 and still further by 2016 (Figure 4.1).[331]

This chapter provides an analysis of these centennial-scale climate oscillations, and the conclusion is very clear. The warming phase revealed by ice cores dating between 1700 and 1940 is a statistically significant outlier in terms of its trough-to-peak temperature rise and the time taken for the temperature to rise. Based on a number of assessment methods, the odds are strong for a switch to a cooling phase. This trough-to-peak then peak-to-trough analysis shows we're long overdue for a switch to a cooling phase, and the higher the temperature sub-peak prior to that switch, the more abrupt the temperature fall will be.

Consider also that a Northern Hemisphere cooling followed all four grand solar minima during the Little Ice Age, and that this cooling was highly correlated with solar activity (see Figures 4.3 and 4.4). Based on this analysis, and if the future climate repeats trends of the Little Ice Age, then the balance of probability is that earth will re-enter a cooling phase during the current grand solar minimum.

When Will This Global Warming Phase End?

To better understand the current climate and the IPCC's global warming story (i.e., since 1880), it is necessary to understand how the global and Northern Hemisphere atmospheric temperature between 1880 and 2016[332] relate to the Greenland ice core temperature data between 5980 BCE and 1960.[333] These data sets overlap each other and permit us to understand that the 1880-2016 trough-to-peak temperature rise is but a fragment of a larger global warming trough-to-peak phase that started in circa 1700 (see Figure 4.1).

By comparing today's temperature only with that in 1880 we are being led to believe that a 1.02^{0}C rise in temperature since 1880 is the highest on record.[334],[335] However, when today's temperature is compared with the Holocene Climate Optimum's temperature 7,980 years ago, then that highest rise in temperature on record actually represents a *decline* of 1.9^{0}C (Figure 4.1). According to climate science experts who specialize in the Holocene Arctic climate, the Arctic temperatures was in general two to four degrees Celsius higher 8,000 to 5,000 years ago than it is today.[336],[337],[338]

The Arctic ice core data, like the Antarctic data, global data, and the Northern Hemisphere insolation data (Chapter 3), tells us emphatically that we have already entered an ice age (8,000 years ago). After the Holocene Climate optimum, the temperature declined 4.86⁰C by 1700.[339] This decline took place in a centennial-scale oscillatory manner. These centennial-scale climate oscillations comprised 39 temperature increases exceeding 0.99⁰C (trough-to-peak phases, range 1–15 decades, mean 88 years), which were followed by temperature declines exceeding 0.99⁰C (peak-to-tough phases, range 1–10 decades, mean 76 years), among smaller temperature oscillations.[340]

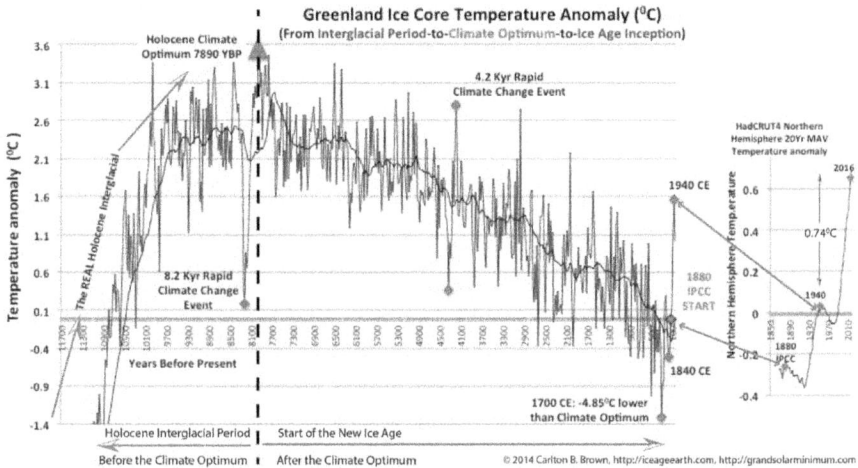

Figure 4.1. A graphic of Greenland's ice core climate reconstruction from 9080 BCE to 1960 CE is positioned alongside a 20-year moving average of the Northern Hemisphere temperature anomaly (1870 to 2018; right hand diagram). This depicts how the modern instrument era-derived Northern Hemisphere temperature data relates to Greenland's ice core temperature data. This juxtaposition of different climate data was done to give an approximate bearing on today's climate relative to the climate optimum. The temperature declined by 4.86⁰C between the Holocene Climate Optimum in 5980 BCE and 1700 CE, then rose by 2.87⁰C between 1700 CE and 1940 CE. The 2016 temperature peak is still 1.9⁰C lower than at Greenland's Holocene Climate Optimum 8,000 years ago. By 2018 the Northern Hemisphere's temperature had declined by 0.27⁰C from its 2016 peak.[341] **Conclusion**: Greenland entered an ice age 8,000 years ago. The lowest temperature since the climate optimum was in 1700, and was reached in a centennial-scale oscillatory manner comprised of dozens of sequential warming and cooling phases. The IPCC's anthropogenic global warming story starts in 1880, and was "grafted on" to the trough-to-peak warming phase that started in 1700 before the anthropogenic influence became significant.

The above-cited 39 trough-to-peak temperature rises exceeding 0.99⁰C between 7,980 years ago and 1960 were extracted from the

Greenland ice core data (see Figure 4.2). This indicates that the 1700-1940 rise of 2.87⁰C was the largest statistical outlier since the Holocene Climate Optimum. These 39 warming phases were stratified into two temperature groupings. Group 1 was comprised of five temperature rises greater than 1.77⁰C, while group 2 comprised 34 rises between 0.99⁰C and 1.56⁰C. Group 1's temperature rises were significantly different from Group 2's, indicating that these were extreme warming phases that *preceded* climate-modifying human activity. You can read the statistical analysis summary in the endnote citation for Figure 4.2 to get the important details.

Based on this comparative analysis of extracted trough-to-peak temperature rises, I concluded that *there is a greater probability the ice core temperature will decline* than continue its rise through the rest of the 21st century.

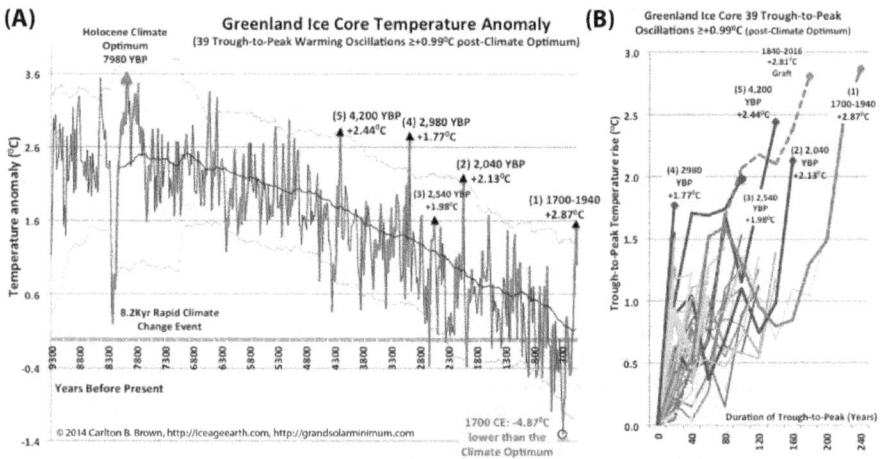

Figure 4.2. A) Thirty-nine trough-to-peak temperature rises exceeding 0.99⁰C (red segments) between 7980 years ago and 1960 were extracted for analysis. To help visualize statistical outliers, upper/lower Bollinger bands (pale grey) are used to highlight the peaks and troughs falling outside two standard deviations (95% confidence limits relative to a 60-period moving average, black line). The 39 trough-to-peak rises were not normally distributed and were stratified into two groups (≤ or ≥ 1.77⁰C) based on goodness-of-fit and outlier tests. The outlier test highlighted that those peaks rising more than 1.77⁰C were significant outliers, and that the 2.87⁰C rise from 1700-1940 was the biggest outlier. The stratification yielded two normally distributed groups that were significantly different from one another. **B)** This figure graphically displays the 39 trough-to-peak warming phases (rebased) plus a grafted peak +2.81⁰C (see Figure 4.1 citation for the rationale of the grafted peak). Group 1 outliers are blue and red (extreme outliers)[342] **Conclusion**: There is a greater probability the climate will switch back to a cooling phase than continuing its rise throughout the 21st century.

Outlier Arctic Warming Peaks Fall Abruptly after a Climate Switch

I wanted to understand if there were differences between Groups 1 and 2 once the climate switched to a cooling phase. My analysis shows that when the climate switches the temperature decline is *deeper and more abrupt* with the Group 1 outliers than with Group 2.

Groups 1 and 2 were compared for their magnitude of temperature decline, and the time taken to reach the first post-peak and the final trough. Group 1 (the big outlier peaks) dropped rapidly to its maximum decline of 1.92^0C within 40 years, whereas Group 2 declined 1.03^0C in a similar timeframe. This difference in temperature decline was statistically significant (see the citation).[343]

Some of the Arctic's coldest periods, biggest glacier advances, and important rapid cooling events since the Holocene Climate Optimum are included in Group 1 (see previous citation's table for the years involved).[344,345,346] Group 1 also includes the 4.2 kiloyear rapid climate change event associated with the extreme drought that precipitated the fall of Ancient Egypt's Old Kingdom, the Akkadian Empire, and the Indus Valley Culture.

The conclusion I drew from this analysis is the bigger the trough-to-peak phase, the *greater the magnitude of temperature drop and the more abruptly it falls from peak-to-trough after the peak (i.e., within 40 years)*. The implication for this current 1700-2016 warming phase is that the climate will switch back to a cooling phase, and the temperature will decline sharply.

Carbon Dioxide Did Not Cause the 1700-2016 Trough-to-Peak Northern Hemisphere Warming Phase

Additionally, Group 1 (big peaks less than the 1700-2016 peak) took an average of 105 years to go from trough to peak, whereas the 2016 peak took 316 years (1700 CE trough start). The implication of this time difference is that we're *two centuries overdue for an abrupt decline in the temperature*. Now, I hear the global warming alarmists among you retort that this delay is due to human activity.

Just so we are clear. This two-century delay in the switch to a cold climate phase is a result of a prolonged grand solar maximum phase.

This two-century delay is not due to the unabated rise in carbon dioxide levels associated with human activity. That carbon dioxide is not the cause of this two-century delay becomes evident when one looks at the Northern Hemisphere temperature data[347] in relation to the data for solar activity and carbon dioxide (see Figure 4.3).

The Northern Hemisphere temperature started to increase in 1713; two years *after* the solar activity data indicated that the sun had entered a solar maximum phase. The use of an 18-year moving average for the Beryllium-10 concentration anomaly (i.e., a measure of solar activity) means the temperature, in reality, lagged behind solar activity by at least one 11-year solar cycle (see Figure 4.3 for an explanation of this lag). The inversely related solar activity and temperature anomaly data have strongly mirrored each other's volatilities (i.e., as one goes up the other goes down), from the start of this 1700-2016 warming phase, as well as since 1406 (see Figure 4.4).

On the other hand, atmospheric carbon dioxide related to human activity did not dominantly feature in 1713 when this current trough-to-peak warming phase began. You can also see that the carbon dioxide concentration has *poorly tracked* the Northern Hemisphere temperature volatility on multi-annual to decade time-scales since 1713. This means carbon dioxide was consequential and not causative of this 300-year climate oscillation. This conclusion is fully aligned with the science that says elevated atmospheric carbon dioxide results from ocean degassing plus human activity, and does not cause or control, the temperature rise.[348,349,350,351,352]

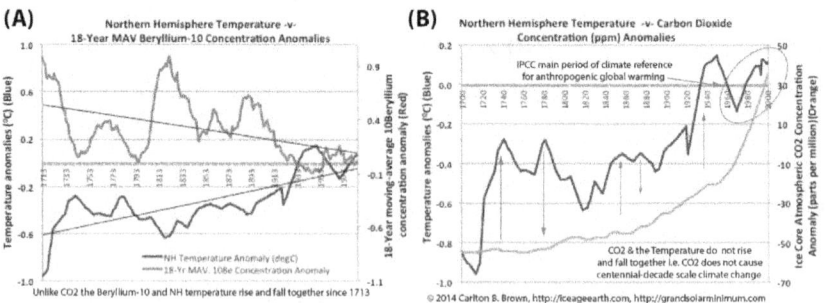

Figure 4.3. The 1700–2000 centennial-scale temperature oscillation represents a global warming "outlier," the cause of which preceded significant human activity by

circa 250 years. **A**) The Northern Hemisphere temperature anomaly highlights a strong inverse correlation with the 18-year moving average Beryllium-10 concentration anomaly ("Beryllium-10", solar activity proxy). In 1711, the Beryllium-10 began its decrease phase (i.e., increase in solar activity), whereas the temperature began to rise two years later from its deep trough in 1713. This relationship tells us the temperature lagged behind the rise in solar activity (viz., a declining Beryllium-10). The use of an 18-year moving average for the Beryllium-10 concentration anomaly means the temperature, in reality, lagged behind fluctuations in the Beryllium-10 concentration anomaly by at least one 11-year solar cycle. This 11-year solar cycle lag (approximate) is composed of 9 years (i.e., half of a 18-year moving average), plus a two year lag in the temperature rise (behind the Beryllium-10 rise), plus one year[353] before the newly produced Beryllium-10 in the upper atmosphere reaches earth's surface where it can be incorporated in ice cores (and thus be measured). These two parameters also tracked one another's variations, putatively supporting a cause-and-effect relationship. **B**) On the other hand, since 1713 the rising carbon dioxide concentration has poorly tracked the Northern Hemisphere temperature volatility on multi-annual and decade time scales (see Figure 4.4), indicating it is consequential and not causative.[354]

Cold Climates Always Follow Grand Solar Minima

Between 1400 and 1900, the Northern Hemisphere was on average about 1^0C colder than in the late 20th century, with this varying on a regional basis. The coldest region during the Little Ice Age was the Atlantic sector of the Arctic.[355,356]

All four officially recognized grand solar minima of the Little Ice Age were associated with troughs in temperature in the Northern Hemisphere (Figure 4.4). These Little Ice Age grand solar minima were the Wolf (1280-1350), Spörer (1450-1550), Maunder (1645-1715), and Dalton (1790-1830) minima. These grand solar minima coincided with the biggest glacier ice advances experienced since the Holocene Climate Optimum (Figure 3.4).[357]

For the years 1406 to 1889, a statistically significant inverse relationship exists between the Northern Hemisphere temperature and the solar activity proxy (i.e., 18-year moving average Beryllium 10 concentration anomaly)—that is, for 484 years of the Little Ice Age period. This relationship is highlighted in Figure 4.4.A. The correlation was maximized using an 18-year moving average of the Beryllium-10 concentration anomaly, over the raw Beryllium-10 data and a 5-year and 11-year moving average. The use of the 18-year moving average is the equivalent of saying the temperature lags behind the solar activity by about one 11-year solar cycle. When all four grand solar minimum phases (viz. increasing 18-year moving average Beryllium-10 concentration anomaly) and associated temperature decline phases were extracted and synthesized into a new data series, one grand solar minimum after the other, the correlation coefficient was further increased (Figure 4.4.B).

The above (together with Figure 4.4) tells us that solar activity was inversely correlated to a high degree with the Northern Hemisphere's temperature during the Little Ice Age, with each data parameter strongly mirroring the movement of the other data parameter, i.e., as one goes up the other goes down. It also tells us that the solar activity-temperature relationship strengthened during the periods of grand solar minima. In other words, declines in temperature during a grand solar minimum always followed a decline in solar activity.

Based on the strong correlation between solar activity and the Northern Hemisphere temperature during the Little Ice Age, if this relationship is repeated during this grand solar minimum, *then the planet will cool.* This conclusion is fully aligned with the consensus conclusion of solar scientists who are experts in climate change (see next).

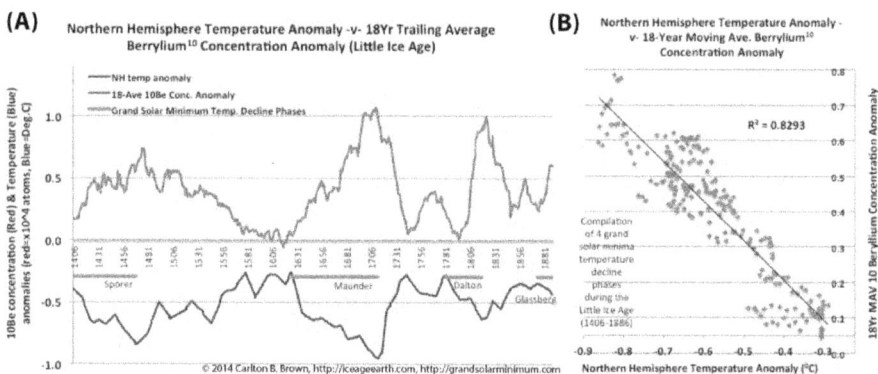

Figure 4.4. A) Time series: A statistically significant inverse correlation exists between the 18-year trailing average of the Greenland ice core Beryllium-10 isotope concentration anomaly (red line) and the annual Northern Hemisphere temperature anomalies (blue line). Both data parameter variations also tracked one another's variations to a high degree, putatively indicating a cause-and-effect relationship. Beryllium-10 is produced in the atmosphere by high-energy cosmic ray collisions with oxygen and nitrogen atoms, and is a well-established proxy for solar activity. High Beryllium-10 ice core concentrations indicate low levels of solar activity, and vice versa. B) All four grand solar minima of the Little Ice Age were characterized by a strengthening of the relationship between the increasing 18-year moving average Beryllium-10 concentration anomaly and the declining Northern Hemisphere temperature anomaly. See the citation for the detailed statistical analysis summary.[358] **Conclusion**: The cold Northern Hemisphere climate during the Little Ice Age lagged behind the decline phase of solar activity. A grand solar minimum in solar activity is always associated with a colder Northern Hemisphere climate.

Solar Scientists Predict a Cold Climate Will Accompany this Grand Solar Minimum

The intention of the following paragraphs is to highlight the fact that there are solar and earth system scientists, expert in climate change, who have made climate predictions linked to the grand solar minimum we are now entering. These leading experts include Habibullo Abdussamatov, Jan-Erik Solheim, Nicola Scafetta, Theodor Landscheidt, Rick Salvador, and Nils-Axel Mörner.

The above experts utilize different solar activity-related climate forecasting methods, all of which have provided a very consistent and consensus set of climate forecasts and predictions relating to this grand solar minimum.[359,360,361,362,363,364,365] Their *consensus view*, at odds with the IPCC, is that there will be a *return to Little Ice Age-like conditions during this grand solar minimum*. They expect that the trough of this grand solar minimum will be reached in the decades ahead, with a range between 2030 and 2060.

Some of the forecasting models utilized by these experts were also able to backward forecast (hindcast) the sun's activity (grand solar minima and maxima) and associated climate change since the Holocene Climate Optimum. This millennial-scale hindcasting ability provides a level of validation for the forecasting methods these solar scientists utilize.[366,367]

These solar scientists' climate forecasting methodologies relate to how solar activity occurs in cycles that have predictable periodicities. These solar activity-related periodicities are also embedded in the climate data in a highly correlated manner. Effectively, solar activity is controlled by the gravitational and angular momentum impact of the giant planets (i.e., Jupiter and Saturn). This planetary impact wobbles the solar system's center of mass and perturbs solar dynamo processes, which in turn are responsible for generating the sun's magnetic field (Chapter 6). The predictable nature of planetary orbits and their impact on solar activity periodicities makes the climate predictable, according to these experts.[368,369,370,371]

CHAPTER 5

Climate-Forcing Volcanism, Grand Solar Minima, and the Climate Oscillator Switch

Volcanic activity and periods of low solar activity were dismissed or ignored by the IPCC in its climate risk assessments provided to governments.[372] This chapter highlights why we should be concerned about climate-forcing volcanism during this grand solar minimum.

Climate-forcing or large magnitude volcanic eruptions have been responsible for some of the worst cold periods, droughts, and famines of the last millennium in countries and regions such as China, India, Europe, Africa, and the Caribbean Basin.[373,374,375,376,377,378,379,380,381,382,383,384] These climate-forcing volcanic eruptions were responsible for the sixteen coldest decades, and the coldest European summers, of the Current Era.[385]

Large magnitude volcanic eruptions have disastrous effects on global agriculture through cold, ice, drought, and flooding. Scientists tell us that if a Laki-like volcanic eruption (Iceland, 1783) occurred today, the global crop production losses would be in the region of *one year's food supply for one-third of the world's population* (see Chapter 7).[386] There is no publicly available plan to manage such a global catastrophe.

Climate experts specializing in climate-forcing volcanism believe the Little Ice Age was triggered by the Current Era's largest magnitude volcanic eruption at Rinjani, Indonesia, in 1257.[387] Rinjani combined with other large magnitude eruptions caused an abrupt and persistent cooling in the late 13th century and in the middle of the 15th century. This cooling coincided with two of the most volcanically active half-centuries since the Rinjani eruption. These cold periods were sustained over multi-decade to centennial timescales, and were associated with centennial-scale glacier ice expansion. This cooling and ice expansion persisted long after the aerosols produced by the eruptions were removed from the atmosphere.[388],[389],[390,391,392,393]

This chapter highlights a number of discoveries and reminders relating to the sun's role in controlling centennial-scale climate oscillations and cooling, and the occurrence of climate-forcing volcanism.

Volcanism and a Grand Solar Minima Caused the Little Ice Age

Volcanic activity has been more frequent over the last 1–2 millennia than during the preceding period back to the Holocene Climate Optimum.[394,395] This increased frequency of volcanic activity during the Current Era, and particularly since the start of the Little Ice Age, paralleled an increase in glacier ice buildup (Figure 3.4).[396],[397],[398]

Large magnitude volcanic eruptions (VEI 4 to 7 on a scale of 1 to 8, with 7 being Rinjani or Tambora-like) are the most relevant to climate change, because of the large quantities of sulphur dioxide they inject into the stratosphere. Once in the stratosphere, the sulfur dioxide is converted into sulphate aerosols.[399,400] These sulphate aerosols disturb the radiative balance of the planet by reducing the amount of sunlight reaching earth's surface. This in turn has a global cooling effect for 1–3 years after an eruption.[401,402],[403],[404,405,406],[407] The cooling impact for the larger volcanic eruptions (VEI 7) can persist for up to a decade.[408]

During the Current Era the global temperature dropped by an average of 0.6^0C with each large tropical volcanic eruption, and by more than 1^0C for Tambora-like volcanic eruptions (VEI 7).[409,410] Large volcanic eruptions in the equatorial latitudes have had the greatest cooling impact over the last 2,500 years, compared with eruptions occurring at more northern and southern latitudes.[411]

The climate impact of large magnitude volcanic eruptions is not simply confined to planetary cooling. Large magnitude volcanic eruptions have also been responsible for some of the worst droughts and famines of the last millennia, in countries and regions such as China, India, Europe, and Africa. The temperature, drought, and rainfall following these large eruptions varied according to the hemisphere in which the volcanic eruption originated, and whether the eruption

45

was at high latitude or of tropical origin. These climate forcings were due to an eruption's impact on moving air circulation systems (see Chapter 7).[412,413,414,415,416] [417,418,419,420,421,422,423]

How Large Magnitude Volcanic Eruptions Cause Centennial- Scale Glacier Expansion

Large magnitude volcanic eruptions trigger atmospheric and ocean circulatory system responses in the years following such an event. These can then induce longer-lived (decade to multi-decade) changes in the Arctic's and North Atlantic's climate. This in turn can have a major impact on the global ocean temperatures for several decades, which can lead to centennial-scale increases in the Northern Hemisphere's glacier and sea ice.[424] [425,426,427,428]

Scientists expert in volcanic activity-induced climate change (see the next paragraph's citations) believe the Little Ice Age was caused by periods of abrupt and persistent summer cooling in the late 13th century and middle of the 15th century. These periods coincided with two of the most volcanically active half-centuries of the last millennium. The Little Ice Age also coincided with four successive grand solar minima, starting with the Wolf minimum in 1280.

A large magnitude volcanic eruption is believed to have triggered the Little Ice Age (Rinjani in 1257, a VEI 7 event),[429] which was then followed by other large magnitude volcanic eruptions, roughly one every decade. These eruptions collectively resulted in volcanic sulfate levels during the 13th century (as revealed by ice core data) that was many times greater than in any other century during the last millennia. The cold periods resulting from this 13th century large-magnitude volcanism had an impact on the climate that was sustained over centennial timescales, and long after the eruptions' volcanic aerosols were gone from the atmosphere.[430] [431] [432,433,434]

Low solar activity-induced alterations of atmospheric circulations are thought to play an important role in glacier and sea ice expansion processes.[435] The North Atlantic Oscillation is a dominant Northern Hemisphere atmospheric circulation, and is the key determinant of the winter climate over the North Atlantic.[436,437,438,439,440,441,442] A prolonged negative phase of the North Atlantic Oscillation was experienced during the Little Ice Age, which was associated with increased ice accumulation during the Little Ice Age.[443,444]

Importantly, the North Atlantic Oscillation is coupled to the upper atmosphere (i.e., the stratosphere) by some complex physical processes,[445,446] and its phase and strength are correlated with geomagnetic activity (i.e., earth magnetism), which is known to be modified by magnetized solar wind.[447,448],[449] The North Atlantic Oscillation is also modified by changes in the loading of the stratosphere with volcanic aerosols.[450,451]

The above paragraphs collectively highlight that climate-forcing volcanism and the North Atlantic Oscillation were instrumental in the Little Ice Age's cold climate and ice accumulation mechanism. This mechanism led to an increase in sea ice entering the sub-polar North Atlantic region from the Arctic (referred to as "sea ice exports"). These sea ice exports in turn weakened the North Atlantic branch of the Atlantic thermohaline circulation (i.e., a salt concentration and temperature-driven ocean circulation system), and reduced warm water entry into the Arctic region. The increased sea ice exports and changes to the ocean circulation system "reinforced" the ice generating process, which led to centennial-scale glacier ice expansion in the Arctic (viz. *glacier ice expansion mechanism*).[452,453,454,455,456]

Are Grand Solar Minima and Maxima the Master Regulators of Glacier Ice Expansion?

One of the key pieces of the climate change jigsaw that is missing in the scientific literature is *the elucidation of a natural oscillatory control mechanism* that is the "master controller" of centennial-scale climate change and glacier ice expansion. Understanding this missing *centennial-scale control mechanism is fundamental to forecasting the 21st century climate*, and in turn to advising governments on mitigating the risks accompanying climate change.

When viewing Figure 5.1, which depicts sunspot numbers between 9104 BCE and 1895 CE, it becomes obvious that sunspot cycles constitute a natural oscillator. This solar activity oscillator is similar to the temperature oscillation evident in the Arctic ice core temperature data (Figure 4.2), albeit the two are not always in phase with each other. The duration of both the larger sunspot number oscillation and the temperature oscillation is typically one to two centuries from peak-to-peak.

A number of analyses were conducted utilizing data from two different volcanic eruption databases. The analysis below assessed if solar activity and climate-forcing volcanism were associated with centennial-scale climate oscillations.

A quantitative filter was applied to the National Oceanic and Atmospheric Administration (NOAA) volcanic eruption database,[457] resulting in the selection of the 73 largest volcanic eruptions over the last 11,000 years. These 73 eruptions caused the greatest climate forcing (i.e., cooling), or reduction in solar irradiance reaching earth's surface through their impact on stratospheric sulphate aerosols. All 73 climate-forcing volcanic eruptions were plotted against the sunspot numbers over 11,000 years, to help visualize their distribution in time and relative to solar activity oscillations (see Figure 5.1.A).

The standout finding of this analysis was that large magnitude volcanic eruptions are concentrated around the major and minor sunspot number peaks and troughs. Seventy-seven percent (56 of 73) of these large magnitude volcanic eruptions occurred at or within a decade of a grand solar minimum or maximum, or a smaller trough or peak of sunspot numbers (see Figure 5.1.B). This relationship is not evident for smaller volcanic eruptions. Three-quarters of these large magnitude eruptions occurred when the 500-year average sunspot number fell below 37.

A similar result was obtained by plotting the 67 total VEI 6 and 7 eruptions from the Volcano Global Risk Identification and Analysis Project (VOGRIPA) database against 11,000 years of sunspot numbers. This analysis showed that 82 percent of all VEI 6 or 7 events occurred at or within a decade of a sunspot number peak or trough. Three-quarters of these VEI 6 and 7 events occurred when the 500-year average sunspot number fell below 34.[458]

(A) Holocene Large Volcanic Eruptions with a Climate Forcing Impact of Lower than <-5 Watts/m2 versus Sunspot Numbers

(B) The Sunspot Peak & Trough Association of Large Magnitude Volcanic Eruptions

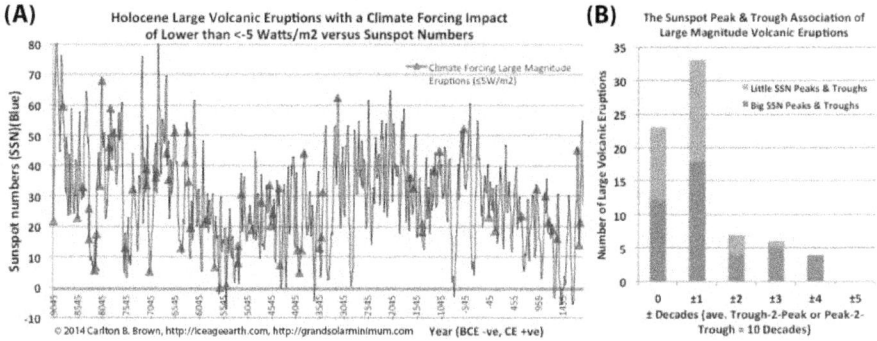

Figure 5.1. A) A quantitative filter (see citation methodology) was applied to a National Oceanic and Atmospheric Administration (NOAA) volcanic eruption database. This volcanic eruption data was derived from Greenland's GISP2 sulphate record and was used to estimate the number of volcanic eruption events. In this manner, the 73 largest climate-forcing volcanic eruptions were selected covering the last 11,000 years. All 73 large magnitude volcanic eruptions were plotted against the sunspot numbers (NOAA provided). This combined data was used to create Figure 5.1.B. **B)** Seventy-seven percent (56/73) of large magnitude volcanic eruptions occurred at or within a decade of a grand solar minimum (i.e., deep sunspot number trough) or grand solar maximum (i.e., sunspot number peak), or within a decade of a smaller trough or peak going into or coming out of a grand solar minimum. This resulted in the skewed distribution of the eruptions in the zero and ±1 decade groups relative to the ±2–5 decade groups. The mean sunspot number trough-to-peak or peak-to-trough duration was ten decades (standard deviation 4.7 and range 3–25 decades).[459] **Conclusion**: Grand solar minima and maxima ±1 decade represent high-risk times for climate-forcing volcanic eruptions.

(A) The Little Ice Age's Large Magnitude Volcanic Eruptions (VOGRIPA/LaMEVE database)

(B) Peak & Trough Association of Large Magnitude Volcanic Eruptions   Peak & Trough Association of Large Magnitude Volcanic Eruptions

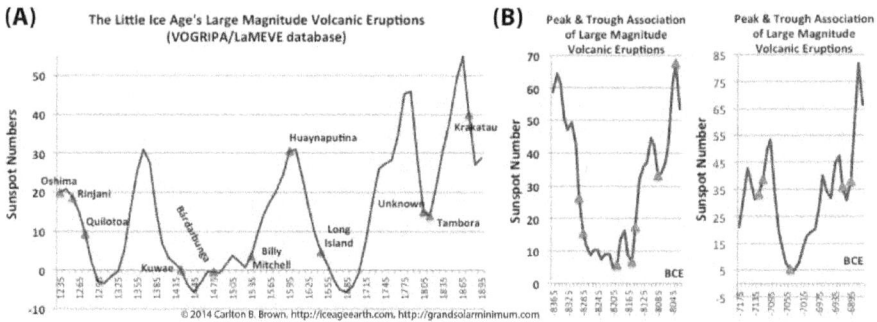

Figure 5.2. A) VOGRIPA's record of large magnitude volcanic eruptions (VEI 6 or 7), associated with the Little Ice Age were plotted against reconstructed sunspot numbers. The Rinjani volcanic eruption occurred at the grand solar maximum just prior to the Wolf minimum and triggered the Little Ice Age. You can see how 5 of the 11 large magnitude volcanic eruptions took place at or near the troughs in these grand solar minima periods. A further 3 of 11 large magnitude volcanic eruptions (VEI 6 or 7) occurred half way into a grand solar minimum, while the remaining 3 of 11 eruptions occurred at grand solar maximum sunspot peaks. **B)** These two figures highlight the association of large magnitude volcanic eruptions with grand solar maxima and minima (i.e., big peaks and deep troughs), as well as with the smaller peaks and troughs. The first of these two figures coincides with the 8.2-kiloyear rapid climate change event, the most abrupt and deepest cooling event in the last 8,500 years, which left its imprint in the climate record around the world.[460,461,462,463] **Conclusion**: Grand solar minima and maxima, and smaller sub-peaks and sub-troughs represent high risk periods associated with the "triggering" of climate-forcing volcanic eruptions.[464]

Since the Holocene Climate Optimum, a statistically significant inverse relationship has existed between the 500-year average sunspot numbers and the 500-year bin totals of climate-forcing volcanic eruptions. This relationship markedly diminished when the period of correlation calculation was extended from the last 8,000 years out to the last 11,000 years. The correlation also diminished when the duration of the 500-year average sunspot numbers and the 500-year bin totals of climate-forcing volcanic eruptions were each reduced to 400 and 300 years. If the solar activity-volcanism relationship is real, then a long-term process involving magnetized solar wind[465] is implicated in causing climate-forcing volcanic eruptions (see Figure 5.3).

A stronger non-linear relationship was demonstrated using the VOGRIPA Large Magnitude Explosive Volcanic Eruption database data, while utilizing the same methodology detailed in Figures 5.3.A and 5.3.B. This non-linear relationship, if real, would seem to indicate that as the 500-year average sunspot number declines below 17 there is an accelerative increase in climate-forcing large magnitude volcanic eruptions (VEI 4–7), i.e., more bang for your low sunspot number buck. However, caution is merited in interpreting this potential non-linear relationship, given that many volcanic eruptions in the more distant past (i.e., before the last millennium) are not part of the scientific record. This can give the impression of an accelerative increase in volcanism during the Little Ice Age.[466,467]

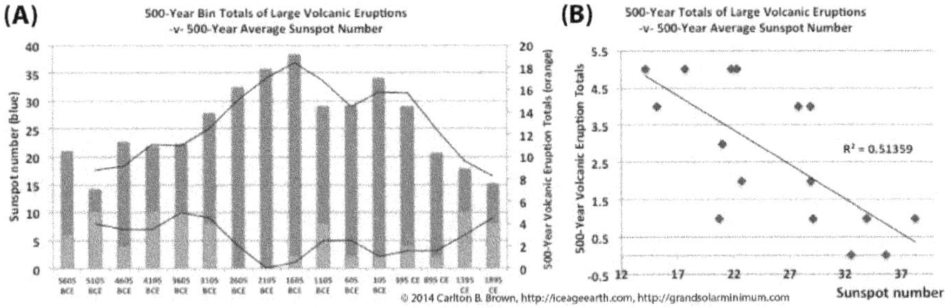

Figure 5.3. A) Five hundred-year totals of large volcanic eruptions were plotted against 500-year average sunspot numbers occurring since the Holocene Climate Optimum. The two-period moving average trend lines highlight an inverse relationship (i.e., one goes up, the other comes down). The last 2,500 years have seen a declining trend in 500-year sunspot numbers, with the 500-year period ending in 1895 having the lowest 500-year sunspot number average in 7,500 years. **B)** A scatter plot presents 5.3.A's data differently, and highlights a statistically significant relationship between long-term

average sunspot numbers and climate-forcing large volcanic eruptions.[468] **Conclusion**: If low sunspot numbers are involved in triggering climate-forcing volcanism, then a long-term process involving magnetized solar wind is implicated.

Figure 5.1 visually demonstrates that the majority of climate-forcing volcanic eruptions occurred at or within one decade of a grand solar minimum or maximum, or at or within one decade of a smaller sunspot sub-peak or sub-trough. This alludes to the potential role of solar activity in triggering climate-forcing volcanic eruptions on centennial timescales. This triggering is associated with the grand solar maximum peaks and grand solar minima troughs, or the change in phase from trough-to-peak or peak-to-trough (smaller peaks and troughs).

At the same time, Figure 5.3 indicates that a long-term declining trend in sunspot numbers (i.e., solar activity) is associated with an increase in the frequency of climate-forcing volcanic eruptions. It is telling that three-quarters of large volcanic eruptions occured when the 500-year sunspot number average fell below 35 (average of NOAA and VOGRIPA data). In other words, as the sun goes into a long-term magnetic sleep cycle, and solar irradiance is in its ice age mode (precession of the summer solstice), the planet cools and accumulates ice at the poles in an oscillatory manner (Figures 3.4 and 3.4.B).

How Solar Activity and Climate-Forcing Volcanism Control our Ice Age Descent

The implications of the data included in Figures 5.1, 5.2, and 5.3 are that grand solar minima and maxima, combined with a long-term declining trend in solar activity, represent high-risk periods for climate-forcing volcanism. Climate-forcing volcanism (linked to solar activity), together with the previously described glacier ice expansion mechanism, are responsible for causing long-term global cooling.[469]

The involvement of magnetized solar wind in triggering climate-forcing volcanism is inferred from the cosmogenic isotope carbon-14 used to reconstruct the sunspot numbers (see the citation for an explanation of this inference).[470,471] This carbon-14 sunspot number data (derived from tree rings) correlates with climate-forcing volcanism (Figures 5.1 and 5.3). Therefore, solar activity-triggered volcanism, mediated via magnetized solar wind, indirectly links the sun to rapid cooling processes (i.e., via sulphate aerosols). This solar activity-induced volcanism operates on a multi-annual timescale and is involved in triggering the previously mentioned multi-decade to centennial-scale glacier ice expansion mechanism.[472,473,474,475,476]

It therefore follows that climate-forcing volcanic eruptions triggered at or near a grand solar minimum trough (Figure 5.2.A, i.e., Spörer minimum) work in concert with the magnetized solar wind climate mechanism to cool the planet and kickstart glacier ice expansion. Sea ice and ocean feedback processes, together with periodic VEI 4 or 5 eruptions, then further propagate this glacier expansion on centennial timescales.

Likewise, it follows that climate-forcing volcanic eruptions triggered at the peak of a grand solar maximum (Figure 5.2.A, i.e., Rinjani in 1257), would rapidly cool the planet (via sulphate aerosols). Such a grand solar maximum-associated eruption would then "regulate" the magnitude of the temperature peak and help realign the temperature oscillation with its climate-leading solar activity oscillator, if the temperature got out of phase (i.e., through other climate system perturbations). Similarly, climate-forcing volcanic eruptions occurring on entry into a grand solar minimum, while the temperature is at a peak (e.g., *as with 2016's temperature peak*) would also rapidly cool the planet.

We should therefore be reminded that one climate-forcing volcanic eruption, occurring at the grand solar maximum, or going into a grand solar minimum, or at the trough of a grand solar minimum, *over which humans have no control*, poses great risks to our way of life. In other words, *the IPCC was wrong to dismiss the important role of volcanism and solar activity* in their climate change theory, climate forecasts, and their risk assessments provided to governments (see Chapter 7).

CHAPTER 6

The Solar Activity Science of Global-Scale Catastrophic Risks

The purpose of this chapter is to highlight two important scientific concepts, linked to magnetized solar wind and its impact on climate and tectonic-volcanic processes. Understanding how magnetized solar wind impacts earth's magnetosphere and its length of day (i.e., rate of rotation) is important because it opens us to *the science of solar activity-related risks associated with global catastrophes.*

First, the sun connects with the climate system in two main ways: via solar irradiance (electromagnetism), and by its magnetized solar wind (magnetism). Magnetized solar wind impacts earth's magnetosphere (i.e., its magnetic shield), earth's length of day, and the earth's atmospheric and ocean circulations, giving it numerous secondary or sub-levers on the climate system.[477,478,479,480,481,482,483,484,485,486,487,488,489]

Second, solar activity-induced changes to solar wind, geomagnetism, and earth's length of day also impact tectonic processes (i.e., earthquakes and volcanism).[490,491,492,493,494,495,496,497,498,499,500,501,502]

The *unseen risks* linked to solar activity include a switch to a cold climate phase and centennial-scale glacier expansion (Chapters 3–5), extreme changes in precipitation (Chapter 7), climate-forcing volcanic eruptions (Chapters 5–6), earthquakes (Chapter 6), and pandemic flu outbreaks (Chapter 14).

Planetary Orbital Forces Make the Sun Wobble, Causing Solar Cycles

The sun physically oscillates around the solar system's center of mass on its journey through galactic space. This wobble effect on the solar system's center of mass is due to the gravitational and angular momentum impact of the giant planets, specifically Jupiter and Saturn.

This wobble effect results in a number of periodic oscillations in the movement of the sun about the solar system's center of mass.[503,504,505]

Physical forces operating between the planets as they orbit the sun also affect the rate at which planets rotate, and the sun's rate of rotation as well. Cycles of differential rotation by the sun are thus established, which then determine the multiple periodicities of the sun's activity. Earth's rate of rotation is also subject to these same planetary forces acting on the sun.[506]

This planetary influence on the sun's motion around the solar system's center of mass perturbs the sun's internal solar dynamo processes. The solar dynamo is responsible for generating the sun's magnetic fields. Cycles of solar activity therefore manifest in sunspots, solar flares, solar irradiance, coronal mass ejections, and the sun's magnetic fields emanating into space (magnetized solar wind).[507]

Sunspot numbers rise and fall over an 11-year cycle (see Figure 6.1), and these sunspots can be observed on the surface of the sun as dark discs. The current Solar Cycle 24 began in January 2008.[508] This is the third 11-year cycle in a row since the peak of Cycle 21 in the late 1980s with diminishing peak sunspot numbers.[509]

These diminishing peaks and troughs of solar activity highlight the influence of longer-term solar cycles that impact the magnitude of the 11-year solar cycle (sunspot numbers), and indicate that the sun is moving into a grand solar minimum phase. These longer-term cycles include the Gleissberg (50–80 and 90–140 year periods) and Suess cycles (170–260 year periods).[510] At this stage of the glacial cycle, the sun spends about twice the time in grand solar minima compared with grand solar maxima.[511]

Yearly Mean Sunspot Number

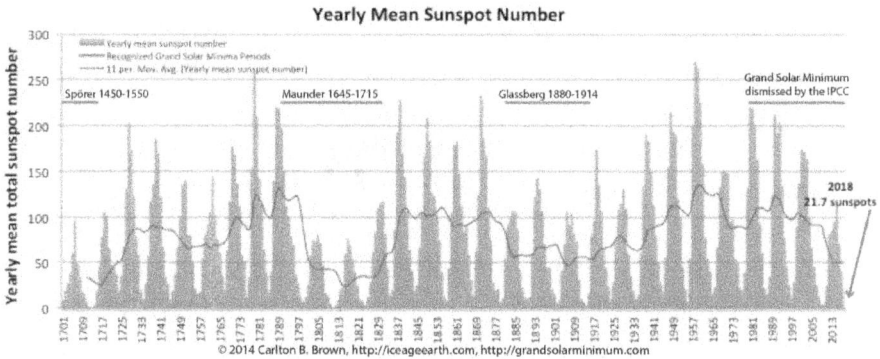

Figure 6.1. Yearly mean sunspot numbers covering Solar Cycles 1-24 between 1700 and 2018. This highlights an approximate 11-year solar cycle duration, and that the peak sunspot number for each 11-year solar cycle vary over longer-term cycles. The peaks and troughs of these longer-term solar cycles are referred to as *grand solar maxima* and *minima* respectively.[512] **Conclusion:** Solar Cycle 24 is progressing toward a grand solar minimum in terms of sunspot numbers. Sunspot numbers during the 11-year solar cycle have been in decline since the late 1980s.

Magnetized Solar Wind Controls Earth's Length of Day

The sun's surface represents a maelstrom of eruptive activity, the result of the sun's internal dynamo processes. We see this surface activity as solar flares, coronal mass ejections, and magnetic loops. This eruptive activity generates solar wind and magnetic fields that emanate from the sun's surface. The solar wind leaving the sun at varying high velocities is composed of streams of high-energy charged particles (i.e., electrons, protons), which are referred to as plasma.[513]

This high velocity solar plasma associates itself with the sun's magnetic field that bursts out from the sun's surface, and drags a small portion of this magnetic field out into the solar system. This magnetic field is then referred to as the interplanetary magnetic field. This surface eruptive activity occurs simultaneously with the sun's electromagnetic emissions, or its solar irradiance.[514,515] However, this magnetized solar wind is ignored or dismissed by the IPCC in its version of the affect of solar activity on the climate.

Magnetized solar wind interacts with earth's magnetosphere,[516] i.e., its cocooning magnetic shield. Solar wind produced during solar flares and coronal mass ejections causes geomagnetic storms, whose impact becomes concentrated at earth's polar regions. We see an aspect of this as the Northern and Southern Lights.

The sun's interplanetary magnetic field interaction with the magnetosphere changes the strength of earth's magnetic field (i.e., registered as geomagnetic activity), alters earth's rate of rotation (i.e., synonymous with the length of day), and modulates the penetration of ionizing cosmic rays into the atmosphere. This is explained below.

The solar wind's interaction with the magnetosphere impacts the ionosphere, or the electrically charged upper atmosphere, as well as the parts of the upper atmosphere comprised of the stratosphere and troposphere.[517] Solar wind is ultimately associated with the atmosphere through various complex coupling processes.[518,519,520]

Geomagnetic storms modify the strength of the stratospheric polar vortex, or a high altitude, low pressure region over both poles, which is ultimately coupled to the lower atmosphere. This stratospheric-atmospheric coupling leads to changes in the pressure distribution in the polar atmosphere and the movement of air mass. This pressure-induced movement of air mass changes atmospheric circulations and the pressure balance over tectonic plates.[521,522,523,524,525,526,527,528]

At a holistic level, the sun and its activity cycles, earth's rotation (and orbit), atmospheric and ocean circulations, and the sea temperature (i.e., acting as a heat sink and source) act together as a dynamic, integrated system. Strong solar wind creates a compressive effect around earth's magnetosphere, which is associated with a reduction in earth's rate of rotation, thus changing the length of day. This change in the length of day perturbs atmospheric and oceanic circulations, as well as earth core, mantle, and tectonic processes.[529,530]

Cyclical variations in the solar wind, driven by solar cycles, create changes in earth's rate of rotation. Significant correlations exist between sunspot numbers and earth's length of day. Variations in earth's length of day are known to occur on 11, 22, and 60-year cycles, as well as other periodicities. These same periodicities are also embedded in solar activity cycles. This confirms the connection between the sun and earth systems.[531,532,533,534,535,536,537]

Solar Activity Modifies the Climate System via Different Mechanisms

While solar irradiance declines in a smooth and gradual manner on the millennial timescale, ice age cooling trajectories can be variable, abrupt, and oscillatory (Figures 3.6 and 4.1). The climate science field still needs to explain the mechanism(s) for controlling centennial-scale climate oscillations, the variable rate of ice accumulation, and the rapid climate change events that occurred after the Holocene Climate Optimum. This chapter aims to highlight the evidence for the involvement of solar magnetism in the control of earth's climate over multiple timescales. In so doing, we can better understand the science of solar activity-related risks.

In reviewing the data contained in Figures 4.3, 4.4, and 6.2 (solar activity and Northern Hemisphere temperature), and in Figures 5.1–5.3 (solar activity and climate-forcing volcanism), as well as the science reviewed below, it becomes clear that the sun has multiple levers on the climate system. These *solar-climate levers* are: (1) millennial-scale climate modulation operating by solar irradiance, which is further modified by changes in earth's orbit (i.e., precession of the solstice); (2) magnetized solar wind driving multi-annual, decade- and centennial-scale climate oscillations (putatively via atmospheric and oceanic circulation systems); (3) grand solar maxima and minima, and long-term trends in solar activity associated with climate-forcing volcanism; (4) geomagnetism and its impact on the entry of cosmic rays into earth's atmosphere, and the modulation of low cloud formation by cosmic rays.

Solar Electromagnetism Drives the Climate System

In Chapter 2 it was mentioned how the Northern Hemisphere's summer temperature decline paralleled the decline in Northern Hemisphere summer insolation, or the amount of solar radiation reaching earth's surface. This millennial-scale decline in Northern Hemisphere summer insolation (irradiance) resulting from precession of the summer solstice,[538,539,540,541,542] *drives our inescapable—and non-delayable—ice age destiny.*

This precession of the summer solstice-mediated decline in summer insolation gradually reduces the level of electromagnetic energy entering earth's atmosphere and climate system. It is this longer-term level of solar irradiance, modified over the short- and medium-term by volcanic and anthropogenic aerosols, natural and anthropogenic greenhouse gases, air and ocean circulation systems, and clouds that creates our climate and weather.

The sun's total irradiance has changed by a paltry 0.1% since modern monitoring of solar activity began in 1978. However, over longer timescales, such as since the Maunder minimum (1645-1715 CE), or with shorter-term fluctuations over the 11-year solar cycle, solar irradiance has varied more significantly.[543]

This variation in solar irradiance could have been by as much as 2–4 Watts per square meter at the upper atmosphere during the Maunder minimum, as compared to today.[544] This solar cycle-related variance is the same as the global radiative forcing impact of atmospheric carbon dioxide concentration,[545] making it significant and not ignorable.

Solar Magnetism Drives the Climate System and Centennial- Scale Climate Oscillations

There are a number of publications cited in the scientific literature highlighting solar activity's relationship to climate. For example, the global temperature and sea surface temperature in different ocean basins have closely tracked sunspot numbers since the mid-19th century. This climate and sunspot relationship coincided with fluctuations in the Gleissberg cycle (80-year solar cycle). The Gleissberg cycle influence increased the magnitude of the sunspots in the 11-year solar cycle through the 20th century.[546]

Similarly, land and air temperatures in the Northern Hemisphere are also known to vary with the length of the eleven-year sunspot cycle.[547] The length of the sunspot cycle reflects the underlying cycles of solar activity, and the influence of the giant planets' gravitational and angular momentum on the sun's center of mass and on its dynamo processes.[548,549,550]

However, these two commonly cited examples of sunspot number and climate correlations *fail to offer* a proven physical means by which the sun connects with earth's climate system. Is the sun's climate system connection, inferred from sunspot numbers, mediated by an electromagnetic or a magnetic mechanism? Only assumptions can be made when attempting to answer this question.

The use of sunspot numbers (recorded on the sun's surface) contrasts with utilizing cosmogenic isotopes (Beryllium-10, Carbon-14) to explore the sun's connections to the climate system. This is because cosmogenic isotopes are recovered from earth repositories (i.e., ice cores and tree rings), and tell us about the physical means of the sun-climate system connection. Cosmogenic isotopes levels are modified by cosmic rays, and therefore indirectly by magnetized solar wind and geomagnetic activity (see below, and the following citation). Cosmogenic isotopes tell us that the means of the sun's connection with earth systems (i.e., Northern Hemisphere temperature, climate-forcing volcanism) is magnetism-based and not solar electromagnetism-based (i.e., for multi-decadal to centennial-scale climate oscillations).[551]

The data contained in Figures 4.3, 4.4, 5.1, 5.3, and 6.2 demonstrate how the sun influences the climate system on multi-annual, decadal and centennial timescales, through processes significantly correlated with magnetized solar wind. The involvement of magnetized solar wind in climate control is inferred from the 18-year moving average Beryllium-10 concentration anomaly ("Beryllium-10"). Beryllium-10 demonstrated a significant inverse correlation with the Northern Hemisphere temperature for the years between 1406 and 1985. Stronger Beryllium-10 and temperature correlations are also evident during grand solar minimum and maximum phases.

The involvement of magnetized solar wind in triggering climate-forcing volcanism is also inferred from the cosmogenic isotope carbon-14 data used to establish past sunspot numbers.[552,553] This carbon-14-derived sunspot number data correlates with climate-forcing volcanism (Figures 5.1-5.3). This climate-forcing volcanism is also involved in the previously mentioned glacier ice expansion mechanism.[554,555,556,557,558]

The Northern Hemisphere temperature closely tracks the Beryllium-10's variability, suggesting a true cause and effect relationship. The relationship between Beryllium-10 and the Northern Hemisphere temperature was strengthened during the Little Ice Age's grand solar minima (Figure 4.4.B), and the 20th century's grand solar maximum (Figure 6.2). These extremes in solar activity were associated with cooling and warming phases, respectively.

The data underpinning Figures 4.3, 4.4, and 6.2 indicate that temperature lags behind the changes in Beryllium-10. This lag is evidenced by the enhanced correlation which is observed between the 18-year moving average Beryllium-10 data and the Northern Hemisphere temperature, compared with the unmodified Beryllium-10 data and the 5-year and 11-year moving averages of the Beryllium-10 concentration anomaly. The use of the 18-year moving average is the equivalent of saying the temperature lags behind solar activity by about one 11-year solar cycle (see Figure 4.3 for an explanation).

When all four grand solar minimum phases and their associated temperature decline phases were synthesized into a new data series, one grand solar minimum after the other, the correlation value was increased (see Figure 4.4.B). The Beryllium-10 and temperature correlations during the Little Ice Age's four grand solar minima and the 20th century grand solar maximum (1889-1985) were similar in magnitude and significance.

Adding further support for a true cause and effect relationship is that the Northern Hemisphere temperature strongly mirrored the variability of the Beryllium-10, i.e., as one went up, the other went down. In other words, declines in temperature during a grand solar minimum always followed a decline in solar activity. Likewise, increases in temperature always followed a grand solar maximum and during times of increased solar activity (i.e., declining Beryllium-10).

This means that solar activity, mediated by a magnetized solar wind mechanism impacting Beryllium-10 levels in the atmosphere, leads both the trough-to-peak (i.e., 1711-1985) then peak-to-trough phases (Little Ice Age's grand solar minima) of the Northern Hemisphere temperature oscillations. This magnetized solar wind relationship is demonstrated on multi-annual, decadal and centennial timescales, indicating it is associated with centennial-scale climate oscillations.

The implication of the temperature lagging behind solar activity by one 11-year solar cycle is that an *indirect* solar-climate regulation mechanism is in play. The magnetized solar wind could be operating through atmospheric and ocean circulation systems (i.e., North Atlantic Oscillation, El Nino Southern Oscillation and others), which would explain the temperature lag. Magnetized solar wind's ability to modify atmospheric circulations via earth's length of day is reviewed in this chapter.

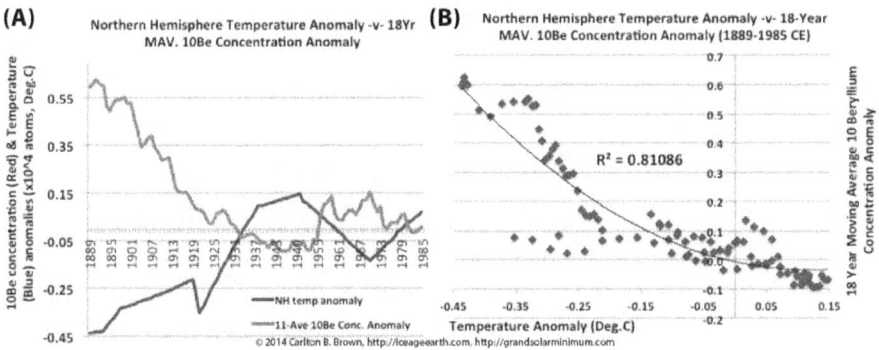

Figure 6.2. A) A time series of Greenland's ice core Beryllium-10 isotope concentration (Be^{10}, red line) and the annual Northern Hemisphere temperature (blue line) anomalies (1889-1985) highlight an inverse relationship. As solar activity increases, the Beryllium-10 isotope concentration anomaly declines and the temperature increases, due to the effects of magnetized solar wind. **B)** A scatter plot of Greenland's ice core Beryllium-10 isotope concentration anomaly and the annual Northern Hemisphere temperature anomaly (1889-1985) is presented. A statistically significant inverse correlation exists between the 18-year trailing average Beryllium-10 isotope concentration anomaly and the annual Northern Hemisphere temperature anomaly.[559] **Conclusion**: The Northern Hemisphere's rising temperature between 1889 and 1985 followed an increase in solar activity. This temperature rise coincided with the 20th century's grand solar maximum.

Earth's Length of Day
Modifies Atmospheric Circulation Systems

The shared periodicities between solar cycles and earth's length of day are also reflected in earth's geomagnetic activity, in various atmospheric and ocean circulations, as well as in the climate data. Shared periodicities in activity across different earth systems highlight the connection between the sun, earth systems, and the climate.[560,561,562,563,564,565,566]

When earth's rate of rotation changes, the modified atmospheric circulations are responsible for most of the short-term variability in the length of day, on an annual or seasonal time scale. On the other hand, over longer-term timescales it is earth's molten core and internal processes that are responsible for mediating most of the multi-decade scale and longer-term changes in the length of day. Glaciation and deglaciation cycles are also associated with earth's rotational acceleration and deceleration, indicating the sun's control over glacial-cycle climate change as well.[567,568,569]

Variations in Earth's climate are also closely linked with atmospheric pressure and ocean circulatory systems, which undergo changes in intensity and location in response to variations in solar activity.[570] The following paragraphs exemplify two of the atmospheric circulations that are important to the global climate and climate change.

The Little Ice Age's grand solar minima were associated with an increase in earth's rate of rotation, or a reduced length of day. This change in earth's rate of rotation had an important effect on the North Atlantic Oscillation. The North Atlantic Oscillation is an atmospheric pressure system driven by the movement of air mass across a pressure gradient between the Arctic and the subtropical Atlantic. Ultimately, this permits cold ocean currents from the Arctic to penetrate further south into the North Atlantic, while redirecting warmer waters into the Mediterranean.[571] The North Atlantic Oscillation is a key controller of winter climate variability over the North Atlantic, Western Europe, the eastern USA, Siberia, and other regions.[572,573,574,575,576,577]

Importantly, the North Atlantic Oscillation is coupled to the upper atmosphere (i.e., stratosphere and troposphere) by some complicated physical processes by means of the stratospheric polar vortex.[578,579] The North Atlantic Oscillation's phase and strength is correlated with solar wind-driven geomagnetic activity.[580,581,582] This confirms the North Atlantic Oscillation's connection to solar activity.

This coupling process in turn explains why the North Atlantic Oscillation is correlated with earth's length of day, global sea surface temperature, and the Aurora Borealis (Northern Hemisphere aurora). All of these earth system responses display a dominant 60-year cycle in their activity.[583] The North Atlantic Oscillation highlights a strong inverse relationship with sunspot numbers since 1880.[584]

Likewise, the El Nino Southern Oscillation (ENSO) results from interactions between the ocean surface and the atmosphere over the tropical Pacific, leading to periodic changes in weather patterns and ocean temperatures and currents. ENSO variations also impact weather systems around the world.[585] The ENSO index of activity has also demonstrated a strong inverse relationship with sunspot numbers since 1880.[586]

Geomagnetism and Cosmic Rays Impact the Climate

Significant correlations also exist between global and regional climates and the strength of earth's geomagnetic field, over both the IPCC's climate reference period (i.e., since 1880) and over much longer periods of time.[587,588,589,590,591,592] How geomagnetism modulates the climate is not fully understood today.

However, what the climate science field understands is that an important relationship exists between cosmic ray intensity, geomagnetism, and low-level clouds. More cosmic rays enter the atmosphere from space when the sun is less active (i.e., a reduced intensity of magnetized solar wind) and with a reduced level of geomagnetic activity. This leads to an increased formation of low-level clouds, which is associated with planetary cooling.[593,594,595,596,597]

The strongest cosmic ray and climate correlations are geography-dependent. The strongest correlations are observed over the North Atlantic, Europe, the Far East, and the Antarctic region.[598] Over the 11-year solar cycle (from peak to trough), the intensity of cosmic rays in the upper atmosphere can vary by as much as 15 percent (globally averaged). Moreover, cosmic ray levels can be 50 percent higher at the poles compared with the equator, thus concentrating their cooling impact in the polar regions and during the troughs of solar activity. These cosmic ray variances can significantly alter the levels of incoming solar irradiance entering earth's climate system. The climate impact of variance in cosmic rays is similar, in magnitude, to a doubling of anthropogenic carbon dioxide emissions from current levels.[599]

Earth's Length of Day Impacts Volcanic and Earthquake Activity

What proof do we have that solar activity impacts volcanic activity?

Knowing the answer to this question can give us a better understanding of how solar activity can trigger climate-forcing volcanic eruptions. In this way, we can better comprehend how the sun's activity and volcanism act in concert to accelerate cooling and ice accumulation processes, and better understand *the risks inherent in the extremes of solar activity,* such as with a grand solar minima.

Earthquakes are explored here because tectonic and volcanic processes are linked, and both are geographically associated with tectonic plate boundaries. There is also a short delay between maximum volcanic activity and maximum earthquake activity, with large earthquakes portending large volcanic eruptions, further supporting the linkage between these processes.[600]

As a general rule and summary, solar maxima or minima portend increased volcanic and earthquake risks,[601,602,603] depending on which fault line you live near. That is to say, different continental plate margins react differently to extremes of solar activity, so the outcome will vary depending on where you live.[604]

Correlations exist between solar activity, volcanism, and earthquakes, indicating that these solar and terrestrial systems are connected.[605] This is not a simple relationship, because the trends in volcanic and earthquake activity depend upon which continental plate this activity is located in, and on the location of events within compression or tension zones of those plates.[606,607]

Scientists researching different types of volcanic explosions and large magnitude earthquakes have identified 11- and 22-year periodicities embedded in the volcanic, earthquake, and solar activity data. These shared periodicities are indicative of the link between solar activity and earth's endogenous processes.[608] These are the same periodicities embedded in earth's length of day variations, atmospheric and ocean circulations, geomagnetic activity, and the climate data (cited above).

Over the last 1,500 years the frequency of volcanic eruptions in Greenland was greater around the 11-year solar minima, with eighty-year periodicities also demonstrated.[609] These cycles correspond with the 11-year solar cycle, and with the Gleissberg cycle.[610] Similarly, in Japan 9 of 11 large magnitude volcanic eruptions during the last three centuries occurred during solar minima.[611]

Since the late 19th century, the frequency of volcanic activity and larva volume has been positively correlated with earth's geomagnetic activity, which in turn is modulated by solar activity.[612] In many of the main earthquake regions around the world, seismic activity is also known to change with variations in geomagnetic activity, both annually and on a daily basis. Thus the occurrence of earthquakes is concentrated around midday and midnight.[613]

Maximum earthquake frequency is also known to occur during solar activity maxima, with high correlations having been demonstrated.[614],[615],[616],[617] Maximum seismic energy is released in earthquakes that occur a few years after the sunspot maxima.[618,619,620]

Changes in earth's rate of rotation are also associated with large earthquakes. Significant correlations have been established by scientists linking length of day with large magnitude earthquakes.[621,622,623,624] Deceleration alone,[625] or both acceleration and deceleration are known to result in an increase in earthquake frequency.[626]

Earthquake event density and energy release are broadly symmetrical around the equator.[627] This symmetry of earthquake events around the equator and the energy released supports a hypothesis that earth's rotational dynamics help modulate tectonic processes (i.e., earthquakes, volcanism). This is because the equatorial regions are subject to greater radial accelerative-decelerative forces than occur at the poles.[628]

The annual energy budgets associated with variations in earth's accelerations and decelerations are an order of magnitude greater than the annualized energy released from earthquakes,[629] thus providing the energy required to disturb tectonic plate boundaries. Likewise, variations in continental plate dimensions and distances between their abutting boundaries can change by at least half a meter through bulging and boundary compression and tension effects during earth's acceleration and deceleration.[630]

Moreover, scientists have also linked earthquakes with variations in high-speed solar wind and the large-scale movement of air masses across perturbed atmospheric pressure gradients and tectonic plates. The science indicates that the annualized energy associated with these atmospheric disturbances is at least three orders of magnitude greater than the annualized energy released by earthquakes.[631,632,633]

CHAPTER 7

The Catastrophic Risks Hidden
by the IPCC's Articles 1 and 2

This chapter describes the catastrophic climate risks hidden by Articles 1 and 2. The climate-specific risks linked to this current grand solar minimum include a long overdue switch back to a global cooling and drought phase, and climate-forcing volcanism with its associated glacier ice expansion. These risks, when they manifest, will place us face-to-face with the brunt of the ice age we entered 8–10.5 millennia ago.

The world is without a risk mitigation plan for anything other than anthropogenic global warming. Under the IPCC's global warming scenarios, the risks vary according to the degree of anthropogenic greenhouse gas emissions.[634,635] The IPCC climate risk assessment ignored numerous eminent climate experts who specialize in solar activity, and who have warned of a return to Little Ice Age-like conditions during this grand solar minimum.[636,637,638,639,640,641,642]

Surprisingly, and almost shockingly, the IPCC's climate risk assessment ignored or dismissed the climate risks associated with the current grand solar minimum, and with climate-forcing volcanism.[643] The prospect of a rapid climate change event was also effectively dismissed.[644,645] The IPCC's assessment reports provide no review of the climate-related catastrophes that occurred during the Little Ice Age,[646,647,648,649,650] including the catastrophic impact of climate-forcing volcanism.[651,652,653,654,655]

The risks the world needs to be planning for (Sections 3 and 4) include a switch to a global cooling phase and its associated extremes of precipitation, i.e., drought, rainfall, and snow (Chapters 3–5, 7). We now live in an era of enhanced risk for large magnitude volcanic eruptions (Chapter 5–6), increased earthquakes (Chapter 6), and pandemic flu (Chapter 14). Centennial-scale Arctic ice accumulation and associated rapid climate change cannot be dismissed either (Chapter 5).[656,657,658,659,660]

Lest we forget, no matter what human activity-related perturbations exist in the earth and climate systems, they are dwarfed by natural phenomena. These natural phenomena proceed unabated, irrespective of Articles 1 and 2, and by their own natural clocks that abide by the laws of the universe.

The IPCC Dismissed or Ignored Volcanism, Solar Activity, Rapid Climate Change, and the Little Ice Age in Its Risk Assessments

The purpose of this section is to help you understand that major climate risks have been ignored, downplayed, or dismissed by the IPCC in the climate assessment reports it has provided to governments since 1990. The citations provided below show you exactly where to look in the IPCC's 1,000–1,500 page documents, so you may see first-hand exactly what they ignored, downplayed, or dismissed. Given the fact that governments have almost unanimously endorsed the IPCC's climate change science and risk assessments, there are consequences to these omissions—omissions that have left *our global society blind to the full scope of climate and solar activity-related risks.*

The IPCC tell us that they have only assessed risks that represent severe impacts relevant to Article 2 of the United Nations Framework Convention on Climate Change.[661] Article 2 concentrates the IPCC's focus on dangerous human interference with the climate system. That is to say, the IPCC ignores or dismisses the climate risks attributable to nature. Article 2 has among its objectives the stabilization of greenhouse gases, while ensuring global food production and sustainable economic development.[662]

The IPCC predicts, with a high confidence, flooding due to sea level rise and increased storm surges. We are told that flooding, drought, and rain extremes will increase the risk of food insecurity, which will vary geographically. Drought is expected to impact water availability, with consequences on food production, water security, and on the survival of populations living in drought areas.[663]

We are told that the displacement of people will increase, as will the risk of violent conflicts.[664] The IPCC informs us of the likely loss of, or damage to, coastal ecosystems due to sea level rise and ocean acidification, which will lead to species extinctions and reduced biodiversity.[665]

The IPCC also tells us that the global risks will vary depending on the level of greenhouse gas emissions reached. Higher emissions are expected to exacerbate all risks identified, especially species extinctions, global food security, and normal human activities. We are told that risks will be substantially reduced under a low emission scenario,[666] which is supposed to help motivate us to reduce our greenhouse gas emissions.

The IPCC's recent assessment reports (in 2014[667,668,669,670] and 2012[671]) were reviewed for their assessment of the risk of rapid or abrupt climate change. The IPCC's review of abrupt or rapid climate change was restricted to a special kind of rapid climate change event, referred to as Dansgaard-Oeschger events that occur in the depths of an ice age after the northern ice caps have formed.[672]

The IPCC's climate risk assessment did not detail the civilization-destroying and population-culling rapid climate change events known to have occurred after the Holocene Climate Optimum. These included the 8.2, 5.9, and 4.2 kiloyear rapid climate change events, the Little Ice Age, and the numerous climate-forcing volcanic eruptions that took place during the Current Era, all of which are highlighted below.

The IPCC's dismissive assessment of Little Ice Age climate history and its associated volcanism means we have not had our "upright attention" drawn to the catastrophic impact of climate change-induced cold, drought, and famine on the populations of China, Europe, Mesoamerica, etc. at that time. During these climate-related crises and catastrophes up to half the population in the affected areas died, and this happened on more than one occasion during the Little Ice Age (see below).[673,674,675,676]

This failure to review rapid climate change risks occurred despite numerous eminent solar scientists expert in climate change warning of a return to Little Ice Age-like conditions during this grand solar minimum.

The IPCC cautioned the reader in concluding that social unrest, political instability, and warfare during the Little Ice Age could occur today based on historical precedent. This caution was justified by the IPCC because the precise causal pathways linking climate change to these societal changes in historical societies are not well understood (see the citation note on *confirmation bias*).[677]

Surprisingly, in the IPCC's 2012 special report titled, "Managing the Risks of Extreme Events and Disasters to Advance Climate Change Adaptation," we are informed the IPCC did not assess the existence of climate system tipping points, and the risks of abrupt and irreversible changes to the climate system.[678] Two years later, in their fifth assessment report, they cited a *low confidence in the science that an abrupt or rapid climate change will occur during the 21st century*,[679] effectively dismissing the possibility.

One very important point to understand is that while the IPCC recognizes that volcanic activity can have a dramatic impact on the global climate (i.e., cooling), their climate forecasts do not include the occurrence of climate-forcing volcanic eruptions.[680] This is in spite of Rinjani and other climate-forcing volcanism triggering and propagating glacier ice accumulation. (see Chapters 5 and 7).

A key word search of the 2012–2014 IPCC assessment reports confirms that no risk assessment of climate-forcing volcanic eruptions or super volcanism was provided, or any review provided of their potentially devastating impact on the climate system and on global food security (Chapter 5 and below).

The role of the sun (solar irradiance), sunspots, magnetized solar wind (i.e., solar flares, coronal mass ejections) and their connection with earth systems (Chapter 6), was dismissed, downplayed, or ignored by the IPCC's assessment report (released in 2014).[681] The term grand solar minimum (or minima) is mentioned only three times in the 1990–1992 and 2012–2014 assessment reports; it is referred to twice without description or explanation, and once in a citation.

The purpose, therefore, of the remainder of this chapter is to draw your attention to the wealth of scientific literature relating to the solar activity, climate, and volcanic related-risks, all of which have been ignored, downplayed, or dismissed by the IPCC. Examine the cited science to see what is being overlooked.

The Little Ice Age and Rapid Climate Change Events

The strongest impact of cooling during the Little Ice Age occurred in the North Atlantic and the northern latitudes of Europe, Asia, and North America. There was either more drought or more rainfall at the lower latitudes, depending on where you lived.

In parts of Europe the Little Ice Age caused more and bigger floods, which coincided with cool periods and lows in solar activity.[682] This contrasted with Spain, where the Little Ice Age was associated with less rainfall, which in turn was directly correlated with low solar activity.[683] In the European Alps colder climates, glacier expansion, and increased levels of snow, rainfall, and flooding were experienced.[684,685,686,687,688]

Prolonged and severe droughts were experienced on the northern Great Plains of what is now the USA, and were in phase with low solar activity.[689] In the northeastern USA, on the other hand, solar minima were associated with wet conditions.[690]

In China the Little Ice Age climate was greatly impacted by the various monsoons moving over the region, which were associated with solar activity.[691,692,693],[694] Depending on where you lived, either colder winters and more rainfall, or drought, were experienced. These climate extremes were concentrated in three main periods—the Spörer, Maunder, and Dalton minima.[695,696]

Across India the Little Ice Age brought a cooler climate, reduced monsoon intensity, and prolonged, severe droughts, with these being more intense during the Spörer and Maunder grand solar minima.[697,698,699,700] Africa also experienced either more drought or more rainfall, depending on the region.[701,702,703,704]

In South America glaciers expanded and there was increased precipitation, evidenced by higher river levels.[705,706],[707,708] A cooler and more drought-prone climate was also experienced in the Caribbean[709,710,711] and on the Yucatan Peninsula.[712]

It's obvious from the published science that solar activity imprinted itself in the paleoclimate record (i.e., historical climate data before 1880), especially during grand solar minima. The climate impact varied by geographical location and the climate system involved.

If the past is representative of the future, then we should expect something similar during this grand solar minimum to what was experienced during the Little Ice Age's grand solar minima, rather than an all-encompassing global warming. This assumes that nature will continue to do as it has done for billions of years and "control and regulate" the climate system via the sun's activity (electromagnetism and magnetism), climate-forcing volcanism, atmospheric and ocean circulations, and through the influence of geomagnetic activity on cosmic rays and low cloud formation.

If you are one of those people who think human activity can cancel out natural forces to the extent that we may avoid a Little Ice Age-like fate, reflect on the following. One large climate-forcing volcanic eruption will rapidly cancel any climate modifying anthropogenic influence, and remind us that natural forces control the climate. Likewise, running out of oil and gas within the coming decades will reduce our ability to modify the natural climate system.

The Little Ice Age Caused Death and Revolutionary Upheavals

Researchers have explored the quantitative aspects of climate change, famine, disease, social unrest, and wars during the Little Ice Age, and particularly in Europe and China. Interestingly, some of the worst human catastrophes of the Little Ice Age took place during grand solar minima.

The big lesson, globally, from the Little Ice Age is that cold phases and drought adversely impacted agricultural production yields, leading to rising food prices, malnutrition, and famines. These in turn, along with other important contributing factors, led to epidemics, social unrest, revolutions, and wars.[713]

In China, most of the peaks in warfare and dynastic transitions took place during cold periods.[714] Likewise, there was a much higher chance of an epidemic occurring during cold periods.[715] For emphasis, and lest we forget, *China lost nearly half of its population* through successive climate-related famines, epidemics, and wars during the 13th and 14th centuries.[716]

Europe was no different. The Great Famine of 1315–1317 was the worst prolonged famine in Europe's history, leading to large-scale mortality (10 percent).[717] Three decades later Europe was ravaged once again, *losing one-third of its population* due to the Black Death, social unrest, and economic hardship.[718,719]

These high mortality events in China and Europe took place during the Wolf minimum (1280-1350), and in the decades after the current era's largest volcanic explosion (at Rinjani, Indonesia in 1257, detailed below).

Fast-forwarding to the longest and coldest phase of the 17th century, Europe's population suffered from high mortality rates, while China lost nearly half its population once again.[720,721] In both cases humans died en masse as a result of famines, disease, and wars. These catastrophes occurred during the Maunder minimum, the deepest and longest solar activity trough of the Little Ice Age.

In Europe a century of successive catastrophes helped precipitate the so-called General Crisis of the 17th Century.[722] During this time, there was a marked decline in harvest yields, and dramatic increases in food prices and famine. Social unrest, population migration (to the New World), and war fatalities dramatically increased.[723]

It wasn't just the northern latitudes that were impacted. The 16th century mega-drought in Central America (1540–1580) coincided with the Spanish conquest and the religious conversion of the Mesoamerican population, and two major epidemics that killed in total between 14 and 17 million people.[724] This horrendous human death toll occurred toward the end of the Spörer minimum (1450–1550).

What staggers me is the sheer scale of human death during times of climate crisis associated with grand solar minima. *Agricultural food production was the critical bottleneck.* However, a review of the episodes of large-scale human catastrophe (i.e., 10 to 50 percent mortality in China, Europe, Mesoamerica, etc.) during the Little Ice Age was not provided in the IPCC's assessment reports.

We need to be reminded that no matter how much technological progress our modern society has made, crops and livestock are still vulnerable to severe changes in the weather. As an example, between late 2006 and 2008 regional crop failures occurred among major food exporters due to drought, and this snowballed into a global food crisis severely impacting many nations.[725,726,727,728,729] It demonstrated how fragile our modern agricultural system and global food supply are (see Chapter 11).

One large magnitude volcanic explosion is predicted to wipe out one year's food supply for one-third of the world—how do you think the world would cope with that?[730] Not as well as it would if it were *forewarned and given time to prepare.*

Rapid Climate Change Events
Terminated Ancient Civilizations

Since the Holocene Climate Optimum the North Atlantic region's climate has experienced a series of rapid cooling events. These took place on multi-decade to centennial time-scales, and each had sufficient global impact to reshape the trajectories of ancient human civilizations.

Three of these rapid cooling events are most frequently cited in the literature, including the 8.2 kiloyear (8,200 years),[731,732,733,734] the 5.9 kiloyear,[735] and the 4.2 kiloyear events before the present[736,737,738739,740,741,742743,744,745,746,747,748,749,750] (multiple citations review the varying geographical impact).

As a general feature, these rapid climate change events were characterized by abrupt (i.e., intra-annual, annual to decadal) and sustained (i.e., over centuries) Arctic and Antarctic coolings. Arctic temperatures declined between 3^0C and 6^0C and there were significant ice sheet expansions. This was associated with a more general global cooling. Aridification (droughts, desertification) also intensified across Northern Africa, the Middle East, parts of Asia, and the tropics.[751,752,753]

These rapid climate change events triggered major famines and decimated ancient human civilizations. The 4,200-year event is well studied and was associated with the collapse of Egypt's Old

Kingdom,[754,755] Mesopotamia's Akkadian Empire,[756] and the Indus Valley (or Harappan) Culture.[757]

In figure 4.2 you can view Greenland's ice core climate data at the time of this widespread collapse of ancient civilizations 4,200 years ago. A deep temperature trough preceded this event, and the temperature then rose from trough-to-peak by 2.44^0C. This was the Holocene's second biggest trough-to-peak outlier in terms of temperature rise. This high temperature peak then abruptly switched to a cooling phase, and civilizations collapsed. This 4.2-kiloyear rapid climate change event was one of the examples of centennial-scale climate oscillation contained within the Group 1 list of outliers detailed in Figure 4.2. Of all the large-scale climate fluctuations since the Holocene Climate Optimum (as revealed by the Greenland ice core data), this 4.2-kiloyear centennial-scale climate oscillation is the one most similar to the 1700-2016 trough-to-peak warming phase.

The Catastrophic Impact of Large Magnitude Volcanic Eruptions

To give you some perspective beyond thinking that it will just get a little bit colder after one of these large magnitude volcanic eruptions (VEI 6 or 7), we must understand that the impact of such events can be far-reaching. At the local level, these large magnitude eruptions can cause devastation, famine, epidemics, and social unrest. At the regional and global levels, these eruptive events can be catastrophic as a result of their effect on the climate system, causing widespread famine on annual to multi-decadal time scales.

The Tambora eruption (VEI 7), which occurred in 1815 during the Dalton minimum, caused disastrous crop failures across the Northern Hemisphere the following year. The year 1816 was dubbed the "year without a summer" (snow fell in June in New England). This led to widespread famine, social unrest, and caused a major human death toll in Europe, Asia, and North America.[758,759,760,761]

The Rinjani eruption (VEI 7) in Indonesia in 1257 took place within one decade of a grand solar maximum immediately preceding the Wolf minimum (1280 CE, see Figure 5.2.A). The Rinjani eruption is noteworthy because it is considered the largest volcanic eruption of the current era, producing a sulphate signature more than twice that of Tambora. Some scientists believe the Rinjani eruption triggered the Little Ice Age.[762] This eruption was associated with climate disruptions in Europe and Eurasia in the ensuing years, causing cold winters and summers, and severe flooding. This led in turn to grain shortages, food price inflation, and famines associated with high mortality over the ensuing years and decades.[763,764,765,766]

The pyroclastic flows (i.e., fast moving streams of hot gas and volcanic matter) from VEI 7 eruptions can travel up to 100 km in any or all directions from the volcano, with few living things surviving these fast moving, superhot flows.[767] Those areas close to the explosion can be covered in pumice and ash tens to hundreds of centimeters thick.[768]

Pyroclastic flows and lahars (i.e., mudflows) are the biggest cause of local deaths and injuries at the time of the eruption. More than 90 percent of fatalities occur within a 30-kilometer radius of an eruption. Famine and epidemics have historically been the biggest single categories of localized human fatality in the aftermath of an eruption.[769,770,771]

The global relevance of such large magnitude eruptions (VEI 7) is their disastrous impact on the world's climate, which in turn impacts agriculture through cold, drought, and flooding. In fact, scientists have modeled the devastating impact of crop loss from some of the world's worst volcanic eruptions over the last millennia, in order to simulate what would happen if such an eruption were to occur today.

The *hard-hitting headline* is if a Laki-like volcanic eruption (Iceland, 1783) happened today, the global crop production losses would be approximately *one year's worth of food for one-third of the world's population.* Smaller-scale eruptions would also cause a major global food crisis.[772]

Are we ready for such an event? How do we prepare? These are among the topics discussed in Chapters 11 and 12.

We should not forget that colossal and globally devastating super-volcanic eruptions (VEI 8) do occur, albeit infrequently. The last VEI 8 eruptions occurred 26,500 years ago (Taupo, New Zealand),[773] 72,000 years ago (Toba, Indonesia),[774] and 84,000 years ago (Atitlan, Guatemala),[775] making an eruption of this magnitude a very rare event.

To provide some perspective, the almost unimaginable Toba erup-tion created a caldera that measured 100 by 30 kilometers, with a depth of half a kilometer, and released 2,800 cubic kilometers of ash and pyroclastic material.[776] At least 800 cubic kilometers of volcanic material was violently exploded out across the Indian Ocean and Southeast Asia, covering millions of square kilometers of the planet's surface in volcanic debris.[777] In fact, Toba is believed to have covered the entire Indian subcontinent in 15cm of ash. For context, one cen-timeter of ash would be sufficient to disrupt regional agriculture and cause widespread famine.[778]

Importantly, Toba occurred during the last ice age and coincided with a transition period associated with rapid ice growth and a falling sea level.[779] This helped to shift the world back into ice age conditions which lasted for 1,000 years.

Grand Solar Minima Portend Large Magnitude Volcanic Eruptions

What are the chances of a climate-disruptive, large magnitude vol-canic eruption (VEI 7), or cluster of VEI 6 events, occurring during this grand solar minimum or in the decades ahead?

Based on evidence from the last four grand solar minima, and if this current grand solar minimum replicates the past, then we can ex-pect two VEI 6 eruptions, a cluster of VEI 4 or 5 eruptions, and a one in four chance that a big VEI 7 eruption will occur (see Figure 5.2).[780]

According to data extracted from the Volcano Global Risk Identification and Analysis Project database (VOGRIPA) hosted by the British Geological Survey, there were 67 VEI 6 and seven VEI 7 large magnitude explosive volcanic eruptions during the last 11,400 years.[781] This equates to, on average, one VEI 7 eruption every 1,600 years, and one VEI 6 eruption every 190 years. Tambora in 1815 CE was the last VEI 7 eruption, while Pinatubo in 1991 was the last VEI 6 eruption.

However, the large-magnitude eruptions (VEI 4-7) in the VOGRIPA database were skewed in their distribution to the current era,[782] and more so since Rinjani erupted in 1257. This skewing makes it hard to know whether we are in an accelerative phase of volcanism compared with the last 11,000 years, or if recent eruptions are simply more likely to have been recorded.

What is a fact, however, is that three of the seven Holocene VEI 7 eruptions, and just under one-quarter of the VEI 6 eruptions from the last 11,400 years have taken place since 942 CE. Theoretically, this provides forecasters of volcanism an absolute number of VEI 4–7 eruptions from which to determine the recent rate of eruptions (i.e., for the last 1,000 years). The rates over the last millennia are more relevant to today and likely reflect a truer frequency of volcanism (for forecasting) than using longer-range frequency data. In other words, the frequency of climate-forcing volcanism during the Little Ice Age is highly relevant for forecasting climate-forcing volcanic activity during the 21st century. When you reflect on the fact that the majority of the Holocene's and Little Ice Age's large-magnitude eruptions occurred at grand solar minima and maxima (see Figure 5.2.A), then forecasters are given a means of improving their climate-forcing volcanic eruption predictions (linked to predictable solar activity cycles).

The above data and that detailed in Figures 5.1, 5.2, and 5.3 can only cause us to conclude that this *grand solar minimum represents a high-risk period that should not be ignored.*

We must also keep in mind the backdrop to this enhanced risk of a VEI 6 or 7 eruption. First, all trough-to-peak phases of centennial-scale climate oscillations change phase to a peak-to-trough cooling phase—*always*. We are in an ice age, at the top of the biggest temperature peak since the Holocene Climate Optimum, two centuries overdue for a relatively abrupt fall in the Northern Hemisphere's temperature (viz., Group 1 outliers highlighted in Chapter 4). The current global and the Antarctic ice core temperatures are also between 1.3 and 3.0°C higher than the average of *all* glacial cycles over the past 2 million years (global) and 800,000 years (Antarctic). If earth systems tend to revert to the mean, then we could be in for a shockingly cold surprise.

Second, if this 500-year low solar activity-linked trend of stratospheric belching eruptions continues, combined with a VEI 6 or 7 eruption triggered by a grand solar minimum, then earth could find itself *facing an accelerated ice age re-entry.*

In my view, simple probability makes the need to prepare for large magnitude volcanic eruptions and a planetary cooling highly pertinent *right now.*

Scientists believe that if a VEI 7 eruption were to occur today, such an event would likely come from volcanoes located in tectonic plate subduction zone.[783] The *majority of the world's volcanic eruption risk* is *concentrated in just six countries* (Indonesia, Philippines, Japan, Mexico, Ethiopia and Guatemala).[784]

Key Themes

Managing the Risks

The Need For

Decentralized Sustainable Development

and

A Rapid Switch of the World's Energy System

CHAPTER 8

Climate Change Risk Mitigation with Decentralized Sustainable Development

Do the concept of human-induced climate change and the need to keep the global temperature below a 2^0C rise over pre-industrial levels motivate you to mitigate your climate impact, or to live sustainably?

Do you feel a sense of urgency to act on your answer to the above question?

This section and this chapter are about *defining solutions to a different climate future and set of risks than those envisaged* by the IPCC (Chapter 7).

We face three major problems in dealing with climate-related risks. The primary problem facing us is that a switch of the climate will happen in the coming decade(s), and we face the prospect of a host of climate related risks en-route (Chapter 3–7). Second, exacerbating our vulnerability to this primary climate problem is the fact that population and economic growth are accelerating the depletion of our finite energy and water resources. Third, a minority of nations controls access to the world's energy resources.

Human activity has already depleted two-thirds of the world's largest aquifers beyond sustainable limits, and increased the periods of time when there is insufficient water to meet human needs available in water basins and major transboundary river systems.[785,786] This water basin depletion eliminates our ability to buffer extreme and prolonged drought, and this will impact our food supply during a severe drought.

World population and economic growth is accelerating the depletion of world energy reserves, such that we only have 50 years of proven oil and gas reserves left.[787,788,789] Compounding this is the fact that 50 percent of the world's oil reserves and 70 percent of the gas reserves are in fact *unproven guesstimates.* This includes shale resources.[790,791] The prospect for new oil and gas discoveries is not as good as the promises made by the oil and gas industry and by governments.[792],[793,794795,796] Exacerbating the situation of dwindling energy reserves is that the ownership of these energy resources is concentrated in a minority of nations.[797]

The above represents a catastrophe waiting to happen, and for which we don't have a plan that anticipates a climate switch or the climate-related risks.

Rapidly *switching the world's energy system* to renewable energy and implementing *decentralized sustainable development* are pivotal strategies for mitigating 21st century climate and resource supply risks—and addressing the three interrelated problems outlined above.

Such a dual strategy will permit a greater degree of self-sufficiency at home and in urban areas, with the aim of reducing our vulnerability during a climate switch and to climate-related risks. A focus on decentralization means that individuals, communities, and cities embrace partial self-sufficiency for securing energy, water, and food. The need for decentralization anticipates that government, corporate, and commodity market actions will restrict resource supply in a crisis event, and so aims to reduce our vulnerability to this risk.

A Minority of Nations Control the World's Oil and Gas Reserves

Ownership of the world's fossil fuel resources is concentrated among a few nations, and state-owned oil and gas companies control most of these resources.[798] The ownership of these energy reserves is detailed in Figure 8.1.A. Thirty-three nations control more than 83 percent of the world's energy reserves. The world's two most populous nations (China and India) hold a further 13.6 percent of global energy reserves, mostly as highly pollutive coal. This means the majority of nations (160+) hold less than 4 percent of fossil fuel reserves.

The world is dependent upon OPEC nations, Canada, Venezuela, Australia, and a small group of exporting nations from among the other major holders of reserves for much of its oil, gas, and coal needs.

For example, just 20 nations supply 90-plus percent of the world's oil and gas exports, while just five nations supply more than half of those exports.[799] Collectively this means that a minority of nations holds the majority of energy reserves, and an even smaller minority exports the lion's share of the world's fossil fuel supplies. This dependence on a minority of exporters will create vulnerabilities for fuel importers when a crisis arises.

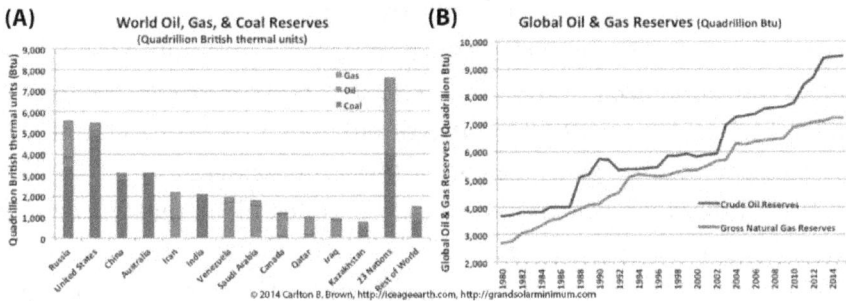

Figure 8.1. A) Twelve nations hold three-quarters of the world's fossil fuel energy reserves. Another group of 23 nations holds a further one-fifth of the world's energy resources. This means the majority of nations (160+) hold only 4 percent of fossil fuel reserves. **B)** The reporting of fossil fuel reserves without backdating volume revisions to the original wildcat discovery creates a straight-line in cumulative reserve growth, similar to that depicted in this figure.[800] This lack of backdating revisions means we miss the plateau that is occurring in the reserve growth rate.[801,802]

Global Fossil Fuel Reserve Revisions Mean 50–70 Percent Are Unproven Guesstimates

The immutable laws of physics limit the maximum level of technical recoverability from a reservoir or field, to less than 50 percent for oil and 80 percent for gas.[803],[804] Once peak production has been reached, or approximately half a reservoir's technically recoverable oil resources have been extracted, then production declines in a well understood manner.[805,806]

Oil recovery happens in phases, and costs increase as technology is deployed. Primary oil recovery extracts the initial 10–30 percent, with secondary recovery methods permitting extraction levels of up to 30–50 percent. Most wells have already used secondary recovery methods for extraction (i.e., gas and water injection), but fewer have employed tertiary or enhanced oil recovery methods, due to complexity and cost.[807,808]

Technically recoverable resources represent the volumes of oil and gas that can be extracted with current technology and know-how. Proven reserves are a smaller subset of technically recoverable reserves with a greater than 90 percent certainty of being extracted. Unproven reserves are technically recoverable but are assigned either a 50 percent (probable) or 10 percent (possible) prospect of recovery, to reflect risk and uncertainty.[809]

From the time production begins, reserve growth typically increases between 1.3- and 4.6-fold over the first 25 years.[810] These recoverable volume increases come from new well additions within a reservoir, increases in the field area, and new reservoir discoveries within the field, as well as the use of enhanced recovery techniques. Revisions to reserve estimates are a frequent occurrence.[811,812,813,814,815]

With knowledge of the above, it becomes clear that *all reserve estimates are subject to uncertainty.* Governments, which own the majority of reserves, can have a significant influence on market perceptions about global reserves, while not being subject, as oil and gas companies are, to stock exchange reporting rules.[816]

One such example of government influence on the market perception of reserve abundance occurred in 2013, when the US Energy Information Agency team increased its technically recoverable global reserve forecasts significantly by 11 percent and 47 percent for oil and gas resources, respectively, only two years after their 2011 forecast of reserves. These significant upward revisions reflected changes to the Energy Information Agency predictions of world unconventional energy resources i.e., shale oil and gas.

These revisions mean that one-third of the Energy Information Agency's world gas and one-tenth of its world oil resource projections are for shale resources. These revisions also mean that *between 50 percent and 70 percent of total conventional and unconventional oil and gas projections are classified as unproven (i.e., guesstimates).*[817] With such a high percentage being unproven, caution is required concerning the Energy Information Agency's optimism.

It is also important to understand the forecasting methodologies and assumptions that are used. The 2013 revisions concerning oil and gas reserves were *based on predictions* involving the application of historic US shale oil and gas recovery rates to foreign petroliferous basins with similar geophysical characteristics. These revisions assumed the same operating context internationally as in the USA.[818] However, using that same operating context elsewhere is a questionable assumption, because the US is not the same as the rest of the world. *Environmentally controversial fracking methods* will be used for the extraction of these unconventional reserves, which adds another level of risk and calls into question the Energy Information Agency's optimism.

The US Environmental Protection Agency (EPA) is unable to assure the public that fracking is not harmful to underground drinking water. The EPA found scientific evidence that hydraulic fracturing activities can affect drinking water resources under some circumstances, while at the same time informing us of their *safety study's significant shortcomings.* The EPA tells us on their website, *"Data gaps and uncertainties limited EPA's ability to fully assess the potential impacts on drinking water resources locally and nationally. Because of these data gaps and uncertainties, it was not possible to fully characterize the severity of impacts, nor was it possible to calculate or estimate the national frequency of impacts on drinking water resources from activities in the hydraulic fracturing water cycle."*[819,820] In addition, enhanced earthquake risk from fracking is well known to science.[821],[822,823,824,825,826,827,828,829]

The above helps us better understand why a number of European countries have officially banned, effectively banned, or will ban fracking, and why major lobbying groups are working to ban fracking in the USA and in other countries.[830],[831]

In 2015, the U.S. Geological Survey attempted to manage market and public perceptions of the status of global energy reserves by significantly revising upwards its global estimates for conventional oil and gas reserves. These revisions were based on new prediction methodologies, assumptions about new technology deployment, and a supposedly greater understanding of reservoirs using desk-based research methods.[832] In other words, these were subjective assessments leading to significant upgrades of estimated reserves, which however lacked any physical validation of the reserves.

Experts remind us that over the long term, government forecasts have a tendency to overestimate reserves, and that reserves may not be adjusted downward despite years of production.[833]

Oil and Gas Discoveries Peaked in the Past

Conventional oil discoveries peaked in the 1960s, with new yearly additions on average below annual production since the early 1980s (Figure 8.2.A).[834] Peak natural gas discovery took place between 1960 and 1980,[835] and we will now spend the decades ahead building pipelines and piping those gas reserves to markets.

In the USA, the lower 48 states peaked in oil and gas discovery in the 1930s, with production peaking in the early 1970s. The North Sea reduced the time lag between peak discovery and peak production from 40 years (i.e., as in the USA) to 27 years, meaning industry has become more efficient at depleting reserves (i.e., over a shorter timeframe).[836]

The largest fields and reservoirs within a given area are typically discovered early on,[837,838] and tend to provide most of the reserve growth.[839,840] No super-giant reservoirs containing more than 5 billion barrels of oil have been discovered in decades.[841,842]

It is also recognized by experts that the easiest measures to increase oil recovery from existing fields and reservoirs have already been taken, meaning that more costly, enhanced oil recovery methods will be needed to increase extractions in the future.[843] This enhanced extraction will require a higher oil price.

Such is the state of oil and gas discovery that only one in four oil and one in three increases in gas reserves are actually coming from new discoveries.[844,845] Companies today are simply not discovering as much in new reserves as they are producing in energy,[846] which means that they are depleting their reserves (see Figure 8.2.A). This is reflected in the world oil reserves-to-replacement ratio being less than 100 percent for 28 of the last 37 years.

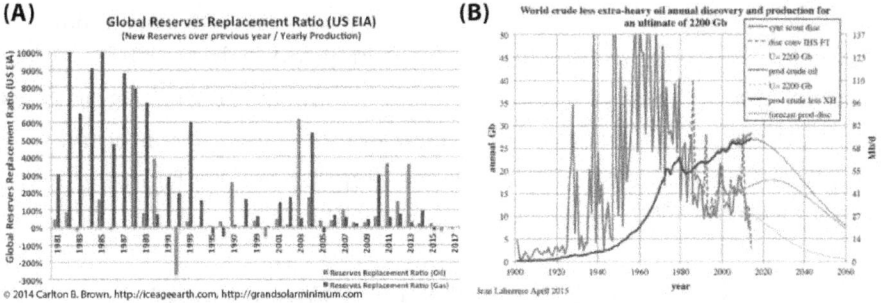

Figure 8.2. A) If cumulative reserve growth was as strong as Figure 8.1.B indicates, then why has the world Reserves-to-Replacement ratio, as depicted here, stood at less than 100 percent in 28 of the last 37 years? In at least three of nine years (2011-2013)[847] with a reserves-to-replacement ratio greater than 100 percent, upward revisions involved unproven reserve volumes, not new discoveries, and were not backdated to the original wildcat discoveries.[848] **Conclusion:** The world is consuming more oil than is being discovered each year. **B)** New oil discoveries (green) versus world annual oil production (brown) shows peak oil discovery occurred in the 1960s.[849] **Conclusion:** The peak in oil discovery is behind us.

A giant field today holds more than 500 million barrels of oil, which is the equivalent of five days of global production.[850] In recent years, the size of discoveries has been significantly smaller than 500 million barrels.[851] The downturn in market price for oil in recent years highlights the sensitivity of exploration to oil price, as new oil discoveries have not been at such low levels since the 1940s.[852,853,854]

New discoveries are increasingly being made at greater depths on land, in deeper sea water subject to larger storms (e.g. Hurricane Katrina), and at greater distances from markets.[855,856,857] These more difficult operating environments and the higher costs involved will need oil and gas prices to recover substantially to make exploration and extraction economically viable.[858]

Unconventional oil and gas resources are touted as the saviors of the oil and gas industry, and are already supplying a significant share of today's production,[859] along with offshore supplies.[860] These unconventional reserves are largely concentrated in North America, Russia, the OPEC nations, China, and South America.[861] Market commentators have cautioned against being overly optimistic about shale prospects, because of the early falloff in production rates at US shale wells. This has necessitated concentrating oil recovery efforts in "sweet spots" until oil prices recover.[862]

When I reflect on all of the above concerning oil and gas reserves, the obvious questions are, first, *has peak-oil discovery and production come and gone?*[863] Is the inclusion of unproven reserves in calculations of world oil supply hiding the fact that peak oil is in fact behind us? And are market commentators' attempts to move the discussion from peak oil and peak discovery to peak demand and the adequacy of supply similarly designed to mask the fact that we are running out of proven reserves?[864] The answer is yes on all counts.

When Will We Run Out of Fossil Fuels?

By dividing the 2013 Energy Information Agency's proven global oil, natural gas, and coal reserves by 2013 levels of production, without assuming any growth or a switch to a colder climate (thus accelerating energy demand), we've got 50 years of proven oil and gas reserves, and 130 years of coal reserves left.[865,866,867,868,869]

By including the more speculative unproven reserves,[870] not included in the Energy Information Agency's reserve data,[871] while assuming a 50 percent certainty of recovery for those unproven reserves, then the reserve timelines can be extended from 50 to 75 years for oil and from 50 years to 117 years for natural gas. These extended reserve timelines do not factor in any population or economic growth, rendering these estimates "optimistic" to say the least.

The Nuclear Energy Agency and the International Atomic Energy Agency indicate that nuclear fuel reserve timelines are similar to that of coal, meaning that coal and nuclear fuels will run out at about the same time in the 22nd century.[872]

With peak oil arguably behind us,[873] with existing "reserves" *comprised, to a great extent, of unproven resources,* and with discoveries of new reserves declining, we need to be very cautious in thinking that there's plenty of oil and gas still out there to be discovered. Clearly, supplier nations will simply not run out of energy resources overnight, but its prospect, when realized, will become a major economic force (i.e., oil price increases) for change in the future supply of energy.[874]

Investing in expensive oil exploration in difficult terrain (i.e., offshore, in the Arctic, deep in the earth), with exploration infrastructure facing more extreme weather (e.g., hurricanes), by an industry that is arguably trying to mask its decline phase, creates a less attractive context for long-term investment. Such a difficult context for oil and gas investment makes renewable energy investments more attractive, and undermines the promise inherent in the inflated, unproven estimates of oil and gas reserves.

Population and Economic Growth Accelerate Energy Reserve Depletion

Now we must consider the prospects for population growth, because growth accelerates demand and more rapidly depletes non-renewable resources.

Between 1950 and 2017 the world's population grew by 5 billion people, and every year the population continues to grow by 85 million people.[875,876] This population growth is unrelenting, with 20-year and 3-year compound annual growth rates (i.e., year after year) broadly the same at 1.2 percent and 1.1 percent, respectively. Since 1980 most of this growth has occurred in Asia and Africa.[877]

Economic growth and development has helped improve access to energy, food, and clean water, and thereby has played a major role in sustaining world population growth.[878,879] Population and economic growth have come at a price: the depletion of finite global resources and increased environmental destruction with little regard for the needs of future generations.

Population growth has tracked the growth in fossil fuel production since the early 1990s (see Figure 8.3), and humans have been steadily increasing their energy consumption per capita during this period, especially for gas, thus accelerating the depletion of energy reserves.

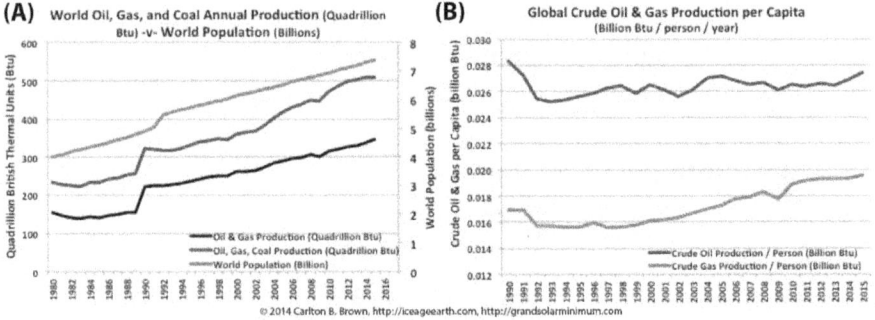

(A) World Oil, Gas, and Coal Annual Production (Quadrillion Btu) -v- World Population (Billions)

(B) Global Crude Oil & Gas Production per Capita (Billion Btu / person / year)

© 2014 Carlton B. Brown. http://iceageearth.com, http://grandsolarminimum.com

Figure 8.3. A) World population growth has tracked energy production growth. The 3, 5, 10, and 20-year compound annual growth rates for oil and gas production have outpaced population growth rates, implying a growth in per capita consumption of energy resources. **B)** Humans are using more energy and especially more natural gas per capita, and this is a multi-decade trend. **Conclusion**: Economic growth leads to increased per capita consumption of fossil fuels, which more rapidly depletes the world's non-renewable energy reserves.[880]

Decentralized Sustainable Development and Switching Energy Systems Are the Solutions

Absent a global catastrophe like pandemic influenza, wars over resources, or a climate switch to a cold-drought phase, a rapid climate change event, or climate-forcing volcanism, the world's population is expected to exceed 9 billion by mid-century. Much of this population growth is expected to occur in Asia and Africa, where nearly 70 percent of population growth is occurring today.[881,882],[883,884] Under the climate and risk scenarios put forward in this book, however, these population growth estimates are *unrealistic*.

Thomas Malthus in 1798 anticipated the population growth crisis of his time in his book, *An Essay on the Principle of Population*.[885] In light of this essay, the term Malthusian crisis was coined to describe the situation in which a population in a given area exceeds its food supply, resulting in famine. Famine would reduce the population until it was once again in balance with the food supply. The technology of the First Industrial Revolution—innovations such as the steam engine, the seed drill, marine chronometer, the spinning jenny, etc.,[886] as well as migration to the New World, for the most part saved 19th century Europe from famine.

Given the climate-related risks detailed in Chapters 3–7 (different from those the IPCC has warned about), unrelenting population growth, and our depletion of resources (i.e., fossil and nuclear fuels, groundwater), we will face a major resource supply problem in the years ahead, *unless we act now*. In a much colder, drought- prone world, and without having switched enough of the world's energy system to renewable energy, today's youth could face the real prospect of a Malthusian-like crisis.

To ameliorate the impact of this problem, the world's top two priorities are to rapidly *switch the world's primary energy source to renewable energy* and transition to living sustainably in a decentralized manner (see below).

But first things first. What is sustainable development?

One of the more widely recognized definitions of sustainable development is contained in the Brundtland Report (1987). This report came out of the World Commission on Environment and Development in 1983.[887] The Brundtland Report defined sustainable development as *"development that meets the needs of the present without compromising the ability of future generations to meet their own needs."*[888]

This report focused attention on the need to promote economic and social development in ways that avoid damaging our environment, such as through over-exploitation of natural resources and pollution. It critically focuses our attention on managing population growth and *ensuring all human rights, especially for women and their basic human right for self-determination and equality.*[889]

From a resource perspective, living sustainably means we must strive to use all energy, water, and food, and other resources sparingly, efficiently, and equitably. We must also maximize residual resource and energy recovery, before generating a minimal amount of waste and pollution. Living sustainably must permeate everything that we do at home, at work, and when we travel. With an impending return to a cold climate, living sustainably will need to become embedded in daily human activity and popular culture, and become a new societal norm, both nationally and internationally.

In order to address the climate switch and its potential risks, as well as the issues of growth and the accelerating depletion of resources, we must develop strategies for rapidly *switching the world's energy system* to renewable energy, and implementing *decentralized sustainable development.* This is how we will prevent a 21st century Malthusian crisis.

Such a dual strategy will permit a greater degree of self-sufficiency for securing energy, water, and food at home and in urban areas, with the aim of reducing our vulnerability to climate switching and other climate-associated risks. Decentralization anticipates that government, corporate, and commodity market actions will restrict resource supply in a crisis event.

Financing the Energy System Switch and Decentralized Sustainable Development

Various researchers have estimated the cost of adapting to and mitigating global climate change to be in the ballpark of 1 percent of world GDP (i.e., gross domestic product), or more than $1 trillion per annum.[890,891,892,893]

One percent of world GDP well exceeds the $100 billion per annum that the signatories of the 2015 Paris agreement plan to spend (starting in 2020) in order to help developing nations adapt to climate change.[894,895] The $100 billion provided by developed nations (but not including the USA) would be additional to funding their own national plans to prepare for climate change.

As things stand, people and businesses in nations belonging to the Organization for Economic Co-operation and Development (OECD) spend the equivalent of nearly 9 percent of gross domestic product (GDP) on insurance premiums per annum, across all insurance categories.[896] Globally, the cost of insurance is about 6 percent of GDP.[897]

The insurance costs cited above provide a crude proxy for the cost of insuring our global way of life and mitigating risks. This way of life will be severely disrupted in a climate switch, or in a sustained climate crisis (i.e., climate-forcing volcanism, rapid climate change), or in the event of a pandemic flu outbreak.

The cost of doing nothing will be far greater than the cost of mitigating climate change and preparing the world for sustainable living. The spending required to prepare for climate change can be viewed as a form of insurance against the potential damage that unmitigated climate change would inflict on the global economy. This cost can also be viewed as an investment in today's youth, who will inherit a resource-depleted planet.

Big Bags of Cash Exist in a Carbon Tax, Fuel Subsidies, and Pension Funds

Someone has to pay for climate change preparation. Governments are fiscally constrained and encumbered with the debt created by the 2008 financial crisis and its followup, the Great Recession. Where is that $1 trillion per annum going to come from?

I will now discuss some of our *big money, new cash*-yielding options.

Carbon markets have emerged at the regional, national, and municipal levels.[898,899] While a carbon tax may not be every nation's cup of tea (as with the USA at the present time),[900,901] other regions like Europe,[902] as well as nations and sub-national governments, have made significant progress in implementing a carbon tax or have it in the planning stage.[903]

Carbon taxation is an important financing option for state and provincial governments and cities, and levies a charge per ton of carbon emitted through the combustion of fossil fuels. This taxation uses emissions trading schemes and direct taxation (collectively, a "carbon tax") to capture revenue.[904,905]

This form of "user pays" taxation is known to incentivize changes in consumption, promote investment in low-carbon technologies, and help direct innovation. If priced right and complemented by well-designed policies, carbon taxation can help expedite the switch to renewable energy and improve energy efficiency.[906]

Already carbon taxation is generating sizable revenues for fiscally constrained governments; it was estimated to be about US$50 billion (five percent of global GDP) per year in 2016. The carbon tax revenues generated so far were only levied over about one-sixth of global emissions in 2015, with the majority of this levy priced at low levels that do not incentivize energy efficiency efforts or switching to renewables. With China implementing its own emissions trading scheme, this will significantly expand the global emissions covered by a tax.[907]

According to one review, a carbon tax priced at $25 per ton would yield about $750 billion in revenues.[908] Therefore a carbon tax would need to be priced at $30+ per ton of carbon to provide the estimated 1 percent GDP equivalent in funding per annum. Theoretically, a carbon tax could also be implemented on an industry- and sector-specific basis to incentivize a switch to renewables and drive energy efficiencies in those industries and sectors. A carbon tax could also be levied on fossil fuel suppliers.

Subsidies for fossil fuel consumption are another big pot of taxpayer cash that could be reduced or redirected to finance the cost of climate change adaptation and to help the poor access affordable renewable energy systems. These fuel subsidies have been estimated to total $500 billion annually,[909,910,911] although this varies from year to year depending on international oil prices.

Reducing fossil fuel subsidies would reflect the real cost of energy generation and make all energy suppliers compete for grid supply contracts on an economic and equal basis. By resetting at a higher level the fossil fuel prices to which incentives are applied, governments can also provide a stronger market signal promoting an energy system switch. A reduction in subsidies, coupled with renewable energy feed-in-tariffs (i.e., higher contract prices paid to renewable energy grid suppliers), could be used in tandem to facilitate a more rapid switch to renewable energy.

Admittedly, some governments will be unable to reduce subsidies, in order to ensure that their citizens and industries have access to affordable energy. Unpopular government decisions have political consequences (i.e., protests, riots, loss of elections, revolutions), and can increase poverty levels and undermine economic development. The key point is that reducing subsidies is part of a package of measures to help governments rapidly induce a switch to renewable energy.

There are also *tens of trillions of dollars* of public and private pension funds in the world financial system which could be put to work on society's behalf. There are two ways pension funds could be harnessed to help fund preparations for climate change.

First, changing the quantitative restrictions (i.e., the upper and lower limits) on asset classes that pension funds can hold[912] is a means by which investments can be directed toward renewable energy and sustainable development projects (see Chapter 13's "Bucket List"). For example, pension fund investments in renewable energy infrastructure projects can be facilitated by increasing or decreasing the share of equities and the types of equities (i.e., listed, non-listed, growth, international) in a fund, or the percentage of bonds and types of bonds (i.e., government, municipal, corporate, green), relative to the other asset classes held by a fund. More specifically, pension fund investments can be facilitated through the issuance of municipal and public infrastructure bonds (i.e., Qualified Public Infrastructure Bonds as used in the USA) coupled with government-backed insurance protection to help de-risk investments for pension funds.

Infrastructure bonds can be used to facilitate public-private partnerships (PPP) in renewable energy and sustainable development infrastructure projects. Such PPPs would improve the scope for pension funds to invest in lower-risk projects, while helping fiscally constrained governments (central and municipal) get big infrastructure projects completed.

Second, what good is a pension fund to retiree or near retiree if they freeze or starve to death in a global climate switch, or during a food or energy crisis? Permitting homeowners to draw down a percentage of their pensions to retrofit their homes for sustainable living and to prepare for climate change would help community development, without governments needing to provide tax-credits (and thus lose tax revenues). This of course assumes the world awakens to the prospect of a climate switch, unanticipated natural climate-related risks, and the diminished prospects for new, large-scale oil and gas discoveries, and then decides to act with a sense of urgency.

CHAPTER 9

Rapidly Switching the World's Energy System to Renewable Energy

Over the last decade, a number of important factors have converged as powerful market forces to catalyze renewable energy's arrival in the mainstream. The renewable energy industry's technological progress, combined with the improving economics of system supply, enabled its readiness for switching the world's energy system. A greater level of climate change awareness, support by governments, the United Nations and its affiliates (IPCC, IRENA etc.), and big business, as well as finance for the renewable energy industry have all helped to catalyze major progress.

The renewable energy industry and its progress in technology development have required decades of nurturing by industry, governments, the United Nations and IRENA, and stakeholders (i.e., utilities, regional transmission organizations, environmental and consumer organizations, etc.). This technological development has resulted in higher energy output, improved efficiency of energy conversion, and lower production costs. This progress has helped bring these technologies to the point of mass-market production and affordability, just when we need them the most.

We have all the renewable energy technologies and energy systems available right now for switching to renewable energy sources across all sectors and market segments. These renewable energy technologies offer society a means for achieving the decentralized supply of energy to our homes, communities, and cities.

If a switch of the world's energy system from fossil fuels to renewable energy ("energy system switch") is to happen, then renewable energy must rapidly replace the three-quarters of the total electricity supply generated today by non-renewables. The big challenge, though, is that renewable energy must also erode a large part of the 63 percent share of fossil fuel energy supplied to industry, transportation, and other sectors for non-electricity uses. This is a mountainous task, given that the non-hydroelectricity share of renewable energy is only about 8% of total electricity generation (see Figure 9.4), and even less of the world's total energy generation (see Figure 9.2.A).[913]

Pivotal to ensuring the world's energy system switch moves ahead rapidly is the need to motivate society and business with the *right messages and economic incentives*. The market message must be *aligned* with the strategic goals of governments, such as moving toward sustainable development and switching from fossil fuels to renewable energy.

The market message must also communicate a *sense of urgency*. A climate switch, climate risks, and running out of oil and gas could together provide that sense of urgency. This energy switch would be greatly assisted by a higher oil and gas price that reflects the message of resource scarcity, and the need to save some fossil fuel resources for today's youth in tomorrow's world.

To more fully comprehend what switching to renewable energy means, this chapter will review the main uses of energy by the industrial and transportation sectors, and the technologies available to switch those sectors from fossil fuels to renewables. This chapter also provides a review of best practices used and promoted by central and municipal governments for catalyzing this energy system switch.

Who Are The Big Consumers of Energy?

A small number of nations consume the majority of the world's energy resources. Four nations—China, the USA, Russia, and India account for half of the world's energy use. If China, the USA, India, and other large industrial nations accelerated their nascent energy system switches, this would have a major impact on extending the lifetimes of world energy reserves.

At the other end of the energy consumption spectrum, some 160+ nations consume 12.6 percent of fossil fuels consumed, while only holding about 4 percent of the world's fossil fuels reserves. These 160+ nations' are obviously dependent on energy imports, which make them vulnerable to a future tightening of energy supplies.

In 2015, electricity delivered to the end-user accounted for 13.5 percent of world energy consumption. Staggeringly, of the 575 quadrillion BTU of energy consumed each year, 72 quadrillion British thermal units (BTU) were delivered as electricity, while 146 quadrillion BTUs (of that 575 quadrillion BTUs) were lost in the production of that electricity at the time of its generation.[914] Put another way, one-quarter of all energy consumed is wasted in the process of generating electricity, due to inefficiencies of converting energy to electricity.

Therefore, of all the initiatives we can undertake, *rapidly switching* electricity generation systems to renewable energy will have the biggest impact on extending the lifetimes of our fossil fuel reserves. At the same time, making improvements in the efficiency of energy conversion into electricity, and the use of heat co-generation technologies in producing that electricity, would also save major amounts of energy.

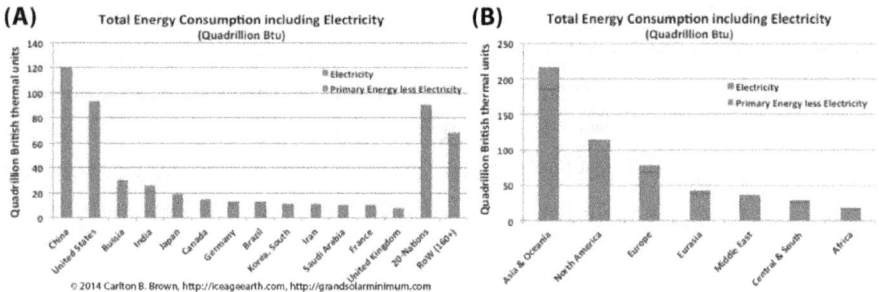

Figure 9.1. A and **B**) Twelve nations are responsible for about 70 percent of the energy consumed, with China and the USA alone accounting for 40 percent. Twenty other nations consume a further 17 percent, leaving the rest (160+ nations) to consume 12.6 percent.[915]

The level and mix of energy use varies by sector and by country, depending on their stage of economic and technological development.[916] At the macro level, liquid fuels, natural gas, and coal supply over 80 percent of energy consumed, with nuclear fuel and renewable energy accounting for the balance, as indicated in Figure 9.2.A.

The global energy market consumes its supply of energy across the industrial, transportation, residential, and commercial sectors. Industrial use and transportation account for 80 percent of the energy consumed globally, making these two sectors important for a switch to renewable energy and the realization of fossil fuel savings.[917]

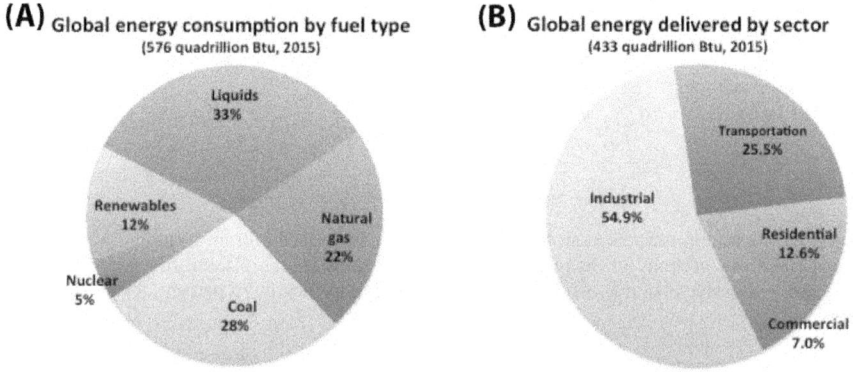

(A) Global energy consumption by fuel type
(576 quadrillion Btu, 2015)

Liquids 33%
Renewables 12%
Natural gas 22%
Nuclear 5%
Coal 28%

(B) Global energy delivered by sector
(433 quadrillion Btu, 2015)

Transportation 25.5%
Industrial 54.9%
Residential 12.6%
Commercial 7.0%

© 2014 Carlton B. Brown, http://iceageearth.com, http://grandsolarminimum.com

Figure 9.2. A) The world relies heavily on non-renewable fossil and nuclear fuels for its primary sources of energy. Fossil fuels supply 83 percent of the world's energy needs. **B)** More than 90 percent of energy consumption is accounted for by industrial, transportation, and residential use.[918]

The industrial sector is the largest global user of energy, consuming more than half of the energy supplied across all sectors. Heat and energy-intensive manufacturing processes consume most of this energy. This energy is used in the manufacture of food, steel and other metals, chemicals, in oil refining, and in pulp and paper production.

Principle areas for industrial action would include switching manufacturing processes to renewable energy for low temperature heating and cooling processes, and for fluid heating and steam generation processes. Switching to renewable energy systems for powering lighting and air temperature control systems inside buildings will also lead to significant fossil fuel savings.[919]

Where heat is generated in industrial and manufacturing processes, energy recovery systems should be utilized to harness the dissipated or waste heat to improve overall energy efficiency.

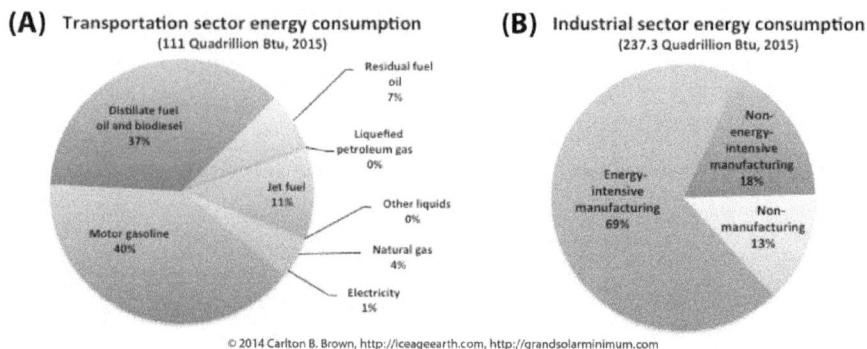

Figure 9.3: **A)** Liquid fuels account for 95 percent of all fuels used in transportation, while natural gas and electricity account for the rest. **B)** Industry is the largest user of fossil fuel energy supplies, with energy-intensive manufacturing processes accounting for nearly 70 percent of industry's total energy use.[920]

The transportation sector is the second-largest user of energy. More than half of the energy used by the transportation sector is in nations belonging to the Organization for Economic Co-operation and Development (OECD).[921],[922] However transportation in non-OECD nations is expected to dominate future growth in fuel use.[923]

Passenger transportation accounts for nearly two-thirds of transportation fuel use, and freight transport for just over one-third. Light duty passenger transportation, air transportation, freight trucks, and shipping are the main users of energy within the transportation sector.[924] All of these means of transport must come under scrutiny designed to find ways to improve their fuel use and efficiency. Reducing energy use and improving energy efficiency will require a general downsizing of engine capacities and reductions in vehicle weights.

Switching transportation to renewable energy systems is a priority, especially for passenger and freight transportation, as well as in cities and on the main intercity routes where most traffic occurs. The two main options for switching to renewable energy sources are reviewed in the last section of this chapter.

The Renewable Energy Switch Must Be Accelerated

In 2015, world electricity generation accounted for 13 percent of the world's fossil fuel and nuclear energy (primary energy) consumption, with a further 25 percent of the world's primary energy consumption being lost in the conversion of that energy into electricity.[925] Hydroelectric and non-hydroelectric renewable energy sources accounted for about one-quarter of the world's electricity generated, with hydroelectricity dominating the renewable electricity supply (i.e., about 70 percent).[926]

If a risk-mitigating market switch to renewable energy is going to occur, then not only must renewable energy replace three-quarters of the total electricity generated (and the energy wasted), but renewables must also replace a large part of the 63 percent share of primary energy supplied to industry, transportation, and non-industrial commerce sectors for non-electricity energy uses.[927]

The early stage progress toward the electricity market switch, and the mountain still to be climbed, are highlighted in Figure 9.4.

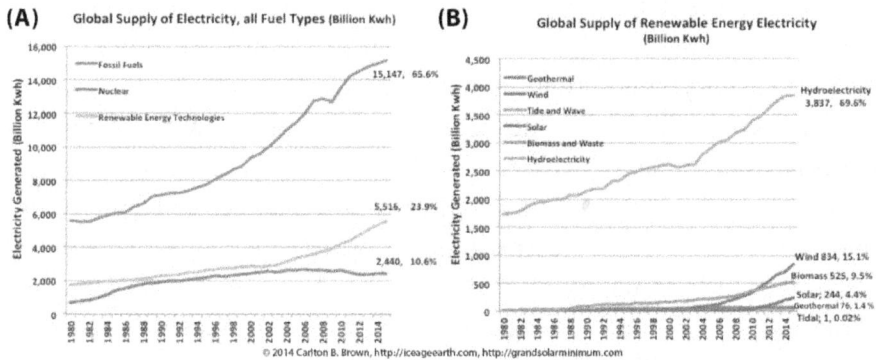

Figure 9.4. A) In 2015 fossil fuels and nuclear energy supplied more than three-quarters of electricity generated. Between 1980 and 2015, these non-renewable energy sources supplied most of the growth in electricity generation (75 percent). In the last decade, non-hydroelectric generation by renewables has grown to account for 7.2 percent of global electricity generation, with hydroelectricity accounting for 16.6 percent. Electricity generation with renewables has shown a 10-year compounded annual growth rate of 5.3 percent, and non-hydroelectric renewables 16 percent. **B)** Wind, biomass, and solar sources of renewable energy are beginning to make an impact in the market.[928]

Technologies Fueling the Renewable Energy Revolution[(RE)]

Hydroelectricity, wind, biomass and waste biogas, and solar photovoltaics (solar PV) account for more than 90 percent of the electricity generated by renewables. These technologies are maturing, and will dominate renewable energy electricity generation for the foreseeable future.

Hydroelectricity accounts for nearly 70 percent of the world's renewable energy electricity supply, with more than 150 countries producing hydroelectricity. China, Canada, Brazil, the USA, Russia, India, Norway, and Japan are the leading hydroelectricity producers.[929] Hydroelectricity typically provides bulk power for supplying base load grid electricity (i.e., supplying the minimum level of grid electricity demand), and offers 24/7 flexibility in its supply.[930]

On a smaller scale, run-of-the-river hydroelectric systems are also used to generate electricity. These systems capture the kinetic energy in rivers without the use of river blocking dams or reservoirs. Small hydroelectric projects (<10 megawatts) and micro-hydroelectric projects (10–100s kilowatts)[931] are used for supplying electricity to small industries and communities.[932]

Small-scale hydropower lends itself well to countries rich in river water resources, while avoiding significant environmental damage associated with new river-blocking hydroelectricity dams. Hydroelectricity generation on this scale can be paired in a complimentary way with other renewable energy sources (wind, solar PV, biomass, etc.) to achieve off-grid or decentralized energy supply.

Wind power has grown over the last two decades to supply more than half of non-hydroelectric renewable energy,[933] and nearly one hundred countries use wind power.[934] Wind turbine technology has advanced significantly, making wind power competitive for grid electricity supply.[935]

A modern wind power plant can have up to 200 turbines, each with three-megawatt output, with rotor diameters exceeding 100 meters. Such wind plants are comparable in capacity to conventional power plants.[936]

Wind power innovation has centered on taller towers, longer blades, and the use of advanced, lighter construction materials, as well as the use of intelligent communications and remote wind sensing. This all helps to increase the energy output and improves its utility in lower wind conditions, enabling wind turbines to be installed on all continents and coastlines.[937]

Offshore wind permits the use of larger wind turbines and blades compared with land turbines because large ships can be used to install them. Offshore locations provide greater and more constant wind speeds,[938] making offshore wind useful for supplying coastal cities with decentralized electricity supply.

Solar PV converts sunlight into electricity through the photo-electric effect using silicon semiconductors.[939] In 2016, solar PV accounted for one-sixth of the global supply of non-hydroelectric renewable electricity.[940]

Solar PV is already an economically attractive option for all scales of market use. Enhanced integration systems and weather forecasting will help make solar PV more competitive with large-scale grid suppliers.[941] Solar PV is also the most important renewable energy technology *for home use* (see Chapter 12).

The other important large-scale solar technology is concentrating solar power. This technology offers utility company-scale supply, with heat storage capability that allows electricity supply to be extended beyond sunlight hours.[942]

Geothermal energy taps reservoirs of steam or hot water from beneath the earth's surface to generate electricity.[943] Geothermal power plants can operate 24/7, and are used to provide base load electricity,[944] as well as grid balancing of electricity supply when large quantities of renewable energy are integrated into the grid.[945] The US, the Philippines, Indonesia, New Zealand, Mexico, Italy, and Iceland are world leaders in geothermal energy supply.[946]

Geothermal, or ground source, heat pumps utilize the outer crust of the earth as a heat source for heat exchange.[947] This type of renewable energy system represents a major source of renewable energy capacity for heating and cooling buildings. A heat pump and ground-coupled heat exchanger are used to move heat energy into the earth (cooling by day) and out of the earth (warming by night). These heat exchange systems are a key component of energy efficient homes, buildings, and greenhouse designs (see Chapter 12),[948] providing free sustainable heat and cooling once the energy system is installed.

Biogas from Agroforestry and Food Biomass, and from Municipal Waste

Biogas recovery is a proven technology and potentially a large source of renewable energy. Biogas recovery from biomass waste is a process widely used in food processing, the processing of farm and municipal wastewater, and also for processing agroforestry residues.[949,950,951] Biogas is generated from biomass using anaerobic digestion processes. It is then converted into electricity, heat, and transport fuels.[952,953] The European Union is a market leader in biogas generation, with biogas already contributing about one-twelfth of the EU's renewable energy supply.[954]

Global agriculture and the food distribution system waste more than one-third of the food that is produced. This waste offers the opportunity for biogas production on the farm, or via larger-scale farming cooperative or commercial ventures.[955] Agriculture also generates massive quantities of biomass, such as forestry byproducts and crop residues, offering farmers a source of renewable energy to offset their energy needs or to provide an additional revenue stream.[956,957,958,959] Biomass recovery from food distribution hubs and retailers, and city-generated biomass and municipal waste, are sources of plentiful (and renewable) biomass waste for the production of biogas.[960,961,962,963,964,965,966]

We Must Plant Billions of Trees and Reforest Our Planet

Without technology, fossil fuels, and renewable energy, how will humans keep warm and cook? Yes, that's right, the same way they did for tens of thousands of years: *with firewood and fire.*

Planting trees today makes good sense for future generations, who will need the wood for fuel[967,968,969,970,971,972] when oil and gas have run out or are being withheld by governments and corporate entities during a global energy or climate crisis.

Enabling Energy Security with Regional Super-Grids and Local Smart Grids

The electricity grid refers to a network of overhead and underground transmission lines, substations, and transformers that deliver electricity from a network of power plants to homes and industries.[973,974] In many countries, the existing transmission and distribution infrastructure for electricity is aging and being stretched to its limits, necessitating the upgrading of national and regional electricity grids.[975,976]

Without upgrading the existing electricity grids and switching to two-way electricity metering, renewable energy will not meet its potential for switching the world's energy system, or permit the decentralized supply of electricity.[977,978]

Integration of more than one-third of renewable energy supply into an electric grid at any one time is achievable with most grid systems and with advanced management methods. To accommodate a large-scale renewable energy switch, transmission and distribution networks will need to handle a much larger supply of renewable energy from a larger network of suppliers, and spread over a much wider geographical area than traditional bulk electricity suppliers.[979,980]

Smart grids are part of the solution to the above quite complex challenge. Smart grids manage the direct interaction and communication between energy consumers and suppliers, and enable the integration of vast amounts of renewable energy into the electricity grid. These smart grids provide a more resilient and flexible grid supply, and are able to accommodate a large number of suppliers and consumers. Smart grids will be integral to a future renewable energy power system.[981,982,983]

Storing energy at times of low prices provides additional grid flexibility to cope with peak electricity demand, or for use in emergency situations.[984,985] The ability to store half a day of energy is considered the sweet spot for large-scale grid suppliers.

Pumped-storage hydropower involves pumping water uphill into reservoirs at times of low price and low electricity demand, and then when the demand and price are higher the energy is "harvested." Pumped-storage hydropower offers large-scale, low-tech grid storage of energy, which helps to make electric grids more resilient.[986] Battery storage of energy at a gigawatt-scale will also be possible in the future,[987] and will further enable smart grids by helping ensure the resilience of the electricity supply when grid supply is unable to meet demand.[988,989,990]

High voltage direct current (HVDC) transmission systems are used for transporting bulk electricity over long distances from power plants to local substations or large consumers and markets,[991,992] including transit of bulk electricity underwater.[993] These HVDC transmission systems are an enabler of smart grids and regional super-grids and represent the best solution for long distance transmission of bulk renewable energy.

As the switch to renewable energy accelerates, supply agreements and transmission links, operating to harmonized standards, will be needed in all parts of the world. This will permit regional networks of interconnected countries with complementary renewable resources to transmit electricity to distant markets and across different time zones. These collective arrangements will also permit the pooling of electricity resources to improve the resilience of electricity supply across multiple countries, when electricity supply in one country is unable to meet market demands.

These regional electricity grids are referred to as wide area synchronous grids or super-grids,[994] and already operate in regions like the EU,[995] North America,[996] Russia, China, other parts of Asia, and North Africa.

Decentralized energy, or distributed energy, is generated at or near the point of use, and is connected to a local smart grid. Decentralized energy supply using renewable sources is also readily scalable.[997,998,999,1000] This type of decentralized electricity supply represents the best energy supply strategy for homes and urban areas, for mitigating the risks of a tightening global energy supply, and in an international energy or climate crisis.

Accelerating the Renewable Energy Switch

What is proposed below, and indeed throughout Sections 3 and 4, may come across as idealistic and unrealistic. You might cite all the reasons why we can't do these things. I am working under the assumption that the near-to-medium term future will not be "business as usual." I assume that a cold climate switch is going to happen, and that risks like climate-forcing volcanism, pandemic flu, and severe drought will eventuate in my lifetime.

I am writing this book for people who realize that we cannot ignore natural climate change, or the extreme global warming temperature peak that is upon us. Nor can we ignore the abrupt fall in temperature that this extreme trough-to-peak outlier portends.

We have three options under the above scenario: (1) continue believing in anthropogenic global warming and going about business as usual; (2) deal with this when the proverbial you-know-what hits the fan; (3) rapidly implement risk mitigation plans, that is, anticipate these potential events and move to reduce our vulnerability. *None of these options is without pain*—the pain is suffered either now at one price, or in the future at a much higher price.

Pivotal to ensuring that a global energy switch moves ahead rapidly is the need to send the right messages about climate change, and provide the economic incentives that will motivate business and society as a whole to make changes. These messages must also convey the reality of our finite and dwindling energy and water reserves, and the damage we are doing (greenhouse gases, pollution) by using these resources. By refocusing the climate and resource messages governments can then engender the right sense of urgency and motivation for people and businesses to act and help switch the energy system to renewables.

Higher oil prices achieved through the perception of energy scarcity and the unveiling of catastrophic risks will be a more powerful motivator for businesses, transportation, and industry generally (i.e., the largest energy users) to rapidly switch the energy system than the perceived need to mitigate anthropogenic global warming.

Governments play an important role in regulating their national energy markets and consumption through a number of different mechanisms. State-based control of energy industries versus promoting a competitive marketplace is the most obvious means of market regulation. Governments also make decisions on what national energy resources are exploited, and when, through government investment and the sale of exploration licenses. Government regulation can come about from the building of hydroelectric dams and geothermal or nuclear power plants, the sale of oil and gas exploration licenses, the use of fuel subsidies, or by promoting renewable energy through the use of feed-in-tariffs (premium-priced supply contracts). Internationally, governments together with the United Nations and its inter-governmental organizations (i.e., IRENA, IPCC) are the key architects of the world's transition to renewable energy.[1001,1002]

Governments are also able to legislate and implement policies, as well as educate and incentivize all levels of society to change their consumption habits. Governments at all levels also have significant budgets, capital expenses, and public sector finances to oversee. That's an awful lot of purchasing power to lead the switch to renewable energy.

In order to switch the world's energy system quickly, major public and private investments will be required for ensuring the required energy infrastructure is put in place. This includes making the investments in renewable energy capacity (at the national and municipal levels), in local smart grids and regional super-grids, in long-distance high voltage direct current transmission links, in biomass-to-biogas and biofuel production, and biomass-biofuel and synthetic fuel filling stations, as well as electric road (vehicle recharging) and rail systems.

These big infrastructure projects need to be financed by both public and private sources. Some of the main financing options were reviewed in Chapter 8. Higher oil prices, as indicated above, would provide a greater economic incentive for switching industries and sectors to renewable energy, because the loss of profits is a powerful motivator.

Feed-in tariffs (i.e., long-term energy supply contracts), regulated by governments, are awarded to renewable energy suppliers at a higher price per kilowatt-hour than the price given to non-renewable electricity suppliers.[1003] This pricing difference reflects the higher electricity supply costs at this stage of renewable energy's technology and market development. Feed-in tariffs have been widely used around the world to promote a rapid expansion of renewable energy capacity by helping incentivize investment in renewable energy systems.[1004,1005] The European Commission concluded that *"well-adapted feed in tariff regimes are generally the most efficient and effective support schemes for promoting renewable electricity."*[1006]

Big city governments are leaders in the promotion of renewable energy, particularly in combination with energy efficiency improvements. Some cities and governments have been forward-thinking in managing and risk-mitigating climate change for their people.[1007,1008,1009,1010,1011]

Progressive government and city leaders use their planning processes and purchasing authority to source national and city energy needs from renewable sources, and to ensure that reductions in carbon dioxide are achieved (i.e., by mandating lower vehicle emission standards, promoting decentralized renewable energy sourcing, etc.). These progressive governments also implement building codes and set energy efficiency targets for buildings. Progressive governments also ensure waste is processed for biogas, while investing in renewable energy public transportation systems, among many other actions. Cities that are investing in renewable energy public transportation systems include Oslo, Bogota, San Francisco, and Melbourne.[1012]

By introducing standards for building energy use and efficiency, municipalities can have a direct influence on new construction and the retrofitting of existing buildings to higher standards of sustainability. Municipalities can also influence the replacement of old equipment that uses a lot of energy with more efficient items. Municipalities and the public sector generally can ensure that public heating systems are switched to renewables.

More than 90 percent of transport energy comes from oil and liquid fuels. To have any meaningful impact on switching transportation energy systems, it will be important for municipal and national governments to facilitate the switch to renewable energy transport systems (i.e., electric rail and road), especially in cities and on the main intercity routes.

In the shorter term, setting stricter fuel economy standards for reducing greenhouse gas emissions and fuel economy standards for small vehicles[1013] and other classes of vehicle transportation will promote an industry-wide downsizing of engine capacities. Stricter fuel economy standards will also promote the development of lighter, smaller, and lower maximum velocity vehicles. Older vehicles not complying with stricter emissions regulations would be phased out over time.

Two main options are available for both central and city governments, as well as commercial enterprises, to move transportation to renewable energy systems.

First, public and private initiatives can be used to switch internal combustion engine vehicles to renewable biomass-waste generated biogas and liquid bio-fuels,[1014,1015,1016,1017] and to synthetic fuels.[1018,1019,1020,1021] These renewable fuel options for combustion engine vehicles are available right now.

A switch to renewable biofuels will be greatly assisted by higher oil prices, once it's realized that peak-oil and peak-discovery are behind us, and once a carbon tax is implemented. An economic imperative is needed to support this biomass-biofuel switch, and a higher long-term oil price will support the business case for investing in biomass-biofuel conversion production capacity and vehicle refueling infrastructure.

Compressed natural gas (CNG) has been in use for decades in some countries and is already powering 20 million-plus vehicles worldwide, offering society a tried and tested renewable fuel alternative.[1022] In New Zealand during the oil crisis of the 1970s, the government incentivized vehicle conversions to CNG, establishing a precedent for government intervention to direct market responses to fuel scarcity. More widespread biomass-biogas use will require new policy initiatives and commercial support. Major investment in refueling infrastructure and biomass-derived biogas production facilities will be required, along with the retrofitting of vehicles using biogas conversion kits.[1023,1024]

Biofuels derived from irrigated crops are not a viable renewable energy solution for transportation, because they would have a major impact on irrigation water use in stressed water basins around the world, and would lead to greater deforestation.[1025]

The second option for switching transportation to renewable energy systems is by electrifying road systems and accelerating the development of the electric vehicle market. With such development of electrification infrastructure, transportation could be moved more rapidly to electric-powered systems.

Electric cars are already a reality and in an early phase of market adoption around the world. The technology for electric vans and buses is at an earlier stage of market development, while electrification for long-range trucks and buses needs further development. The electrification of road infrastructure has started, but this will require major public and private investment before it becomes widespread.[1026,1027,1028]

The emergence of fuel cells using hydrogen for transportation is also an interesting area of development. Hydrogen generated for fuel cells offers scope for providing motive power for small cars, buses, trucks, and specialty vehicles.[1029] In the long run, the integration of renewable energy-driven electrolysis systems for splitting water to release the hydrogen fuel will make this technology fully renewable. This water-splitting process involves the use of sunlight and specialized semiconductors (i.e., a photoelectrochemical process), and requires further development.[1030]

CHAPTER 10

Ensuring Water Supply during a Grand Solar Minimum

Archeology teaches us that great civilizations were strategically located next to big rivers, and that extreme drought played a major role in the downfall of many great civilizations. For example, the 4.2-kilo-year rapid climate change event (see Chapter 7) and the severe drought this caused contributed to the end of Egypt's Old Kingdom,[1031,1032] the Akkadian Empire,[1033] and the Indus Valley Culture.[1034]

Yes, it is true that the technology-driven world that we live in today is very different from the world of 4,200 years ago or the Little Ice Age. However, with 7.5 billion people on earth today, and with groundwater resources having been depleted in many places, we are highly vulnerable to extreme drought. Cocooned in our cities and cozy lives, we may feel invulnerable to Mother Nature, but we should remember that the crops and animals that feed us remain vulnerable to climate change.

Humans, with their proclivity for living unsustainably in order to fuel profligate growth, have already depleted two-thirds of the world's largest aquifers beyond sustainable thresholds, and pushed the other one-third into significant stress.[1035] We've also dammed, diverted, or over-extracted water from the majority of the world's largest rivers[1036,1037] to meet the demands of agriculture, industry, and urban growth.[1038,1039] In consequence, coastal outflows of major rivers have markedly declined.[1040]

Globally, agriculture accounts for between 70 percent and 90 percent of freshwater extractions,[1041,1042,1043] and as such has a major impact on water basin stress. This water basin stress and its associated transboundary river stress, as well as the over-extraction and depletion of aquifers, constitute an ever-worsening global problem,[1044,1045,1046,1047] particularly in Asia and Africa.[1048,1049] These regions are important to global agriculture and the global supply of food (see Chapter 11).[1050,1051]

Asia and Africa also account for 70 percent of the world's population,[1052] and will account for the majority of future population growth; they are also rapidly undergoing urbanization.[1053,1054] In fact, almost three-quarters of the world's megacities, each home to more than 10 million people, are located in Asia and Africa.[1055]

The conclusion I draw from the above is that the unrestrained use of water by agriculture, combined with economic and population growth and growing urbanization, will create an ever-widening gap between freshwater supply and demand. This will worsen already existing water stress and scarcity.[1056,1057,1058] This ever-widening gap between supply and demand will be exacerbated by low rainfall and drought, particularly in Asia and Africa, during this grand solar minimum (see Chapter 7).

The objective of this chapter, therefore, is to explore global water supply and demand, and review the different methods, technologies, and innovations that are available for supplying water and alleviating water scarcity.

Three main topics are reviewed: (1) integrated water resource management, particularly artificial groundwater recharge; (2) renewable energy desalination; and (3) the use of pipelines to transport bulk water. These are in addition to the remedies available to agriculture and industry that would allow them to reduce their demand for water.

The Growing Gulf between Water Supply and Demand

Where does the water that humans depend on come from? A review of the water cycle helps us answer this important question, and provides a basis for understanding the options for managing water scarcity.

About 2.5 percent of the world's water is fresh water, with the oceans comprising the rest. Of that 2.5 percent, two-thirds is stored as glaciers and ice caps, while one-third is stored as groundwater. Only a small percentage of the world's fresh water is actually found on earth's surface.[1059]

The sun is the driver of the world's water cycle and evaporates water from the oceans, lakes and rivers, while evapotranspiration (i.e., from plants and irrigation) adds to atmospheric water vapor, which is then transported by circulating air currents. Over land, the mountains intercept these vapor-enriched circulating air currents, forcing them to rise, after which the vapor condenses into clouds.

This process yields the rain and snow that falls over the mountains and water basins below. The rain and melt water infiltrates the ground and causes runoff into streams, as well as recharging groundwater stores. Streams merge into rivers, and rivers flow down through the valleys, merging en route into even bigger rivers that flow out across the open flood plains and end their journey at the sea.[1060]

Nearly half of rainfall becomes surface runoff, eventually flowing into the rivers and lakes, while the other half recharges groundwater stores and the deeper aquifers. Rivers and floodplains absorb rainfall and release it over time, protecting us from floods, while preserving water supplies in times of drought.[1061]

On reviewing the various elements of the water cycle, water flows, and water interactions within the water basin, we can appreciate that there are two main sources of water for human use.

First, there is water runoff, which is captured in rivers, dams, lakes, and reservoirs. Second, there are deeper ground water stores and aquifers (i.e., nature's underground water tanks), placed there long before humans began extracting water from these deep reservoirs.

Population and Economic Growth
Accelerate the Depletion of Aquifers

Water basin stress, transboundary river stress, and the depletion of groundwater and aquifers constitute an ever-increasing problem, especially in semi-arid and arid regions. This growing water stress and depletion contribute mightily to global water scarcity,[1062],[1063,1064] which is further exacerbated by population and economic growth and climate change.[1065],[1066,1067]

USAID, a US government-operated international humanitarian and development organization, tell us that nearly 30 percent of the world's population lacks access to safe drinking water,[1068] and more than half live with water scarcity.[1069,1070]

The ground water that humans rely on today accumulated over long periods in vast underground aquifers.[1071] Today groundwater is an important source of fresh water to over one-quarter of the world's population. Human activity has depleted two-thirds of the world's largest aquifers beyond sustainable thresholds, and pushed one-third into significant stress.[1072]

This overexploitation of groundwater and aquifers is a growing phenomenon,[1073] and is happening in regions important to agriculture, especially in Asia and North America.[1074,1075,1076] These regions are important to global food security (see Chapter 11), and were impacted by severe and prolonged drought during the Little Ice Age's four grand solar minima (see Chapter 7).

On the surface, human activity's impact is no different than it is below the surface. More than 60 percent of the world's largest rivers have already been dammed,[1077,1078] with more than 40,000 large dams in existence today.[1079] Many more thousands of hydroelectric dam projects are also in progress or are in the planning stages.[1080]

River outflows have markedly declined in major river systems because of the over-extraction of water, especially for agricultural use.[1081] Asia, Europe, and North America are the biggest continental users of irrigation, while India, China, and the USA are the three top country users of irrigation.[1082,1083]

Rivers also unite nations and communities, with more than 260 large rivers crossing more than one country. Their combined water basins cover nearly half the planet's land surface area.[1084] These transboundary river systems highlight big problems, with an estimated one-third to one-half of transboundary river populations already impacted by water stress and scarcity.[1085]

As such, transboundary river systems are potential sources of conflict. The already high level of water stress associated with transboundary river systems will be exacerbated by ongoing economic and population growth, and by future droughts. Increasing water stress combined with a lack of international, cooperative water agreements[1086,1087] means we are facing increased risks to our global food security, and will face the prospect of conflict during times of extreme drought.[1088,1089]

Depleted Water Basins Have Limited Drought-Buffering Capacity

Agriculture Is the Thirstiest of All Sectors

Agriculture is the main global user of land, with 1.5 billion hectares used for crops and 3.5 billion hectares used for grazing livestock.[1090] Agriculture is also the biggest user of freshwater resources by far, accounting for between 70 percent and 90 percent of freshwater extractions.[1091,1092,1093]

Human food and non-food crops account for seventy percent of agricultural water use, while animal production consumes the balance, mostly for growing animal feed crops (i.e., maize and soybeans).[1094]

Forty percent of global crops are produced using irrigation, on less than one-fifth of the world's cropland. Six nations, led by India, China, and the USA, account for half of the irrigation water used.[1095] This concentration of food production in just a small number of stressed water basins, in areas that were impacted by the Little Ice Age's grand solar minima, presents a challenge to global food security as we enter this grand solar minimum.

The world's staple food crops, including wheat, rice, maize, barley, sugarcane, soybeans, and ten other vegetables, account for more than half of global crop production and two-thirds of crop irrigation.[1096] On the other hand, non-food crops like cotton, tobacco, coffee, tea, cocoa, spices, and rubber, while not the largest crops, are thirsty water consumers on a per ton yield basis.[1097]

Another important consequence of irrigation is that large quantities of energy are used for delivering irrigation water. In the USA up to one-sixth of the energy used for crop production is used for pumping irrigation water. This reflects the fact that two-thirds of irrigation water is being pumped from increasingly deeper underground aquifers, due to shortages of surface water.[1098,1099]

The relatively high cost of treating and delivering water has led governments to subsidize water for agricultural use, as well as subsidizing the cost of implementing efficient irrigation systems. Paradoxically, however, this use of subsidies incentivizes greater water use and increases water basin depletion. More land gets irrigated and farmers change to higher value but thirstier crops, and less water is returned to groundwater storage.[1100,1101]

In times of drought and increasing water scarcity, we should remember that animal-related products are also thirsty consumers of scarce water resources. In fact, the water requirement for producing one kilogram of meat is some 20–50 times higher than for producing the equivalent nutrition in crops for direct human consumption.[1102]

The important lesson here is that in times of severe drought affecting already stressed water basins, *crop rotation away from non-food crops, biofuels, grain-fed livestock, and super-thirsty rice* makes good water-saving sense. Switching staple foods will no doubt represent a cultural challenge, especially in Asia where rice is a primary staple, and among the affluent who consume large quantities of meat.

Sustainable Industrial Use of Water

Industrial use of water varies according to stage of economic development. In China and the USA industrial water use accounts for about one-fifth of total water extractions, whereas in Europe it's about 40%, while at the global level average industrial use is about 5%.[1103,1104]

Industry's move toward sustainable water use would be facilitated by policies directed toward ensuring that companies interact with their water basin and environment in a sustainable and non-polluting manner.[1105,1106]

Water is crucial to many industrial processes. The biggest industrial user of water is the energy and electricity production sector,[1107] with mining and quarrying, construction, and manufacturing (i.e., food, beverages, chemicals, and paper) also being big water users.[1108] Water is generally used in industrial processes to wash, rinse, heat, cool, dilute and mix substances, and to drive turbines.

Principle areas for industrial action include improving efficiency of water use by reengineering products and processes. Reengineering aims to eliminate non-essential water use, while reducing the use of water in manufacturing and other industrial processes.[1109] Ensuring that manufacturing process water and wastewater is purified and recycled is a key part of sustainable industrial development, as well as preventing the pollution of groundwater and rivers. Industry, like agriculture, must ensure its leaky pipes are fixed to reduce losses.

Because energy production uses more water than any other industry sector, and industries are the largest consumer of electricity, other industries can help the energy sector reduce its water footprint, as well as their own, by switching to renewable energy. Vast quantities of fresh water are used annually in the inefficient production of energy and its conversion to electricity. This water is used mostly for extraction, production, conversion, and cooling processes.[1110,1111,1112] Renewable energy technologies eliminate water use during the generation of electricity, with the exception of concentrating solar power in which small quantities are used for cleaning reflectors.

Supporting Urbanization and Megacity Growth with Sustainable Water Resources

Domestic water consumption competes with the demands from agriculture and industry, and accounts for about 4 percent of total water extractions.[1113] Big cities located in stressed water basins and arid and semi-arid regions will be vulnerable to drought in this grand solar minimum.

Giving this water supply vulnerability some context, in the space of 50 years after 1950 there was a five-fold increase in the number of world cities exceeding one million inhabitants.[1114] Similarly, the population of the world's 100 largest cities grew to more than 6 million inhabitants per city on average.[1115] This means that more than half of the world's population currently lives in cities, rising to over two-thirds in some regions, such as Europe.[1116,1117,1118]

Rapid urbanization and overall population growth are relentless, particularly in regions such as Asia and Africa.[1119] They present fundamental challenges to city planners' abilities to manage ever-scarcer water resources.

The second big urbanization trend is that coastal population growth outpaces non-coastal population growth. Between 30 and 40 percent of the world's population live within 100 kilometers of a coast.[1120,1121]

In fact, two-thirds of cities with more than 5 million inhabitants as well as most of the world's megacities are ten meters or less above sea level. This grouping of populations near the coast occurs in the majority of the world's nations. While this proximity to the coast poses risks to cities (i.e., coastal flooding), important opportunities are also afforded for the decentralization of municipal freshwater supply through renewable energy desalination.[1122,1123,1124,1125]

Mitigating Water Supply Risks during a Grand Solar Minimum

The conclusion I draw from the above review is that the unrestrained agricultural and industrial use of water, combined with population and economic growth and increasing urbanization, *create an ever-widening gap between freshwater supply and demand.* This gap will be exacerbated by low rainfall and drought, especially in arid and semi-arid regions, and particularly during this grand solar minimum. This collectively leaves us vulnerable to extreme drought and exposes populations to higher levels of disease, malnutrition, and famine, leading to higher mortality rates. Water scarcity could also precipitate human conflict between communities and between nations.

Three main topics considered for mitigating water supply risks are reviewed below. First is integrated water resources management, especially the need for artificial groundwater recharge. Second is renewable energy desalination. Third is the use of water pipelines to transport bulk water to regions and cities in need. These three topics of discussion are in addition to the remedies reviewed above for the agricultural and industrial sectors.

Integrated Water Resources Management and Artificial Groundwater Recharge

Integrated water resources management actively promotes the sustainable use of water resources and their equitable supply among different consumer segments (i.e., agriculture, industry, municipalities). Integrated water resources management is coordinated by government and integrates key stakeholders (industry, consumer groups) involved in the management of water catchment, flood and drought control, and the environment generally.[1126]

Integrated water resources management takes a holistic view of water basin resources, from the mountains and valleys to the coastal river mouth. Integrated water resources management embraces flood control and drought management and maximizes groundwater storage while ensuring the treatment of wastewater and the prevention of pollution.[1127,1128,1129,1130,1131,1132]

At the water basin level, it will become increasingly important to establish controls on water extractions by large water users, such as agriculture and industry. Policy implementation and the use of volume quotas and full economic pricing can be used to limit extractions, while promoting efficiency and innovation in water use. Price controls are not always easy to implement, because they have political consequences (in terms of voting, protests, etc.).[1133,1134] However, when water is scarce and drought becomes extreme, compromises and solutions must be found.

Water flow and groundwater storage in a water basin are modified by dams high up at the headwaters, and by construction at the flood plain level. [1135] Floodplain construction includes channelization, which creates strategically located waterways (i.e., canals, causeways) to control flooding and improve land drainage, and floodplain reclamation (i.e., landfill) which increases the usable land area, permitting greater urbanization and other changes in land use (i.e., roads, industrial parks).[1136]

The downside of flood prevention and floodplain reclamation is that they undermine the natural connection between a river and its floodplain. By eliminating the river-floodplain connection, ground water storage and the water basin's ability to buffer drought are adversely impacted. Reconnecting rivers to flood plains and enabling water to enter natural waterways offers a means for increasing groundwater storage.[1137,1138]

Water basin managers, municipalities, and both large-scale and smallholder farmers are increasingly using *artificial groundwater recharge to increase water stores*.[1139,1140,1141] For large-scale groundwater recharge projects, expert decisions must be made concerning the best location.[1142,1143,1144]

A variety of surface structures are used for groundwater recharge. These improve surface water retention and its infiltration. They include low-level river and streambed dams, contour bunds (i.e., low level earth embankments following a land's contour to hold back or slow water runoff), low level dams in gulleys, reservoirs, storage ponds, irrigation and drainage canals, as well as the terracing of hill slopes. Woodlands and riverside trees and vegetation (i.e., riparian buffers) also assist groundwater recharge.[1145,1146,1147]

Artificial groundwater recharge is increasingly being used for short- and long-term underground water storage. Artificial recharge utilizes permeable surface soils, trenches, or specially located well shafts to directly inject water into the aquifers.[1148,1149] Groundwater recharge utilizes river and wastewater, as well as desalinated water, for the recharging process.[1150]

Forests and woodlands provide a natural means of flood control, especially in the upper water basin and floodplains. Forests regulate storm discharge into rivers and control river flow rates, while increasing the duration of stream flow after rains cease. Forests also help recharge groundwater. Thus forests, and reforesting, provide many benefits—such as a source of renewable energy, a means of flood control and groundwater recharge, and ecosystem benefits (i.e., increasing biodiversity, oxygen production, carbon storage, etc.). These big benefits mean we must make *reforestation a priority.*

Flood control and prevention involves increasing the number of strategically located waterways to direct water away from urban and other areas, and the use of physical flood defenses to protect important areas. Flood defenses include physical structures like flood control dams in the upper catchment areas,[1151] weirs, flood banks, flood protection dikes, contour bunds, water detention basins, and reservoirs. Urban flood protection involves the use of green infrastructure to reduce water runoff and prevent flood control systems from being overwhelmed.[1152,1153,1154]

Decentralized Water Supply Using Renewable Energy Desalination Systems

Do we really have a water shortage? Do we need to face water scarcity and shortages of supply?

We have options for meeting our water needs that go beyond the water cycle, water basin and river systems, and aquifers. With 97 percent of the world's water supply residing in the oceans, and our ability to pipe water thousands of kilometers, the simple answer to the above questions is *No* in both cases.

In an ideal world, we would just need to separate the fresh water from seawater or brackish (i.e., estuary or river mouth) water, using renewable energy desalination systems, and then transport it to where it is needed using canals and pipelines. Pumping would be done with pumps powered by renewable energy.

With a sizable portion of the world's population and the majority of the world's largest cities located close to the coast, renewable energy desalination makes compelling sense. Renewable energy desalination offers coastal municipalities and industries a means for decentralizing their water supply while mitigating water scarcity.

Municipalities account for 70 percent of desalinated water use, while industries use about one-fifth.[1155] Desalination is now an economically viable option for water supply, and is becoming more cost-competitive with other water sources, particularly where large-scale production facilities are located in proximity to renewable energy sources.[1156,1157]

Desalination is presently used to supply about 1 percent of global fresh water, with renewable energy desalination supplying only a small fraction of this. Middle Eastern countries such as Saudi Arabia, the United Arab Emirates, and Kuwait,[1158] as well as the USA,[1159] Spain,[1160,1161] and China[1162] are some of the largest users of desalination technologies. Some examples of small to medium-sized desalination plants are cited here for reference purposes.[1163,1164,1165,1166]

Seawater desalination accounts for most of global desalination capacity worldwide, with river and wastewater also being utilized in desalination systems.[1167,1168] Brackish and river water desalination is a lower cost option compared with desalinating seawater, due to brackish water's lower level of salinity.[1169]

Desalination technologies mainly utilize semi-permeable *membrane desalination processes* such as reverse osmosis, or *thermal desalination processes* using evaporation and distillation processes, such as multi-stage flash and multi-effect distillation.[1170],[1171]

Solar photovoltaic and wind-generated electricity are well suited to reverse osmosis processes, while concentrating solar power, geothermal, fossil fuel, and nuclear electricity and heat co-generation are useful for thermal desalination processes that vaporize the feed water to produce a purified water condensate.[1172,1173,1174]

Solar photovoltaic reverse osmosis systems are highly scalable using bolt-on system additions to scale up water production, making them ideally suited for both small and large towns, and small cities.[1175,1176] Solar stills and solar ponds offer a good solution for areas of low volume and low demand, such as small or remote coastal communities, and for use at home (see citations for homemade designs and design principles).[1177,1178,1179,1180]

The shortcomings associated with renewable energy desalination (caused by the variability in the supply of energy) have made these desalinating systems less attractive when compared with fossil fuel-powered systems. However, advanced management practices now provide desalination plant operators with the ability to buffer the fluctuations in the supply of renewable energy, while extending the hours of daily water production.[1181]

Bulk Water Transportation by Long-Distance Pipelines

It never ceases to amaze me how humans can with ingenuity and innovation find solutions to big problems or needs, if given sufficient time to prepare. Freshwater pipelines and cross-country canal systems transporting bulk water supplies from resource-rich regions (i.e., coastlines, rivers, mountains, and aquifers) hundreds or thousands of kilometers to where they are needed, would be an example of such ingenuity and innovation translated into action.

In China, the South-to-North Water Diversion Project will theoretically transport 45 billion cubic meters of water a year (i.e., the annual volume of the river Thames in England) across a waterway network of nearly 4,500 kilometers. This waterway network comprises giant canals, pipelines, aqueducts, and pumping stations, and crosses the Yangtze, Yellow, Huai, and Hai Rivers.[1182]

Equally ambitious, the Trans-Africa Pipeline Project aims to deliver potable water for up to 30 million people in eleven African Sahel (i.e., the southern Sahara desert margin) countries. Two coastal solar power desalination plants in Mauritania and two on the Red Sea, plus land-based wind turbines, will pump 400,000 cubic meters of desalinated seawater per day inland along an 8,000km pipeline.[1183]

The Great Manmade River of Libya is another of these grand projects undertaken by a government on behalf of its people. This pipeline transports 3.7 million cubic meters of water daily over 2,800 kilometers to Tripoli from deep underground aquifers. Likewise, the California Aqueduct transports bulk water from the Sierra Nevada Mountains and valleys over 640 kilometers away to Southern California, where it is badly needed.

These projects demonstrate great vision born from absolute necessity, as well as the sheer scale of what governments, stakeholders, and their industrial partners are prepared to take on to help their people. They demonstrate that the bulk transportation of water over great distances is eminently feasible. These projects also highlight the different sources of bulk water supply (oceans, rivers, aquifers, and mountains) that can be used to extract and then transport water using massive pipelines hundreds or thousands of kilometers in length.

The combination of coastal seawater and renewable energy, plus industry's proven capability to build pipelines thousands of kilometers long, means we can solve water scarcity problems for coastal and inland cities, industries, and agriculture. These innovations could be used to eliminate human vulnerability to the water cycle and the solar phenomena that will influence it during this grand solar minimum.

CHAPTER 11

Mitigating Climate-Related
Food Supply Risks

What do you think poses the greatest global food security risk?

Is it all bets on the IPCC's Articles 1 and 2 and its global warming story that ignores nature's undismissable risks? What about a global economic model based on growth, resource depletion, and environmental destruction? Could it be growing urbanization and the fact that we've forgotten how to grow food at home? What about the real prospect for government trade restrictions during a food crisis? What about low oil prices, which are helping to prevent a switch to renewable energy? Could pandemic flu trump all other risks?

My top two food security risks are as follows. First, realizing that the Holocene Climate Optimum is behind us, while a switch to a colder climate lies ahead of us, I think Articles 1 and 2 represent by far the biggest unrecognized global food security risk (see Chapters 1–7). Second is the high degree of urbanization, which in turn has caused us to forget how to grow food at home. These two food security risks represent catastrophes waiting to happen.

How do we improve the resilience of our national and global food supply? How do we mitigate the big food supply risks? How do we supply food in a global food crisis?

This chapter's purpose is to provide answers to these questions, and share the major options available to governments, municipalities, and individuals for securing food should a global food supply crisis occur.

The Big Risks to Global Food Security

Weather-related shocks have a major impact on food production because crops and animals are sensitive to weather extremes.[1184,1185]

With the Holocene Climate Optimum behind us, and the likelihood of a switch to a global cooling phase (see Chapters 3–6), the UNFCCC and IPCC's Articles 1 and 2 focused on global warming represent the biggest unrecognized global food security risk by far (see Chapters 1, 2, and 7).

Our "business as usual" global economic model, based on growth and resource depletion with little regard for the environment, also poses major food supply risks. Under business as usual, it is expected there will be 9-plus billion people living in the world by 2050. Most of the population growth from current levels is expected to be in drought-prone, aquifer-depleted, rapidly urbanizing, and malnourished[1186] Asia and Africa,[1187,1188,1189,1190] which depend heavily on fossil fuel imports to achieve economic growth.

Under the business-as-usual scenario it is expected that global food supply will need to increase by between 60 and 100 percent by the mid-21st century.[1191,1192] The downside of growing urbanization is that there are fewer food producers relative to the number of consumers,[1193] something which increases vulnerability to food supply risks. In a food crisis, trade restrictions imposed by exporting nations will be the norm. Corporate food suppliers in nations that do not restrict trade in food will also likely divert food supplies for profit, causing local food price inflation. This will undermine access to food for people in urban areas—especially the poor and, more generally, those living in developing nations.

We face another big risk to future food security in the form of higher oil and gas prices (see Chapter 8) as these resources become increasingly scarce. Ultimately, food price inflation will be the consequence of this. Higher food prices will for the most part impact the poor and developing nations, making access to food more costly and contributing to an increase poverty.[1194,1195]

Our Fragile Global Food Supply

Today nearly one-quarter of global agricultural food production, or more than half a trillion dollars' worth, is traded internationally.[1196,1197] Eighty percent of people now live in countries that must import food, with 10 percent of the world's population living in countries importing more than half of their food supply. This reliance on food imports is most notable in the Middle East, Asia, Africa, Central America, and Southern Europe.[1198]

Twenty food-exporting nations account for the majority of global food trade. The leading agricultural food exporters include the European Union, the USA, Brazil, China, Canada, Argentina, Indonesia, Australia, Malaysia, India, Russia, Ukraine, and Kazakhstan.[1199,1200,1201]

Three main crop types, wheat, maize and soybeans, account for half of exported food crops.[1202] Five nations account for between two-thirds and 80 percent of wheat, maize, and rice exports.[1203,1204,1205] This dependence on a small number of crop-exporting nations creates huge vulnerabilities in our global food security. When crop yields in the major food producing and exporting nations are adversely impacted by extreme weather, import-dependent nations will become vulnerable to trade restrictions imposed by exporting nations, and will therefore be exposed to massive food price inflation.

For example, the food supply crisis of 2008 was the result of a price spike that resulted from drought-related crop failures among major exporters in late 2006, along with other market factors. Prices of the main staples increased by more than 50 percent, and in the case of rice prices tripled. This resulted in food riots in some countries, following the imposition of trade restrictions by some exporting nations. Export disruptions necessitated the introduction of import subsidies by the governments of importing nations, for the purpose of quelling public unrest.[1206,1207,1208,1209,1210]

All of the above demonstrates the high level of interdependence between nations when it comes to food, the vulnerability of the global food system to climate change, as well as its vulnerability to price inflation. This high level of interdependence makes global trade, government restrictions on trade, and global commodity markets important influences on global food security.

How Do We Improve Food Supply Resilience?

Given the fragility of our global food supply to climate disruption, and the high level of dependence of many nations on food imports, how do nations improve the resilience of their food supply?

The simple answer to this question is that each one of us is a part of the solution. A number of important levers are available to people, communities, cities, food importers, governments, and organizations involved in developing climate-adapted crops (see citation)[1211] to improve the resilience of our food supply, both globally and nationally.

The most obvious solution is for more nations, municipalities, and homes to become almost self-sufficient in food supply, so that we may provide for a higher percentage of our own food needs. This will make us less dependent on food imports and on food supplied by corporations. This self-sufficiency underpins decentralized sustainable development. Decentralized sustainable development can protect us against food trade restrictions by governments, the impact of large food suppliers diverting food for profit, and commodity market speculation, all of which hinder the supply of food and drive up prices in a food crisis.

For countries that rely on food imports, increasing the number of supplying countries, suppliers, and the number of crop varieties imported can help them diversify their food supply.[1212] Diversifying the number of countries from which food is imported, and whose weather is controlled by different air circulation and monsoon systems, will also protect against the impact of low solar activity and volcanism (see Chapter 7). Trade agreements can be implemented to ensure food supply in times of crisis, with corporate suppliers prioritizing trade agreements over general exports. In times of a food crisis or short supply, consumers can also diversify their palates and reduce their reliance on any one type of food, such as, for example, water-thirsty rice.

Governments hold grain reserves for emergencies or for stabilizing prices in a food supply crisis. For countries with food stockpiles, these can be used to make good any food supply deficits for a number of months while food is being grown. However, food stockpiles have declined in recent decades, for a number of reasons.[1213,1214] National food stockpiles should be reassessed in preparation for a climate switch and accompanying risks such as climate-forcing volcanism or a pandemic flu outbreak. Industrial-scale and automated greenhouses, high-tech indoor and vertical farming, as well as the large-scale manufacture of single cell proteins, are all proven climate- and sunlight-independent food production technologies (reviewed below). These food production systems could be used to provide emergency food supplies in a crisis.

In the event of a climate-induced food crisis, switching from non-food crops such as those grown for biofuels (e.g., woody crops), beverages (e.g., coffee, cocoa, tea), and fibers (e.g., cotton) to food crops is an option for increasing available food crop acreage.[1215] Croplands changed from non-food crops, as well as livestock pasturelands, can be planted with short-cycle crops such as potatoes, sweet potatoes, millet, pulses, maize, wheat, barley, etc. Diverting grain from livestock production to human consumption will also help expand food supply.

Mitigating Climate Extremes with Climate-Adapted Crops

The world's main food staples are wheat, rice, maize, barley, sugarcane, soybeans, and ten other vegetables. Other important food staples include millet, sorghum, rye, barley, oats, roots and tubers (potatoes, cassava, etc.), as well as animal products (meat, eggs, milk, fish).[1216],[1217]

Climate-adapted maize, wheat, rice, beans, and potatoes are available, both commercially and through the plant breeding programs of international crop development organizations. These organizations include the International Maize and Wheat Improvement Center, International Rice Research Institute, The Pan-African Bean Research Alliance, International Crops Research Institute for the Semi-Arid Tropics, and the International Potato Centre (collectively, "crop stakeholders").[1218]

These crop stakeholders maintain gene banks, seed collections, and breeding programs for the development of climate-adapted crop staples. Crop stakeholders have also assembled global networks to help the farmers of developing nations grow climate-adapted crops in a sustainable and equitable manner. In the future, once a climate switch and other climate risks are recognized worldwide as distinct possibilities, then crop stakeholders will need to increase their efforts to develop cold- and drought- adapted crops. The sooner crop stakeholders change their focus the better, given that it takes years to introduce new climate-adapted crops to farmers in developing nations.

Crop Staples Adapted for Harsh Winters and Cold Climates

A cold temperature threshold must be crossed for a minimum amount of time before plant damage occurs. The species and variety of plant, its stage of development, soil conditions, and climate factors associated with the freeze (such as wind chill factor) influence this temperature threshold. Inadequate acclimation of young plants in the autumn, and the duration and intensity of sub-zero temperatures, determines how well crops survive the winter cold or frosts.

Vernalization is a natural adaptation mechanism by cereal crops growing in harsh winters or short growing seasons. This adaptation ensures flowering occurs in the spring and seeds mature before the next winter.[1219] Winter cereals must be planted before the end of the winter's intense cold phase for vernalization to occur, whereas spring cereals will flower soon after their spring sowing without the need for vernalization.[1220]

Wheat is a very important global crop in Asia, Eurasia, and North America.[1221] Wheat has the broadest adaptation of all cereal crops, and has good cold tolerance.[1222] Spring and winter wheat varieties have been bred, and the most cold-tolerant wheat varieties are killed at just below −20°C.[1223] With adequate cold acclimation in the autumn, winter wheat can withstand freezing temperatures for extended periods,[1224] making wheat a versatile winter and cold climate crop.

Winter rye has historically been the national crop in colder lands, such as northern and central Russia and northern Europe.[1225] Rye is the most cold-tolerant and drought- tolerant of the cool season grass crops, followed by wheat, barley, and then oats.[1226,1227] The most cold-tolerant rye varieties are killed at about −30°C, which gives hope that it can be a staple crop during a climate switch.[1228]

Rice is an important crop, especially in Asia, [1229] and is widely cultivated between the mid-latitude regions, and at up to 3,000 meters in altitude. The Japonica varieties have a higher degree of cold tolerance than the Indica varieties.[1230,1231,1232] Further efforts to improve rice's tolerance for the cold are currently in progress.[1233]

Maize is also a very important global crop, grown in most tropical and temperate latitudes and at altitude. Early sowing of maize increases yields and helps avoid late summer drought, but this early sowing requires cold tolerance traits. Selective breeding of maize for cold tolerance has resulted in varieties able to withstand cold spring temperatures and short-term frosts,[1234,1235,1236,1237] which has improved maize's ability to survive in a colder climate.

Drought-Tolerant Crop Staples Need More R&D

Conventional maize breeding has resulted in improved grain yields under drought conditions. This selective breeding has targeted such traits as increased plants per hectare, ears per plant, seeds per ear, and seed weight.[1238,1239]

The International Maize and Wheat Improvement Center (CIMMYT) develops drought-resistant maize, and supplies half of the world's maize varieties.[1240] The CIMMYT's drought-tolerant varieties are products of conventional plant breeding, and these have provided improvements in yield of up to 50 percent.[1241] Progress has also been made with genetically modified maize varieties that minimize drought impact.[1242]

More than half of the wheat acreage in developing countries utilizes CIMMYT- developed varieties.[1243] The reality, though, is that progress in breeding drought-tolerant wheat varieties has been slow,[1244] and has so far provided insufficient yield improvements to feed a growing population in a more drought-prone world.[1245]

This limited progress in breeding drought-resistant wheat is the result of the fact that other plant stresses such as high temperatures, solar irradiance, and nutrient deficiencies typically accompany drought. These stressors complicate plant breeding selection by multiplying the number of variables to study in the selection process.[1246,1247,1248]

Integrating genetic material from distant wheat ancestors or drought-tolerant rye would be a quick route to improving yields under drought conditions.[1249,1250,1251] Hybrids (i.e., wheat-rye) provide fast-track development options to more rapidly improve wheat's drought tolerance, as opposed to using traditional plant breeding methods.

Given Asia's high dependency on rice that is highly vulnerable to drought, it will be necessary to improve the drought tolerance of rain-fed rice. Two main options are offered for improving the drought tolerance of rice.[1252,1253]

First is the transfer of genetic material from upland drought-tolerant rice varieties to lowland rice varieties vulnerable to drought and grown in drought-prone regions.[1254] The second option is the development of genetically modified drought-tolerant rice, which currently is an area of active research.[1255] Alternatively, and in the face of a worsening drought, changing the palates of Asians and Africans to less thirsty wheat and maize, which require half the water needed by rice, would provide another water-saving solution.

Drought-tolerant cover crops such as millets,[1256] sorghum,[1257,1258] pigeon pea,[1259,1260] and cowpea[1261] help farmers survive drought, especially during the hunger months of the dry season when food supplies run low. Pearl millet and sorghum are among the most drought-tolerant of all the main staples, and are important crops in arid and semi-arid regions.[1262,1263]

Millets are small-grain grasses, and are the most drought-tolerant of the summer annual grass crops. They will germinate when very little moisture is available.[1264] The fast growth and maturation of millets make them well suited to intensive cropping systems in semi-arid and arid climates, particularly as drought progresses and limits the potential use of irrigation.

At the other extreme of water supply, rice normally dies within days of its complete submergence, making rice in flood-prone regions (like parts of India and Bangladesh) vulnerable to extreme rainfall. Progress has been made in the development of submergence-tolerant rice that is able to survive completely submerged for up to two weeks.[1265,1266] Regional differences in the impact of grand solar minima make rice's ability to tolerate submergence important for parts of Asia, which could experience more monsoon rainfall while other areas experience more drought.

How Do We Produce More Food Sustainably?

Promoting Sustainable Agriculture and Supporting Smallholder Farmers

Yield increases alone are unlikely to meet the projected 60–100 percent food supply increase required to feed 9 billion-plus people by mid-century, meaning cropland area will need to increase.[1267,1268] Without a switch to sustainable farming, more cropland will be required, to the detriment of biodiversity and the environment.

Sustainable farming means increasing the intensity of food production, improving the resilience of food production systems, and doing both with less environmental impact. These things move the focus from a resource-intensive system (i.e., heavy use of water, fertilizer, pesticides, and energy) to a knowledge-intensive system, with a key focus on managing environmental impact and biodiversity.[1269,1270]

The most practical and sustainable way to increase food production yields is for farmers to close the yield gap—in other words, the gap between the potential yield using best practice farming methods and currently realized farm yields. A wide variation in crop yields exists within and between different countries, for many reasons. By closing the yield gap, more food can be produced without increasing cropland area. Potential yields depend on seed genetics, the judicious use of irrigation water and fertilizers, optimal soil quality and management, pest management, farmer knowledge and production practices, and also on the local or regional climate.[1271,1272]

More than 1.5 billion rural people worldwide live on more than 500 million small farms, each averaging less than two hectares in area. Three-quarters of these smallholder farms are in Asia and 10 percent are in Africa.[1273] These smallholder farmers account for more than half of food grown for domestic consumption in low- to middle-income countries. In Africa, smallholders are the mainstay of domestic food production.[1274,1275,1276]

The smallholder farmer is therefore critical to achieving food security in most nations,[1277] and particularly in Asia, Africa, and Latin America. In addition, smallholder farmers are responsible for feeding most of the world's poor.

When smallholder farmers adopt sustainable agriculture, average crop yields increase significantly.[1278,1279] Climate change, particularly drought, is one of the greatest risks to smallholder farmers and the poor whom they feed. Therefore, the interests of governments and crop development organizations[1280]are best served by helping smallholder farmers move to sustainable farming systems—by ensuring that irrigation water is made available to them, and by helping them gain access to climate-adapted seeds.

Up to one-third of global food production is wasted.[1281,1282] Food waste is higher for fresh fruit and vegetables than for grains. More food is wasted between the farm and retail outlets in developing nations, whereas more waste occurs in the retail outlets and at home in developed nations.[1283,1284]

By reducing the net food waste from farm to plate, the effective food supply can be increased (i.e., not wasted). Optimizing food storage, processing, and times of transit, and educating consumers about food waste will all be important for reducing food waste and increasing the effective food supply.

Aquaculture Produces More Protein from Less Food than Livestock Production Systems

We cannot expect to sustainably harvest more fish from the oceans, because we have already pushed most fish populations to or beyond their sustainable limits.[1285] Therefore, in order to produce more fish for consumption, sustainable aquaculture will be required. Aquaculture already produces half of the world's fish supply,[1286,1287] with Asia (principally China) producing 90 percent of global supply. Carp, tilapia, shrimp, marine molluscs, and catfish dominate aquaculture production.[1288,1289,1290] Aquaculture utilizes lakes, ponds, canals, tanks and cages, and a wide range of feed types and production technologies.[1291,1292]

When we consider meeting the animal protein needs of 9 billion-plus people, farmed fish has three big advantages over grain-fed livestock. Less grain, less fishmeal, and less water basin depletion is required per kilogram of fish protein produced.

Fish are among the most efficient converters of food into high quality protein, and require only one-quarter and one-third of the grain required by cows and pigs, respectively, to produce one kilogram of fish protein.[1293] This is because fish do not rely on food to make body heat as mammals and birds do, or use muscles to stand upright, utilizing buoyancy instead. This in turn lowers their food requirements for producing body mass as compared with terrestrial livestock.

The impediment to market growth for aquaculture is the amount of fishmeal and oils required to produce farmed fish. On average, for every kilogram of fish produced in aquaculture systems, 0.7 kilogram of wild fish is required (i.e., non-edible fish, fish by-products). Much more fishmeal is needed to produce carnivorous fish (i.e., salmon, trout), and much less for herbivorous or omnivorous fish (i.e., carp, tilapia, milkfish, and catfish).[1294,1295]

Reducing aquaculture's dependence on wild fish is therefore seen as a key priority by the fish feed industry. Soybeans, maize, meat by-products, yeast, and microalgae are being used as substitute nutrients in fish feeds, to make fish farming more sustainable now and in the future.[1296,1297,1298]

Another big plus of fish protein production via aquaculture, as against grain-fed livestock protein production, is the fact that the poultry and swine meat industries are the world's largest consumers of fishmeal for animal feeds (i.e., a source of protein and oils).[1299] This makes fish production more sustainable than livestock production when it comes to maintaining ocean fish stocks.

Traditional aquaculture rearing systems use little or no fishmeal. In fact, most of carp and two-thirds of tilapia production worldwide do not use professionally manufactured fish feeds.[1300] Tilapia and carp aquaculture is well suited to smallholder farmers and for urban fish supply, given that homemade diets composed of rice, beans, sweet-corn, supplements, etc. can be fed to fish instead of using commercial brands of fish food.[1301]

Outside of China and the rest of Asia, in regions such as Africa, aquaculture is not yet a dominant source of fish supply. Depletion of ocean fish stocks, reduced fish catch in coastal Africa, and Africa's increasing urbanization justify the further development of aquaculture in Africa.[1302,1303]

For aquaculture to become more sustainable, the industry and stakeholders need to expand the use of herbivorous-omnivorous fish aquaculture (i.e., tilapia, carp, shellish), reduce aquaculture's dependency on fishmeal and oils, develop multi-species systems to increase productivity (i.e., salmon or trout plus shrimp or mussels) and provide environmental benefits (i.e., biological waste treatment with hydro-ponics), while minimizing aquaculture's environmental impact.[1304,1305]

Urban Agriculture Provides Food Security and Feeds the Poor

For almost the entire duration of the Holocene, right up until the Green Revolution really got underway in the 1950s and '60s,[1306,1307] humans produced most of their food close to home. Sixty years on, most city-dwelling humans have forgotten how to grow food, or have no facility for growing food at home. With more than half the world now living in urban areas, the challenge will be to feed urban residents during a sustained food supply crisis.

On a worldwide basis there are perhaps 200 million urban farmers, with two-thirds of these food providers being women.[1308,1309] Urban agriculture is already an important component of urban food supply to the world's poor, and would play a key role in helping provide food security during a serious food crisis.[1310]

To put a global food crisis in perspective, under normal conditions the world's poor spend the majority of their income on food,[1311] and would therefore be highly vulnerable to food price inflation during a food crisis. With an estimated 1 billion poor people subsisting in urban slums, particularly in Asia, Africa, and Latin America,[1312] urban agriculture has a key role to play in creating urban food security.

Urban agriculture involves the production of food at home or in a variety of urban land locations and building types. Urban agriculture utilizes gardens, greenhouses, balconies, walls, rooftops, and plots of peri-urban land (i.e., land surrounding cities) up to tens of kilometres from the city boundary. Food production is achieved using traditional land garden-based methods and technology-enabled systems such as vertical gardens, greenhouses, covered crops, hydro- and aqua-ponic systems, and aquaculture systems (see below and Chapter 12).[1313,1314,1315,1316,1317]

Typically, urban and peri-urban agriculture supplies food all year round, generally consisting of perishable, short-cycle crops such as vegetables (leafy greens, tomatoes, potatoes, cucumbers, peppers, etc.), herbs and fruits, medicinal crops, and fish and small livestock.

Urban and peri-urban food production is complementary to rural food supply, which ensures perishable products are grown or raised close to their point of consumption. This reduces food waste, particularly where cold storage is limited.[1318,1319,1320]

In some countries the level of urban food production is high, demonstrating that with the right municipal support significant food quantities can be grown in or around cities. For example, in Asia, Africa, Latin America, and Eastern Europe urban agriculture supplies about one-third of urban household food, rising to over half among the poor.[1321] For certain foods such as greens, grains, milk, and eggs, urban and peri-urban farming can fully provide a small city's food supply.[1322,1323,1324]

What is very clear is that if city governments wish to decentralize a significant share of a city's food supply in order to achieve improved food security, then they must make urban food production a key part of city planning. This includes supportive policies, land designation, and the provision of infrastructure and services (i.e., irrigation, waste management, compost, markets, and storage), advisory help, and educating children about urban agriculture.

Urban Food Production without Soil

The common denominator between hydroponics, aeroponics, and aquaponics, is that food can be grown efficiently using nutrient-rich water instead of soil. These production systems are well suited to growing vegetable greens, vine crops such as tomatoes, cucumbers, peppers, squash, courgettes, and green herbs.[1325,1326]

Hydroponics utilizes a small fraction of the water resources used in conventional land-based methods, while bathing plant roots in nutrient-rich solutions.[1327,1328] Aeroponics on the other hand utilizes misting systems to deliver nutrients directly to plant roots, and thus drastically reduces resource inputs.[1329]

Aquaponics couples freshwater aquaculture with hydroponics to produce fish (e.g., tilapia, carp), as well as fruit, vegetables, and herbs. Plant roots suspended in flat tanks or in racks of horizontal PVC-tubes that can be stacked vertically filter the nutrient-rich fish wastewater.[1330,1331]

Hydro-aero-aqua-ponics are important technologies for commercial and private urban food production, and can be deployed indoors, on rooftops, and in greenhouse settings, and on different production scales from small to commercial.[1332,1333,1334] While expensive and knowledge-intensive during setup, once operational food production is cheaper than by conventional farming methods.[1335]

These systems eliminate the need for fertile land, utilize a fraction of space compared with traditional food production, and can be stacked vertically using various methods. They permit year-round food production, resulting in increased crop yields.[1336,1337] All of the above qualities make hydro-aero-aqua-ponic systems among the most promising technologies for sustainable urban food production.

These systems can be made more sustainable by integrating them into energy efficient buildings, and employing renewable energy systems for lighting, water pumps, and water and air heating. Recovering energy from biomass (i.e., gas) or converting that biomass to body mass by feeding it to poultry or goats will help make these systems fully sustainable.[1338] Rainwater harvesting and recycling water and nutrients will improve the sustainability of hydro-aero-aqua-ponic systems.

Emergency Food Production Systems

How will our climate-dependent agricultural system cope with a climate switch, prolonged and extreme drought, a rapid climate change event, or worsening extreme weather events? How will we cope with a climate-forcing volcanic eruption (VEI 7) wiping out half a continent's agriculture, while blocking out the sun and sharply cooling the planet for a couple of years to a decade afterwards?

The limited risk assessment contained in IPCC Articles 1 and 2 means our governments don't have a publicly available plan for the above agriculture-disabling climate risk scenarios.

Climate- and Sunlight-Independent Indoor Farming

Large-scale greenhouses are already a well-developed means for creating a fresh supply of urban food. These highly automated, controlled environment agriculture systems are well suited for intensive food production in peri-urban locations. The integration of large-scale greenhouses with renewable energy systems will help control operational costs and improve the profitability of commercial greenhouses.[1339,1340,1341,1342]

Commercial indoor and vertical farming operations are sprouting up around the world and operating 24/7, growing a wide range of crops like fruits, vegetables, and herbs.[1343,1344] Indoor farms utilize optimized modular hydroponics and aeroponics food production systems, which control the main plant-growing conditions (i.e., temperature, light, nutrients, carbon dioxide) and permit automation and system monitoring. These production units are stacked vertically inside high-rise and purpose-built buildings.[1345]

In order to improve the economic viability of indoor farming, it will need to be employed on a very large scale, while utilizing or adapting existing greenhouse-controlled environment and automation technologies and systems.[1346,1347]

Indoor farms (and greenhouses) have high operating costs, particularly as regards the supply of energy. They can be made more sustainable and profitable by housing them in energy efficient buildings and integrating them with renewable energy systems such as solar and wind, ground-coupled heat exchangers, and biomass energy recovery, as well as using renewable energy desalination systems for their water supply.

Indoor vertical farms can be located in and around cities, in the desert, on wasteland, in proximity to renewable sources of energy, and close to water. They can also be located close to where the food grown is consumed, ensuring high quality food supply while limiting transportation costs and reducing food waste.

Wealthier Middle Eastern and industrialized nations, countries rich in renewable energy resources, and large cities could all benefit from this type of climate-independent farming. It would provide decentralized food supply, emergency food supply, as well as improving overall food security. This type of farming, coupled with physical food reserves, could be used to provide the means of *weathering a major food crisis.*

Emergency Human Food Supply and Single Cell Proteins

Single-cell proteins are produced by an array of microorganisms, and have been used as protein supplements in human foods and animal feeds for decades.[1348,1349] Products for human consumption include the old standby brewer's yeast and commercial brands like Pruteen, Torula, and the Quorn™ range of products.

Manufacturing processes for single-cell proteins use biomass and petrochemicals as starting ingredients. Production yields, orders of magnitude greater than for plant proteins, are rapidly achieved. This makes single-cell proteins highly suitable as emergency food.

Various microorganisms, including yeast, fungi, bacteria, and algae, have been used to produce single-cell proteins. These microorganisms can utilize a variety of starting materials like agroforestry and industrial waste, fossil fuels, and alcohols, to make food.[1350]

Single-cell protein food results in high quantities of uric acid in the blood, which is not good for human health in the long term because it usually causes gout. This medical risk has limited the use of single-cell protein in human foods. As emergency food on a short-term basis, this is probably less of an issue.

The relatively high cost associated with single-cell protein production has had an impact on the broader use of single-cell proteins in animal feeds. Grains and soya supply most animal feed protein at a fraction of the cost.

Single-cell proteins could be repositioned as sunlight- and climate-independent food for complementing emergency food stockpiles, or even as a partial replacement for current food stockpiles. Regulating the animal feed industry for a period of time (i.e., this grand solar minimum), while supporting the single-cell protein production industry financially, would help ensure that an economically viable industrial capacity for emergency food production is developed ahead of a climate or food supply crisis. This animal feed capacity could then be borrowed for emergency human production in times of food crisis.

Regulating the animal feed industry could ensure minimum quantities of single cell proteins were used in animal feeds, in lieu of irrigated crop and fishmeal resources. Financial support for the single-cell protein industry could be justified on the basis of investing in a "just-in-time" food stockpile production capability.

CHAPTER 12

Living Sustainably at Home in Cold Climates and Climate Extremes

What will it mean to live sustainably at home during a switch to a colder climate with extremes of precipitation (drought, rainfall, snow)? How can we mitigate energy, food, and water supply risks during a climate or food crisis, in the knowledge that governments will restrict trade and corporate suppliers will seek to profit even more when resource supply tightens?

In Chapter 8, the principles for living sustainably were reviewed. In essence, this review highlighted that we must strive to use all energy, water, and food resources (as well as other resources) sparingly and efficiently. We must also maximize residual resource and energy recovery before generating a minimized amount of waste and pollution. Living sustainably should permeate our way of life in everything that we do at home, work, and when we travel.

Decentralized sustainable living promotes a level of self-sufficiency for supplying your own renewable energy, water, and food—at home. That way you will have solutions at the ready in a time of crisis. This will help you manage the risks associated with a tightening of resource supply by corporate suppliers and by trade-restricting governments, and help you avoid becoming a victim to hyper-price inflation. Decentralized sustainable living places you in control of your and your family's fate, and reduces your vulnerability to climate-related risk.

A sense of urgency is required to prepare for a climate switch and its associated risks, and to begin living sustainably. We're more vulnerable to climate switching and its associated risks, (i.e., cold climate, extremes of precipitation, pandemic flu) because we've depleted our oil, gas, and water reserves more than is generally realized. This resource depletion will impact energy access, food supply and prices, and water access. This is not a transitory situation we face, but the future long-term reality for our species. The Holocene Climate Optimum and, therefore, the start of the current ice age are already 8–10.5 millennia behind us. When the climate switches or a climate-forcing volcanic eruption occurs—welcome to our Ice Age return.

To that end, this chapter reviews best practice principles for building or retrofitting your home in order to reduce your energy needs, operate your home energy-efficiently, and power it using renewable energy. Practical advice is given on how to minimize water use in the home while efficiently using, re-using, and harvesting water, and recharging groundwater.

Also reviewed are best practice methods concerning how to grow climate-adapted food at home using a variety of urban and climate-adapted methods, while ensuring you have food stockpiled, a seed bank, and food growing systems ready to go.

Efficient Renewable Energy Use At Home

Residential use of electricity consumes 30 percent of electricity supplied globally.[1351] This makes house design for new buildings and the retrofitting of existing homes pivotal to living sustainably. The home design principles for living sustainably include reducing your energy needs at home, improving efficiency, and installing renewable energy and water heating systems to supply your needs.[1352,1353]

House Design Principals for Minimizing Energy Use

For new buildings, in order to optimize the natural energy gained during the day, the house should be oriented to maximize the sun's free heat for as much of the daylight hours as possible. Thus, in the Northern Hemisphere houses should be south facing, while in the Southern Hemisphere houses should be north facing. This orientation will also be optimal for solar photovoltaic and solar water heating systems.

Existing homes will pose challenges for ensuring natural energy gains are maximized while minimizing energy losses. A home energy audit can help you understand where your home is losing energy, what structural changes you can take to improve it, and what renewable energy systems you can install.

Consider the following energy-saving and efficiency principles, including home remodeling, before installing a renewable energy system.[1354,1355] You can apply the principles as best you can to your existing house design and budget, either by having professionals do it for you, or by doing it yourself.

Ensure your house has an airtight seal and utilizes maximum quantities of insulation in the walls, roof, and floor (but see the ventilation caution below). This will minimize heat losses during cold weather and keep the house naturally cool during the summer. Seal and weather-strip all cracks, joints, and large openings to the outside. Insulated cover boxes can be used for attics and non-utilized fireplace entrances.

An "airtight" seal should not compromise adequate house ventilation, to ensure healthy air quality. A heat recovery ventilation system will be required to eliminate the risk of poor air quality with an airtight house seal. Of course, if you don't have the budget for a heat recovery ventilation system, then seal the house but permit some natural ventilation that can be mechanically controlled during extreme cold.

Further minimization of heat loss can be achieved by installing double-glazed windows, low-emissivity coating windows (which reflect heat back inside), or storm windows. Outside window shutters can also be installed, providing a cover over the window space to completely close it off in extremely cold weather. The use of a vestibule inside the main doorways (i.e., double doors) will minimize heat loss as well.

Two types of heat-exchange systems, ground-source[1356] and air-source,[1357] can be utilized to efficiently heat and cool your house. A ground-source or geothermal heat pump and exchange system is used to move heat energy below the frost line, or about two meters below the earth's surface, where the ground temperature is ambient and constant all year round. During the winter and on cool nights the house is heated, and during the summer and in the daytime the house is cooled.

If your home remodeling budget is limited, you can consider making one room in the house or basement very warm and energy efficient for emergency use. This way you have a solution for keeping warm during a climate and/or energy crisis. Other emergency options are discussed below.

Renewable Energy Electricity and Water Heating

Once steps have been taken to minimize the house's energy requirements, then consideration should be given to renewable energy system(s) that can be used to meet your electricity and heating needs.

Rooftop solar photovoltaic (PV) systems are the most common renewable energy systems used for providing home electricity. Solar PV systems are usually connected to the local electricity grid, thereby ensuring a reliable and continuous supply of electricity during sunlight and non-sunlight hours. Energy generated above your immediate needs is transferred back to the electricity grid, paying all or part of your monthly electricity bill. Install more electricity generating capacity than you think you will need, or ensure you can increase the system capacity in the future.

Solar PV with a battery storage system can be installed if your budget and space permits. Ensure your system can be upgraded to include a battery storage system, if it is not installed upfront. A battery storage system will allow you to be grid-independent 24/7, including during blackouts or in an energy crisis when electricity prices will rise.

Battery storage also allows you to take advantage of variable grid pricing tariffs, and use stored energy during peak electricity times when electricity prices are higher, such as in the evening and morning when electricity demand rises.

Don't forget that there are other sources of renewable energy for your home, depending on where you live and your budget. If wind resources are available in your location, a small wind electric system[1358] or hybrid solar-wind electric system[1359] become options. If you have a river or large stream running through your land, then a microhydroelectric power system[1360] becomes an option.

Hybrid systems will protect your electricity supply during variable weather and across different seasons. For example, during the summer there is more sun, which is ideal for solar PV, whereas during the winter there is generally more wind and rain (i.e., increased river flow) and less sunlight. Hybrid systems with a battery storage system better enable you to go off-grid without experiencing power outages. Likewise, in the event of a climate-forcing volcanic eruption that blocks out much of the sun's light, wind and microhydroelectric power will give you electricity-generating options, as solar PV would be limited.

Alternative home sources of energy have been with us for centuries, so keep these in mind for emergencies. The Handbook of Homemade Power[1361] is a gem for learning more about do-it-yourself solar heated house designs, homemade electricity generation, solar water heaters, parabolic solar cookers and ovens (for outside use), homemade biogas production, and ramjet pumps for pumping water up small heights without electricity. This is the do-it-yourself book for those with a small budget and good practical skills.

If you have spare land, then plant plenty of trees for future firewood. You may never need or use that firewood, but others might need it at some point in the future.

Water heating can account for about one-quarter of your electricity usage, so it makes good sense to install a rooftop solar hot water heating system to supply this energy need. These systems include solar heat collectors with a well-insulated storage tank, and can be active or passive systems. The system you install will depend on, first, whether the temperature in your area falls below zero degrees Celsius; second, on your budget; and third on the available space.[1362] Other options exist, including air and ground heat pump water heaters.[1363]

A hot water backup system is useful in conjunction with a solar hot water heating system for high demand times and for cloudy days. You can also lower the thermostat on your water tank, reducing your standby heat loss. Insulating your hot water tank and pipes will improve hot water heating efficiency. Insulating the hot water pipes also means heated water is available instantly, saving you from having to pour precious water down the drain.

Heat recovery from wastewater or effluent water can be achieved by using drain-water heat exchangers. These heat exchange systems can be used with showers, bathtubs, sinks, dishwashers, and clothes washers. This heat can be stored for later use.[1364]

Energy Efficiency at Home

The majority of electricity in the home is used for heating and cooling, lighting, water heating, and running high-energy appliances like washers and dryers, dishwashers, refrigerators and freezers, and televisions.[1365,1366]

A house utilizing the latest energy-efficient electrical appliances and heating systems uses less energy than a house utilizing old, low-efficiency electrical items. However, a house equipped with energy-efficient technologies will fail to achieve its potential energy performance and savings if you ignore how and when to use electricity.

Paying close attention to temperature settings and timings used for space heating and hot water saves energy. Upgrading heating and cooling systems to more energy efficient systems will also save energy.

If you don't live in a very cold region, do you need to use a home heating system when just putting on warmer clothes will keep you warm? Should windows or doors be open when the heat is on? Do you need to heat the entire house if only one or two people are at home?

Changing conventional light bulbs to low-energy, light-emitting diodes is an obvious energy-saving measure. Likewise, turning lights off in rooms not in use, or installing movement sensors to do it for you automatically, will save electricity.

Consider if the fridge needs to be so cold. Do you need a hot wash for the washing machine? Filling up the dishwasher and washing machine fully before putting it on saves electricity and water. Why not dry your clothes outside using sunlight and fresh air rather than an electric dryer? All these things can save energy and money.

Sustainable Water Use at Home

Why is it that a California urbanite uses significantly more water per day than an Aussie urbanite?[1367] It's about changing water use through reeducation, paying the full cost for water, and implementing policies that induce water conservation and efficient use.

What we need to remember is that most of our homes are located in water basins with depleted groundwater resources, and that water is piped into our homes at great cost. On top of that, most water immediately goes straight down the drain after we open the tap.

Water conservation, and the efficient use and re-use of water, remain the most important ways of saving water supplies.[1368]

Reducing the flow rate of water from house taps and shower-heads will mean less water is used, and less wasted hot water will go down the drain. Replace tap fittings and showerheads with water-efficient fittings that have flow restrictors, aerators and pressure-limiting valves.[1369] If you can't afford these fittings, then do it the old-fashioned way by only partially opening taps to reduce the flow rate.

Every time we take a shower, wash the dishes, do the laundry, use sink taps, or use water frivolously, most of that water becomes wastewater, commonly referred to as greywater. Greywater accounts for more than two-thirds of household wastewater.

With the right technology (i.e., filtration, drain water heat exchanger),[1370] suitable house and garden plumbing, and municipal permission, we can re-use this grey water to flush the toilet, supply an underground irrigation system, and recharge groundwater.[1371] Utilizing greywater for garden irrigation will amount to substantial water reuse, and will save a lot of precious fresh water.[1372] However, in many developed nations it is likely that municipal regulations and permission processes will need to change to accommodate more home greywater use.

We can also use smart, low-pressure, targeted irrigation systems (i.e., avoid sprinklers), and irrigate early in the morning or evening. Choosing the right garden plants adapted to the climate (i.e., native plants), covering the ground with mulch to reduce evaporation, and ensuring proper lawn care (i.e., not cutting too short or too frequently) will also save on water use for irrigation.[1373]

Have you fixed those leaky pipes yet? Remember, one drop of water leakage per second wastes 27 liters per day, or 10,200 liters per year. That's a lot of water![1374]

If you live in a drought-prone area, you might wish to consider your emergency water supplies. A large water storage tank for potable water will provide drinking water during a severe drought. If you live near the coast, you might also consider having or making a solar still for purifying brackish water or seawater.[1375,1376,1377,1378]

Growing Food at Home in Climate Extremes

In today's world, if a global food crisis were to occur, we would face a major vulnerability in managing our food supply because of our high degree of urbanization and our reliance on purchased food sourced from many countries around the world. By working for a salary and then buying all or almost all of their food, many people have forgotten how to grow and store food at home so that it is available during the dry and cold months.

I lived in a poor indigenous community in Guatemala for nearly five and one-half years, and I marvelled at how these impoverished people survived with little or no money. Their youth have basic survival skills that most developed nations' kids (and adults) don't have. They know how and when to grow food, and how the weather and seasons impact food production, and they understand the value of planting trees.

When I look to a future that will be marked by climate and food crises, one Biblical saying stands out above all, *"Blessed are the meek, for they shall inherit the earth."* It is the poor, indigenous people, and those connected more with the land and less with the dollar who are the meek, for they still know how to grow food. Knowing how to grow food is an invaluable skill, and it's very rewarding.

If you mock this point of view, then be reminded that during severe famines of the Little Ice Age cannibalism was an innate quality of people and in communities all around the world. This served to feed the survivors and depopulate communities.[1379,1380,1381,1382,1383,1384,1385]

This chapter therefore reviews methods for growing food at home during climate and weather extremes, and is aimed at helping you discover best-practice methods for regaining or developing the ability to grow food, and giving you semi-independence from the global food supply system.

Growing Winter- and Cold-Adapted Vegetables

For summer crops, the most important thing that we can do to grow food in colder climates is to select food types and varieties that will mature within the available frost-free period for a particular region.

The best bet for a homegrown staple crop in cold climates is the trusty potato. Potatoes can be planted directly after seeding, and are hard to beat when it comes to yields and energy content. If you grow them in costales, bags, or containers, you can carry them indoors at night if it's really cold outside. In times of a food crisis this will also ensure your crop is not stolen in the night by the starving.

Turnips, parsnips, and carrots are definitely worth considering, especially in light of their energy content. These vegetables grow well in colder climates, and even in below freezing temperatures. The cold autumn and early winter temperatures actually help them sweeten, because the increased sugar content the cold induces makes them less prone to freezing, i.e., plant sugars are a natural antifreeze.

Cold-adapted maize, climbing beans, and winter pumpkins (squash) make for a three sisters garden, typical of North America, providing cold-tolerant crops with a natural resilience to climate risk. Collards (cabbage, broccoli, and kale), spinach, leeks, asparagus, brussel sprouts, and radishes are well adapted to cold weather, and some varieties are able to survive varying degrees of frost.

Hardy and resilient winter crops like potatoes, beets, carrots, cabbage, and onions have all proved well suited to the cold winter climate of Russia and Siberia, so keep these vegetable staples in mind for a colder climate.

These hardy vegetables can be planted four to six weeks prior to local frost-free dates, or can be grown through the autumn into the winter, depending upon the crop. Look online for growing information specific to your particular region, and pay attention to what the seed packets tell you.

Cold Climate Greenhouse Design Principles

Greenhouses are a cost-effective way of extending the growing season and growing food all year round. Winter greenhouses are a specialized form of greenhouse used to start seedlings in the late winter and early spring, and for growing food during the winter.

Winter greenhouses are designed to capture as much light and heat from the sun during the daytime as possible, storing heat during the day, and radiating stored heat at nighttime. Greenhouses are best oriented in an east-west direction, with the longer, glazed side facing south if you live in the Northern Hemisphere, or facing north if you live in the Southern Hemisphere. This maximizes the amount of sunlight entering the greenhouse during the shorter days of winter, thus maximizing heat gain.

The back wall on the surface of the greenhouse, or the wall opposite the main face of entry for sunlight, is best left unglazed, and can be made of a heat-absorbing material such as rocks, concrete, or mud. The back wall can also be insulated and conjoined with your house, a garden building, or a wall.

Maximum solar heat gain is achieved when the glazing is nearly perpendicular to the angle at which the sun's light enters on the winter solstice, and when the greenhouse's length-to-width ratio is at least two to one. A double-door vestibule at each end is ideal to prevent heat loss during entry and exit.

Heat storage can be utilized using black-colored 55-gallon drums filled with water and placed along the back wall. When the air temperature drops in the evening, the stored heat is then radiated back out into the greenhouse, thus helping keep the inside warmer for longer.

The smart money for winter greenhouse design also utilizes a geothermal heat pump coupled with a heat exchange system to reduce cold weather heating costs. This could utilize the same system as your house, or have its own dedicated system. Specially designed greenhouse heat exchangers are called air-to-soil heat exchange systems, earth-air tunnel systems, or ground tube heat exchangers. The cold interior greenhouse air is slowly pumped underground through large-diameter tubes where the air is warmed, and then circulated back into the greenhouse to impart its warmth.

Winter greenhouses can also be covered at night with insulating blankets, rigid foam sheets, or plastic sheets, to reduce heat loss and keep plants a few degrees warmer than outside. Renewable energy sources like electric heaters, biogas burners, or wood pellet biomass heaters can also be used to warm the greenhouse interior.

Another cold climate design option is an underground greenhouse, or walipini. This is dug six feet or more into the ground, where the average ambient earth temperature remains constant throughout the year, and is much warmer than the ground surface temperature. Underground greenhouses utilize a heat exchange with the earth on five of its surfaces, whereas with a conventional greenhouse heat exchange happens only at ground level.

During the day, the walls and floor of an underground greenhouse absorb heat, acting like a heat storage battery, with the aforementioned five surfaces providing radiant heat during the night to warm your food crop. You can also place an insulating cover over the roof at night to reduce heat loss.

These greenhouse design principles have been used by Siberian, Russian, and Canadian farmers and others who must endure cold winter and spring climates. Their methods are well proven for growing winter food.

Growing Food in Colder Climates

This section reviews various means for reducing frost damage and winterkill. It also adds to the previous discussion on greenhouses.

In the Northern Hemisphere, land with a southern-facing aspect is the best choice for early crops, as south-facing slopes warm up earlier in the spring and gain more solar energy. A higher elevation site is much better than low-lying land, because it warms up quicker. Low-lying areas and valleys tend to be more frost prone because cold air pools there, and so should only be planted after the frost-free date.

In the spring, you can start vegetable seeds indoors, in a greenhouse, or under a cold frame, or by planting them directly into black plastic or organic mulch. Plastic mulch helps warm the soil and keeps it warm longer. Raised growing beds will also help warm up the soil quicker, and these too can be covered with black plastic.

Where crops are planted early in the spring, in order to minimize frost or freeze risk, there needs to be a mix of crop types and sowing dates (different maturities). In that way, only a portion of a crop is susceptible to frost at any one time.

Choosing later-flowering crops (to avoid late frosts) can be a double-edged sword, and result in lower yields if your area is drought prone. It is therefore important to know the expected flowering and harvesting dates for each crop relative to your local growing season's climate, and plant accordingly.

Winter vegetables need to be planted by late summer so a root system can be developed before the autumn frosts and cold weather. Plants can be hardened to withstand frost by exposing the seedlings to varying temperatures and conditions.

Watered plants are more frost resistant; watering improves the conduction of heat stored in the ground. Likewise, application of proper amounts of nutrients can also help maintain plant vigor in cold conditions.

Gardeners and food growers throughout the centuries have learned to use available materials to produce crops earlier in the spring, maintain production well into the fall, and harvest crops throughout the winter. The key is to insulate the crop from the cold as much as possible with your available resources.

A variety of structures are used to extend growing seasons in the spring and autumn, by protecting the crops from frost and extremes of cold. These include low and high tunnels, cold frames, floating row covers, and frost blankets, or covering plants with organic mulch (leaves, straw, bark, etc.). You can even place buckets or big containers over your plants to protect them at night. These structures and covers give plants a few extra degrees of frost and cold protection.

Growing Food during Drought

The only sure method to prevent fruit and vegetable crops from being drought-stressed is to use irrigation and to select crops that are more drought-tolerant. *A crop that needs fewer days to mature needs fewer days of irrigation before harvest.*

If you own peri-urban land, then drought-tolerant cover crops can be used. Drought-tolerant cover crops provide food and help to manage soil fertility (i.e., fix nitrogen) and soil quality (release nutrients, provide organic biomass), and reduce surface water loss. The more drought-tolerant cover crops include legumes that can fix nitrogen (i.e., pigeon pea, cowpea, peas, beans), or non-legume cover crops like cereals (i.e., millets, sorghum, rye, and wheat).[1386] For garden use, planting short-cycle crops like potatoes, sweet potatoes, and drought-tolerant maize and pulses (i.e., pigeon pea, cowpea) will provide you with high energy-yielding food. Other short-cycle crops include leafy greens, tomatoes, cucumbers, and peppers.

There are things that can be done to reduce the risk of drought stress. For example, you can start plants off in a greenhouse. Earlier planting helps avoid late summer drought. When transplanting seedlings, ensure they are planted into deep beds with a good quantity of organic material mixed in. This permits their roots to grow quickly down to where the water is stored before the drought season begins.

Covering the soil surface with plastic or organic mulch can be used to conserve soil moisture and reduce weed growth. Using mulch cover increases the interval between irrigations during the dry season. Weed control is very important, in that it reduces competition for water and nutrients. If you are using raised beds, you can line these with perforated black plastic to better retain water and reduce irrigation requirements.

If you live in a drought-prone area, then harvest rainwater at every opportunity and use house greywater for irrigation.

Growing Food during High Rainfall

If your home is in a flood-prone area, there are things you can do to make growing food easier. Three main options are considered here. These are improving soil drainage, covering your food crops, and the use of vertical gardens. If you can't beat the rain, then join it by raising fish (aquaculture) and coupling this with hydroponics to create an aquaponics system (see Chapter 11).

If you wish to avoid rain or flooding, then greenhouses, hoop houses, and covered beds, with suitable ground drainage, are your best bets. Raised beds are useful for managing flood-prone gardens that lack covering.

A number of things can be done to improve soil drainage. You can increase the soil's organic matter content using compost, bark, leaves, or use sand and gravel. You can also dig to twice the normal depth when preparing ground, to improve the drainage of compacted soils that drain poorly.

If your garden is constantly waterlogged, then small canals, underground drainage tubes, and drainage trenches can be used to drain the water away. Land space permitting, floodwater or excess rainfall can also be diverted to a pond, reservoir, or swale. Minimizing concrete and other impermeable surfaces will limit runoff in your garden and improve drainage.

Vertical gardens are a simple and cheap way to adapt to climate change. Vertical gardens keep plants out of the water during flooding and excessive rainfall. This type of gardening is ideal for wherever space may be limited or unsuitable for ground-based food production.

Vertical gardening is suitable for growing vegetables, fruits, herbs, and other crops, giving you more food production per square meter of ground than normal food growing. Containers can be made from large costales (nylon sacks), plastic or metal drums, or wooden, wire, or bamboo frames. Alternatively, you can make vertical structures to plant your crops in. Check out Pinterest for ideas.[1387] Soil is pre-mixed with organic material (i.e., leaves, compost, old crops, etc.), and a variety of manures, and you can add some worms for releasing nutrients, aerating soil, and improving drainage.

Long-rooted vegetables such as potatoes can be grown on top of large vertical containers. Small holes can be cut into the sides of these vertical containers where short-rooted vegetables can grow. So far, I've managed to grow potatoes, sweet potatoes, pumpkin, watermelon, beans, small stature maize, tomatoes, onions, garlic, carrots, cauliflower, broccoli, herbs, sweet peppers, spinach, chards, and strawberries in vertical gardens.

Raising fish in ponds also makes sense, if stream flow or water is plentiful and you have the space. Hydroponics can also be used to grow food indoors and outdoors using wall surfaces (see Chapter 11).

Preparation for a Food or Climate Crisis
In an ideal world, to be able to survive a global food crisis during a high-risk period, each household should have a three-month stockpile of food at home. Such foods could include sacks of rice, barley, dried maize, beans, peas, oats, grains, dried fruit, tinned foods, milk

powder, etc. You can create a system whereby you replace this stock-pile on a rolling basis to ensure food does not spoil.

As part of your preparation for a climate switch and living sustain-ably at home, you should have a plan for growing food. Ideally, you should have the growing systems already in place and ready to go. Teach your children this more self-sufficient way of obtaining food, just as rural people in developing nations do with their children.

Practice using the different growing methods described above, and growing and storing food like potatoes, beans, peas, maize, pumpkins, and gourds. Practice growing crops (see below) and then harvesting and replanting their seeds. Practice growing food in costales, especial-ly potatoes. In a climate crisis or a freezing northern summer you can then take your food indoors and keep it safe at night.

Keep a seed bank at home that contains short-cycle and cli-mate-adapted vegetable seeds, and ensure these seeds yield plants that are not sterile. Keep seed potatoes at the ready in a dark, dry place, and replace these as required all year round. Develop a calendar for planting seeds to help organize your growing seasons.

The following are seeds to consider for your seedbank. (1) Cold-Adapted Crops. Potatoes, beets, cabbage, onions, turnips, parsnips, carrots, maize, climbing beans, winter pumpkin (squash), broccoli, kale, spinach, leeks, asparagus, brussel sprouts, and radishes. (2) Drought-Tolerant Crops. Millets, rye, drought-tolerant maize, sor-ghum, rye, wheat, pigeon pea, and cowpea. (3) Short-Cycle Crops (3-4 months). Potatoes, sweet potatoes, leafy greens, tomatoes, cucumbers, peppers, and radishes.

What to Do in Response to Climate, Volcanic, Earthquake, and Pandemic Flu Emergencies

Use Google Translate Web to translate hyperlinked emergency-re-lated website pages into more than 100 languages (hyperlink).[1388]

In New Zealand our government via the Ministry of Civil Defence and Emergency Management provides excellent information and plans on preparing for extreme weather events (i.e., storms, floods, tsunamis, landslides) and volcanic and earthquake emergencies (hy-perlink).[1389] Please click on the following hyperlinks for important information on a "Household Emergency Plan" (hyperlink),[1390] "How to Get Ready" (hyperlink),[1391] and "Emergency Survival Items and Getaway Kit" (hyperlink).[1392]

The US government also has one of the best public emergency information websites covering almost every natural disaster, and gives great practical advice and links to information and plans for cold-snow, drought, flooding, storms-hurricanes, and volcanic and earthquake events, etc. (hyperlink, hyperlink).[1393]

If you live in a volcanic region, a gas mask could be a wise investment. Information on volcanic ash impacts, and what to do in the event of a volcanic eruption, can be found at one of the following websites: The New Zealand Crown Research Institute (hyperlink); Auckland Engineering Lifelines Project (hyperlink); the US Government's Volcanic Ash Impacts & Mitigation website (hyperlink); and the International Volcanic Health Hazard Network (hyperlink).

Important emergency information pertaining to the heightened risks for a pandemic flu outbreak and the vaccine debacle that's basically just waiting to happen during this grand solar minimum are detailed in the next chapter.

For a pandemic flu emergency plan, your first port of call should be WHO's Public Health Preparedness website,[1394] which is a must for pandemic-related information. This site will provide you with the current global situation and planning information for government and municipal levels of disease control.

Likewise, the US government's Centers for Disease Control and Prevention website and the US Government's Ready website (hyperlink),[1395] and New Zealand Government websites for pandemic flu preparedness (hyperlink) are useful sources of public information.[1396]

What are you going to do if there is a sudden freeze and deterioration in the overall weather and the power goes out? The US Government has a great website for helping you understand your best options (hyperlink).[1397] A sudden freeze is when having a special warm room in the house will be especially useful. Make sure everyone has thermal clothing and warm hats, gloves, socks, shoes, and sub-zero temperature-rated sleeping bags.

Have water filters, a water-purification system, and a means for treating water at hand. Remember too that homemade water stills can be cheaply made to purify dirty water (i.e., greywater or water from streams and ponds), and for desalinating sea water if you live on the coast.[1398,1399,1400,1401]

CHAPTER 13

Revolution in a Bucket List

This book is titled Revolution(RE) and it is meant to serve as an information bomb that will shatter your illusions about climate change before the climate switch occurs. Revolution(RE) reviews the catastrophic climate-related risks that lie on the horizon, and provides best practice solutions to help you mitigate these risks.

This book calls for four peaceful and cooperative revolutions to bring about change before it's too late: (1) a scientific revolution that allows us to clearly perceive the natural climate change risks; (2) a renewable energy revolution that will help us avoid the coming energy crisis; (3) a pandemic flu vaccine revolution, linked to pre-pandemic immunization, that will prevent tens of millions of needless deaths in a pandemic outbreak; (4) a voting revolution that elects leaders who will help us to prepare for a climate switch and its associated risks, and who will undertake sustainable development and projects facilitating a switch to renewable energy on our behalf.

The book's title is also a reminder of what happens when governments get it wrong and people starve and die en masse, just because nature and the best science were ignored.

Besides preparing your home, family, and community for living sustainably during a climate switch and its associated catastrophes—what can you do?

You have voting power. "Forewarned is forearmed."

You have the ability to vote into power leaders who will take charge of preparing your country, city, and community for a very different world than is currently envisaged and that definitely lies ahead.

This chapter, and Section 3 overall, give you the essential, practical things we all can do to move toward living sustainably, and to switch the world's energy system to renewables. It's a bucket list of things we need to accomplish—*before we die.*

If your leaders are not talking about these issues at election time, then don't vote for them. Find someone to vote for who will get these jobs done for your community and your nation.

And if you want to know what's really going on with the temperature, instead of being manipulated by the media every time there's a hot week or an out of control fire started by arsonists—then look at the temperature data.[1402] The previously cited climate data is provided by the UK government without fearmongering or hype, and gives you the global, northern and southern hemisphere, and tropical temperatures—and yes, the temperature is already in decline (from 2016).

A Bucket List of Jobs Your Elected Leaders Need to Get Done

What follows is a condensation of all of the best practice ideas from Section 3 organized into a bucket list—the things we must do to switch over energy systems to renewables, to live sustainably, and to prepare for the world that the youth of today will inherit.

What it means to live sustainably: that we use all energy, water, food, and other resources sparingly, efficiently and equitably, maximizing residual resource and energy recovery, while generating a minimal amount of waste and pollution. This philosophy should permeate everything we do at home, at work, and when we travel, and be the norm for our urban, rural, and national ways of life.

Financing. Implement a global carbon tax priced at a minimum of $30 per ton. Deploy more renewable energy feed-in-tariffs (i.e., premium-priced electricity supply contracts) to incentivize the installation of more renewable energy capacity. Partially redirect fuel subsidies to help the poor gain access to affordable renewable energy systems. Change pension fund regulations to better enable pension funds to invest in renewable energy infrastructure and sustainable community development projects (promote public-private partnerships). Accelerate the Paris Agreement's $100 billion annual financing and start project investments. Make adequate funding available for pre-pandemic flu immunization and, as a matter or urgency, immunize the willing throughout the world.

Renewable Energy and Energy Efficiencies. Invest in massive quantities of renewable energy (centralized and decentralized). This will take the form of more efficient hydropower and pumped-storage hydropower, on/off-shore wind, solar PV and concentrating solar power, geothermal, and energy from biomass waste. Install renewable energy heating and ground-source heat exchange systems wherever possible. Invest in efficient fossil fuel co-generation and heat recovery systems. Implement massive reforestation projects (i.e., for energy and groundwater recharge). Technology R&D: Invest in high-yield (i.e., fusion, fission, and the free energy concept-promise) and alternative energy systems.

Upgrade the Electric Grid. Install regional super-grids and local smart grids, massive quantities of high voltage direct current transmission interconnections, and gigawatt battery storage (i.e., to improve energy system resilience).

Transportation. Incentivize fuel efficiency and vehicle weight reductions, as well as biomass-waste derived biofuel and synthetic fuel conversions. Invest in electric road, rail, and public transportation systems. Avoid biofuels made from irrigation-dependent crops.

Sustainable Water Supply. Ensure that integrated water resource management becomes the regional and municipality norm. Establish transboundary river agreements and full economic pricing of water. Implement groundwater recharge by reconnecting rivers to floodplains, increasing artificial recharging of groundwater, and undertaking massive reforestation projects. Invest in massive quantities of renewable energy desalination systems and bulk water pipelines. Implement policies and pricing to ensure agriculture and industry conserve and efficiently use water resources. Fix the leaky pipes.

Food supply. Prepare large-scale agriculture for climate switching with cold- and drought- adapted crops and methods ready to be used in a climate crisis. Ensure agriculture is sustainable (i.e., reduce yield gaps, reduce deforestation, minimize chemicals and waste fertilizer). Support smallholder farmers with sustainable farming methods and access to climate-adapted seeds. Reduce food waste from field to plate. Decentralize food supply for towns and cities by implementing urban and peri-urban agriculture. Invest in urban high-tech indoor farming. Develop sustainable aquaculture (urban and rural). Support home food production. Promote urban soil-less food production (i.e., hydroponics, aquaponics, aeroponics).

Emergency food stockpiles and climate-independent food supply. Reassess food stockpiles. Ensure municipal seed banks are well stocked (no sterile seeds). Invest in peri-urban industrialized greenhouses, high-tech indoor urban farming, and bulk-scale single-cell protein manufacture. Invest in R&D for low-cost, renewable energy LED lighting systems for crop growing (for both commercial and home use).

A Shortlist for Living Sustainably in Climate Extremes

Municipal and Government Support. Support will be required in urban planning processes, education, and funding. Bylaw amendments may be required to enable greywater reuse and permit home and garden modifications with minimal red tape.

Home design principles. Reduce energy needs by ensuring house airtight seals and using maximum insulation. Install double-glazed windows, low-emissivity windows, storm windows, and double doors. Install a heat recovery ventilation system. Have one emergency warm room ready to go (if you live in northern latitudes and at high altitudes).

Renewable Energy Systems. Install a rooftop solar photovoltaic system, or other systems depending on your local renewable resources available, such as a wind electric system, a hybrid solar-wind electric system, or a microhydroelectric power system. Install a battery storage system or at least ensure that it can be installed in the future. Utilize ground-source and air-source heat pumps and exchange systems. Plant plenty of trees in and around your community. Install solar water heaters, or air and ground heat pump water heaters. Insulate your water tanks and pipes, and lower the water tank thermostat. Utilize drain water heat exchangers.

Be Energy Efficient. Learn how and when to use electricity efficiently while operating your space heating system, hot water, and lighting. Utilize efficient electrical appliances and heating systems. Use low-energy light-emitting diode light systems. Turn off lights when not in use or install automated movement sensors. Avoid washing half loads of dishes and clothes. Use the sun and breeze to dry your clothes.

Water conservation, efficient use and re-use of water. Reduce tap and showerhead flow rates by using water-efficient fittings with flow restrictors, aerators, and pressure-limiting valves, or only partially open taps. Utilize house greywater (after filtration and energy recovery) for in-house use (toilet), garden irrigation, and groundwater recharge. Avoid sprinklers and use low-pressure targeted drip irrigation systems, together with mulch and ground cover. Harvest rainwater, utilize large water storage tank(s), and get a solar still for purifying dirty water and seawater for severe drought. Fix your leaky pipes and taps.

Growing Food in Cold Climates. Utilize a winter or underground greenhouse, maximizing its solar heat gain while storing heat. Utilize a geothermal heat pump and heat exchange system, or an air-to-soil heat exchange system in your greenhouse. Use renewable energy heaters, and use greenhouse covers at night. Utilize cold structures such as low and high tunnels, cold frames, floating row covers, and frost blankets. Use elevated land and raised growing beds. Start seedlings indoors, and use local and climate-adapted seeds (non-sterile). For summer crops, ensure a mix of crop types, with sowing, flowering, and harvesting dates within the frost-free period.

Growing Food in Drought. Harvest rainwater. Use house greywater for irrigation. Cover ground with plastic or organic mulch, and control weeds. Use efficient low-pressure drip irrigation systems. Start plants off early under cover. Use crop types that take fewer days to mature, or short-cycle crops (see below).

Flood Prone Cropping. Cover your crops with greenhouses or hoop houses, and use covered beds. Use raised beds, vertical gardens, hydroponics and aquaponics. Improve soil drainage using sand, organic material, stones, double-depth digging, and underground drainage tubes.

Seeds to consider for your seedbank: (1) Cold Crop Food. Potatoes, beets, turnips, parsnips, carrots, maize, climbing beans, winter pumpkin (squash), cabbage, onions, broccoli, kale, spinach, leeks, asparagus, brussel sprouts, and radishes. (2) Drought-Tolerant Crops. Millets, rye, sorghum, wheat, drought-tolerant maize, pigeon pea, and cowpea. (3) Short-Cycle Crops (3-4 months). Potatoes, sweet potatoes, leafy greens, tomatoes, cucumbers, and peppers.

SECTION 4

Key Themes

**The Risk of a Pandemic Is High and Bordering
on "Red Alert"**

**Immunizing the World before a Pandemic Happens Is
Technologically Feasible, and Eliminates the Need to
Stockpile Refrigerated Vaccines**

**Insure against a Food Crisis with Pre-Pandemic
Immunization**

CHAPTER 14

The Worst Time to Immunize a Population Is after Pandemic Influenza Emerges—but That's Our Plan

This chapter takes a look at why we are seeing more highly lethal influenza-A virus transmissions from animals to humans, and also looks at the increased risks for a pandemic flu outbreak after a climate switch (i.e., cold) during this grand solar minimum.

These increased risks make it a priority to pre-immunize the world's population against high-risk, potentially pandemic H7N9 and H5N1 influenza-A viruses. These two influenza-A strains are knocking loudly at our species' door; they have killed between 25 percent and 50 percent of infected humans (the percentage may be even higher).[1403,1404,1405,1406,1407] A highly lethal pandemic flu outbreak could kill between one and three percent of the world's population. This was the mortality rate during the 1918–1919 Spanish flu pandemic.[1408]

For all diseases that can be prevented by vaccination, it is both *absolutely imperative and the norm to immunize people before a disease emerges.*[1409,1410,1411] However, under the existing paradigm of pandemic influenza immunization and supply, governments and WHO wait until a pandemic emerges before calling for the manufacture of a pandemic flu vaccine to immunize the population. Under this paradigm, sufficient quantities of vaccine cannot be manufactured quickly enough[1412,1413] to protect the world's population before the peak of a pandemic's mortality is reached.[1414]

Since the 2009 swine flu pandemic, vaccine technology advances offering immunological protection against ongoing viral mutation have made it possible to immunize people ahead of the outbreak of a pandemic (i.e., pre-pandemic immunization).[1415] An upgraded flu vaccine technology would permit a broader immunity against emerging pandemic flu strain mutants, and enable us to better protect the population before the outbreak of a pandemic.[1416,1417,1418,1419,1420,1421,1422,1423,1424] Why haven't we upgraded the flu vaccine technology and implemented pre-pandemic immunization? *Read on to find out.*

Developing nations produce a significant share of the world's food. Look at the top five grain exporters in the following citation and you will see the importance of developing nations to grain supply and global food security.[1425] Developing nations will be hit hardest by vaccine supply inequities, because their people will not be immunized in time under the current plan for supplying pandemic flu vaccine.[1426,1427,1428] People living in developed nations will be hard pressed as well, unless their governments have stockpiled sufficient quantities of a pre-pandemic flu vaccine (unlikely).

There is clearly a need for change, what I term *a pandemic flu vaccine revolution.* With a pandemic flu vaccine revolution our governments can *ensure* we are protected, and our fragile global economy and the global food supply are *insured* against the pandemic flu threat. Pre-pandemic immunization makes good sense before a climate switch and the increased pandemic flu risks it portends.

Why Are More Animal-to-Human Influenza-A Viral Transmissions Occurring?

Why have more animal-to-human influenza-A viral transmissions been occurring since 1997? Is this just because of improved influenza-A viral surveillance and detection, or does it represent a genuine trend?

Human influenza-A viruses have their origin in avian and swine species, and for a new pandemic flu strain to emerge, a number of successful viral mutations and transmissions must first take place. First, transmission of a novel influenza-A virus from an animal to a human must take place. Next, viral transmission between humans must occur, followed by transmission from one human to another in a sustained manner. The flipside of this influenza-A virus mutation and transmission process is that a highly pathogenic influenza-A virus does not usually adapt particularly well to its human hosts in the first instance, because it tends to kill them too quickly for sustained transmission to occur.[1429]

This latter point makes the difference between sporadic animal-to-human infection killing 25 to 50 percent or more of its human victims (i.e., H7N9, H5N1) and a full-blown pandemic killing a lower percentage of infected people. For example, in the 2009 swine flu pandemic the mortality rate was estimated to be 0.002-0.009 percent,[1430] while at the other extreme the 1918–1919 Spanish flu had a mortality rate estimated at between one and three percent of the world's population.[1431]

Conventional epidemiological thinking highlights important factors potentially associated with this increase in animal-to-human influenza-A infections. For example, the growth and intensification of global poultry and swine production combined with climate change increases the stress on animals and makes them more susceptible to infections and disease.

China, where pandemics have historically originated,[1432] and other parts of the world still utilize open markets for livestock trade in support of home-based rearing systems. These rearing systems often mix poultry, swine, and humans in the same areas, which facilitates the emergence and spread of new viral mutants. The intensification of corporate food production systems worldwide (typically, half a million birds per flock) also increases the risk of new viral strains emerging and being transmitted from birds to humans working in close proximity to them.

Most of the influenza-A viruses being transmitted to humans have resulted from viral gene re-assortments (in swine) and viral mutations (both avian and swine). Swine can be co-infected with all three swine, avian, and human influenza-A virus strains at the same time. This yields new, mutant viral strains through viral gene re-assortments as the viruses replicate together inside the cells of the pig.

Many migrating birds, on the other hand, fly to the Arctic Circle every summer, and this brings many bird species into contact with one another, with millions of birds often closely packed together. This summertime mixing occurs while the birds, their eggs, and their offspring are bathed in extremes of geomagnetism and ionizing radiation, such as cosmic rays (which are putative factors involved in viral mutation).

Death by H7N9 or H5N1 Viral Infection Would be Horrific

The influenza-A viruses *we really have to worry about are highly pathogenic avian influenza-A H7N9 and H5N1.* Since 1997 other animal influenza-A viruses have also killed humans, and these continue to pose risks.[1433,1434,1435,1436]

H7N9 is killing between 25 and 40 percent of humans infected.[1437,1438] Animal-to-human infections emerged in China in 2013, and grew year by year to a total of 1,600 animal-to-human infections reported by 2017.[1439,1440,1441] Specific viral mutations that facilitate human infection have since emerged,[1442] *meaning human-to-human transmission is next.* The situation is similar with the H5N1 virus, which has killed more than 50 percent of humans infected.[1443,1444,1445,1446]

Pandemics have historically spread rapidly throughout the world, and up to half the human population is typically infected.[1447,1448,1449,1450] Pandemic flu viruses that kill a high percentage of their victims do so because they cause a high incidence of severe pneumonia and multi-organ failure. This requires intensive hospital care, with the availability of hospital intensive care a potential bottleneck.

The 1918–1919 pandemic flu virus caused acute swelling of and bleeding from the lungs, and *people who were infected typically suffocated within one to two days.* The second wave of the pandemic was responsible for the most deaths, due to an unusually severe hemorrhagic pneumonia. H5N1 victims today experience similar pathologies to those of the 1918–1919 pandemic, with acute respiratory distress syndrome occurring in 50 to 75 percent of infections.[1451,1452]

Likewise, since 2013 more than 90 percent of humans dying from H7N9 infection suffered from pneumonia, respiratory failure, or acute respiratory distress syndrome. Most of the infected people who were hospitalized were admitted to an intensive care unit. With ongoing viral mutations of H7N9 known to improve human viral transmission,[1453] *this is a very worrying virus indeed.*

Historically, in the 1918 and 1957 flu pandemics, the second and third waves of the outbreak were worse than the initial wave of the disease.[1454,1455,1456]

Cold Climates and Extremes of Solar Activity Portend Pandemic Flu Outbreaks

The odds of a pandemic flu outbreak are small, and yet pandemics happen more frequently around 11-year sunspot minima and maxima, and during cold climate phases (see Figure 14.1).

According to the scientific literature, there were 24 pandemics and 29 major or regional influenza epidemic outbreaks between 1500 and 2009.[1457,1458,1459,1460,1461,1462] There were three pandemics during the 20th century—in 1918, 1957, and 1968, and one in the 21st century (2009)—*so far, that is.*

To investigate pandemic flu outbreaks and their association with climate change and solar activity, I conducted my own epidemiological study. This study utilized Greenland ice core and Northern Hemisphere climate data, and influenza pandemic and epidemic outbreak data compiled from the scientific literature between 1500 (or 1700) and today.

The standout finding across the different solar activity, cosmic ray, and climate-related data studied (collectively "solar activity and climate parameters"), was that 76 percent of all influenza-A pandemic and major regional epidemic outbreaks during the Little Ice Age took place at a peak or trough in these solar activity and climate parameters, or within a year of one.

Half of influenza-A pandemics and epidemics (22/45) between 1610 and 2000 occurred when both the Northern Hemisphere temperature and total solar irradiance anomalies were negative, which corresponded with the troughs of the Little Ice Age's grand solar minimum periods. In other words, grand solar minima pose increased risk for pandemic influenza-A outbreaks.

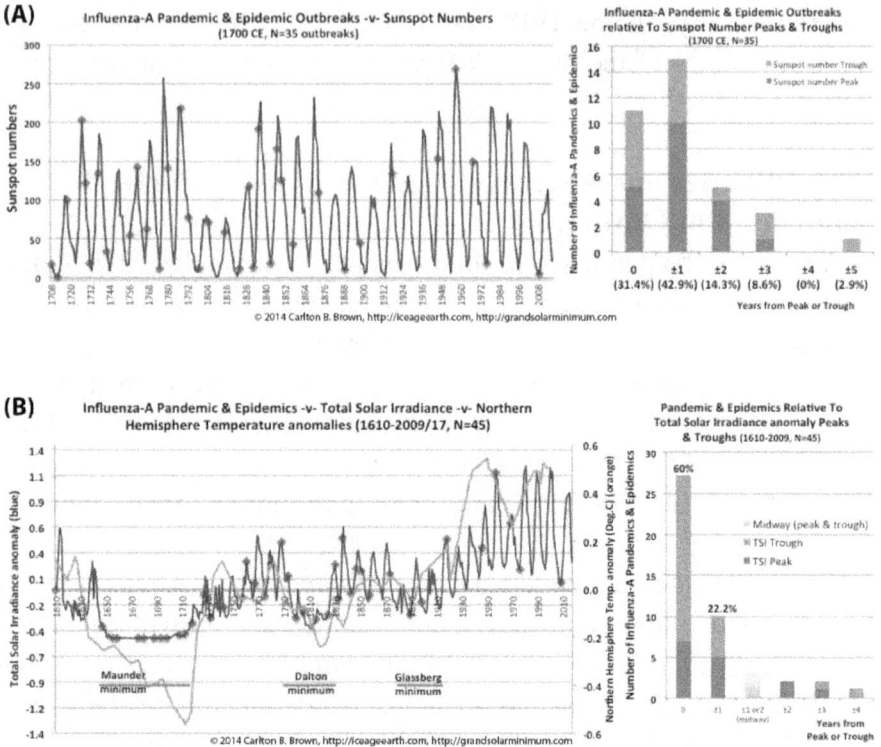

(A)

Influenza-A Pandemic & Epidemic Outbreaks -v- Sunspot Numbers
(1700 CE, N=35 outbreaks)

Influenza-A Pandemic & Epidemic Outbreaks
relative To Sunspot Number Peaks & Troughs
(1700 CE, N=35)

© 2014 Carlton B. Brown, http://iceageearth.com, http://grandsolarminimum.com

(B)

Influenza-A Pandemic & Epidemics -v- Total Solar Irradiance -v- Northern
Hemisphere Temperature anomalies (1610-2009/17, N=45)

Pandemic & Epidemics Relative To
Total Solar Irradiance anomaly Peaks
& Troughs (1610-2009, N=45)

© 2014 Carlton B. Brown, http://iceageearth.com, http://grandsolarminimum.com

Figure 14.1 (A-B). Historical pandemic and epidemic influenza-A epidemiological data used in Figures 5.1 A-B and citations C to F were extracted from six scientific publications reviewing the history of influenza (see citation), providing a general consensus on pandemic flu outbreaks (and major regional epidemics) back to 1500. Climate and solar activity data started in either 1500 or 1700 and the end dates varied, meaning the number of pandemic events varied (N = 35 to 53). **A)** Seventy-four percent of influenza pandemics and epidemics (26/35) since 1700 CE occurred at or within one year of the peak or trough in sunspot numbers, increasing to 89 percent (31/35) within two years. The average sunspot number for pandemics occurring at sunspot number troughs was 12 (18 for pandemics occurring within one year of a sunspot number trough). The 2018 sunspot number was 22.[1463] **Conclusion A**: Based on sunspot numbers, *we are approaching a high-risk period* for pandemic flu. **B)** Between 1610 and 2000, eighty-two percent of influenza pandemics and epidemics (37/45) occurred at or within one year of a peak or trough in the total solar irradiance anomaly. Sixty-four percent (29/45) of influenza pandemics and epidemics occurred during a negative

Northern Hemisphere temperature anomaly. Half of outbreaks (22/45) occurred when both the Northern Hemisphere temperature and total solar irradiance anomaly were negative, which corresponds with the trough of grand solar minimum periods. Negative anomalies resulted when the temperature or irradiance value was less than the 1610-2000 average for that parameter. **Conclusion B**: Grand solar minimum periods associated with a colder climate pose increased risks for pandemic flu outbreaks.

(C)

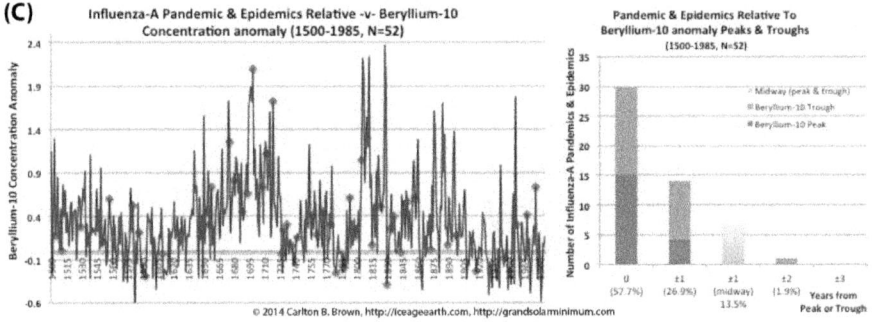

Influenza-A Pandemic & Epidemics Relative -v- Beryllium-10 Concentration anomaly (1500-1985, N=52)

Pandemic & Epidemics Relative To Beryllium-10 anomaly Peaks & Troughs (1500-1985, N=52)

© 2014 Carlton B. Brown, http://iceageearth.com, http://grandsolarminimum.com

(D)

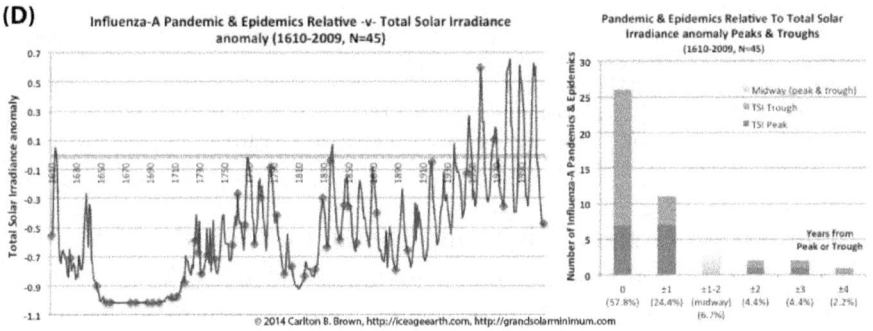

Influenza-A Pandemic & Epidemics Relative -v- Total Solar Irradiance anomaly (1610-2009, N=45)

Pandemic & Epidemics Relative To Total Solar Irradiance anomaly Peaks & Troughs (1610-2009, N=45)

© 2014 Carlton B. Brown, http://iceageearth.com, http://grandsolarminimum.com

(E)

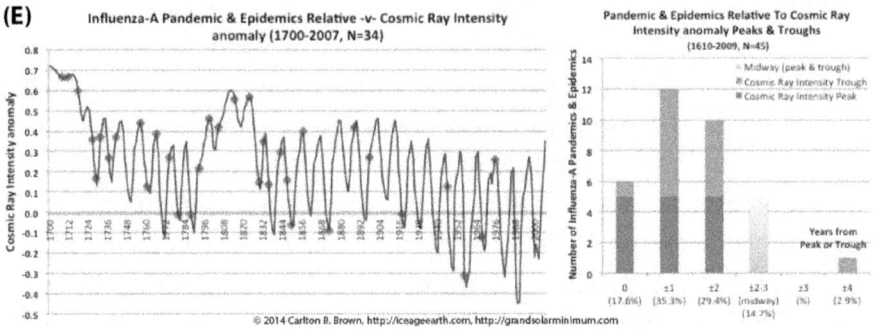

Influenza-A Pandemic & Epidemics Relative -v- Cosmic Ray Intensity anomaly (1700-2007, N=34)

Pandemic & Epidemics Relative To Cosmic Ray Intensity anomaly Peaks & Troughs (1610-2009, N=45)

© 2014 Carlton B. Brown, http://iceageearth.com, http://grandsolarminimum.com

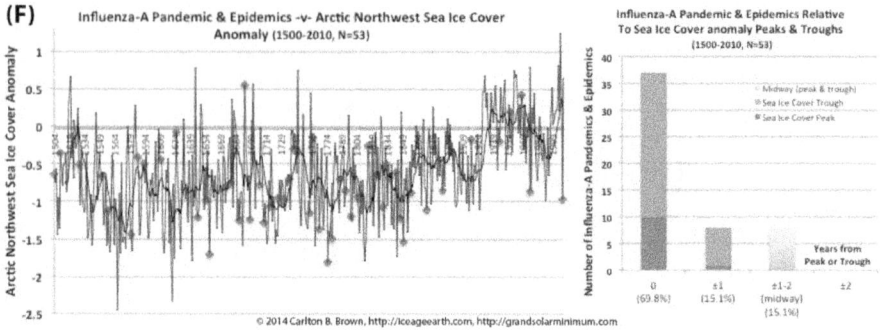

Figures 14.1.C to 14.1.F. The percentage of influenza pandemic and epidemic outbreaks occurring at or within one year of the peak or trough are: **C)** Beryllium-10 concentration anomaly, 85 percent (44/52) since 1500. **D)** Total solar irradiance anomaly, 82 percent (37/45) since 1500. **E)** Cosmic ray intensity anomaly, 53 percent (18/34) since 1700. **F)** Sea-ice cover anomaly, 85 percent (45/53) since 1500. Arctic algal growth declines when increasing sea-ice blocks the sunlight reaching the ocean floor. Anomalies were calculated relative to the 1961-1990 average for total solar irradiance, cosmic ray intensity, and sea-ice cover, whereas the anomaly is calculated relative to the 1960-1985 average for the Beryllium-10 concentration.[1464]

The Arctic Circle, Migrating Birds, and Viral Mutation

There is a small amount of scientific literature available on influenza-A viral mutation and solar activity, and their links to pandemic flu outbreaks.

My research summary for pandemic flu outbreaks presented above indicates that there is a range of climate- and solar-related parameters associated with normal population health i.e., fewer pandemics. Extremes of these parameters appear to be associated with more pandemics and major regional influenza epidemics.

Using similar methods, the above research findings broadly replicate other researchers' data for pandemic flu outbreaks (sunspot numbers). One researcher determined that the relationship between pandemic flu outbreaks since 1700 and the peaks and troughs of sunspot numbers during the 11-year cycle (± 1 year) was *statistically significant* compared with other times of the 11-year solar cycle.[1465] My research builds on this researcher's data with a broader array of solar activity and climate-related parameters, all showing the same peak and trough relationship with high frequency (see Figure 14.1.A-C).

According to other publications, influenza pandemics,[1466,1467] influenza-A viral mutation,[1468] and Ebola hemorrhagic fever outbreaks in Africa[1469] were all associated with the peaks and troughs of the 11-year sunspot cycle.

Interestingly, some of the earliest mutations of the surface proteins of influenza-A viruses, which helped create today's family of viral strains, coincided with two grand solar minima during the Little Ice Age. According to scientists using advanced methods of genetic analysis, these influenza-A viral mutations took place between 1672 and 1715 CE (Maunder Minimum), and between 1825 and 1868 CE (Dalton minimum).[1470] This adds further support to the hypothesis that grand solar minima represent high-risk times for influenza-A virus mutation and pandemics.

Influenza-A viral genomes are known to be responsive to earth's magnetism (geomagnetism). This geomagnetism is known to alter influenza-A viral gene expression (switching on and off) and protein synthesis.[1471] High-energy cosmic rays are guided by earth's magnetic field into the polar regions,[1472] thus concentrating their effects in the Arctic, which migrating birds visit annually. This would putatively provide a source of concentrated ionizing radiation and magnetism to which migrating bird populations and their vulnerable egg embryos are exposed. The ioninizing radiation and magnetism could in turn help mediate viral mutation processes.

This research data is *suggestive of a causal relationship*, and highlights environmental risk factors and potential biological mechanisms that could contribute to avian influenza-A virus mutation. Drug regulators, government, and WHO could modify their perception of risk based on this epidemiological information. This could then help them make the decision to implement pre-pandemic immunization for the willing, at risk public (see the discussion below).

We Can't Immunize the Population before a Pandemic Peaks in Mortality

The number one goal in infectious disease prevention is to immunize your target population *before* a disease outbreak, not afterward. As a rule of thumb, across all vaccine-preventable infectious diseases a minimum rate of 70 percent immunization is targeted in the population before a disease outbreak. This rate of immunization then

imparts a "herd immunity effect" able to protect the whole population.[1473,1474,1475] This herd immunity effect halts or blunts an epidemic
in its tracks. However, in the influenza vaccine field *humans are only
immunized after a pandemic emerges*, though this now need not be
the case.

In the 2009 swine flu pandemic, China and the USA were able
to supply a vaccine within six months of the outbreak. However,
substantial quantities of vaccine only became available *after the pandemic had peaked*. Only 100 million doses had been delivered in the
USA nine months after the pandemic emerged,[1476] while in China
only 90 million doses had been administered nearly a year after the
pandemic emerged.[1477]

Therefore, the biggest shortcoming of the response to the 2009
swine flu pandemic was the failure to supply enough doses of vaccine
before the pandemic peaked.[1478] Industry, government, and WHO
failed to come close to achieving the ideal 70 percent vaccination rate
(see Figure 14.2).

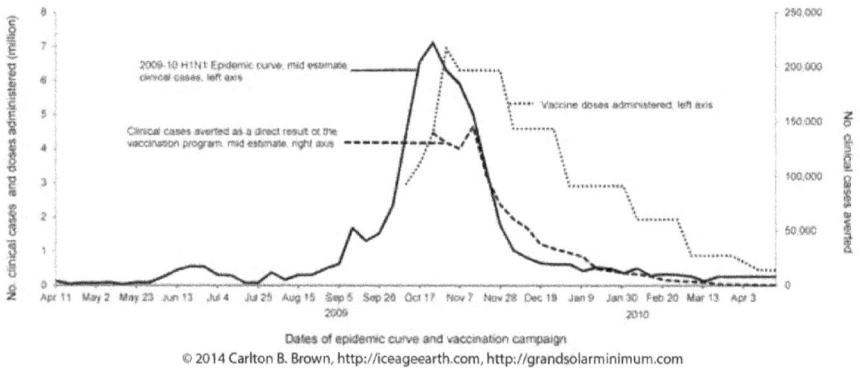

Dates of epidemic curve and vaccination campaign
© 2014 Carlton B. Brown, http://iceageearth.com, http://grandsolarminimum.com

Figure 14.2. A plot of the weekly number of clinical cases of pandemic influenza A (H1N1) virus
infection and the number of pandemic flu vaccine doses administered. This highlights that the first
vaccine doses were only supplied just before the peak of the pandemic (i.e., after approximately
six months), and that by the pandemic's peak only about 4 million doses had been supplied.[1479]
Conclusion: Waiting until after a pandemic emerges before manufacturing a H5N1 or H7N9 pandemic
flu vaccine is *a recipe for human catastrophe, when in fact this need not be the case*. Note: Unlike in
2009, when most humans had preexisting immunity to swine flu H1N1 (developed after a lifetime of
seasonal flu H1N1 infections and immunizations), a H5N1 and H7N9 pandemic will likely be associated with high mortality rates. This higher mortality rate will result because we have no preexisting
immunity to H5N1 or H7N9 influenza-A strains that would normally provide partial protection
against these viruses.

At its root, this 2009 *vaccine supply debacle reflected two main issues.* The most important problem was that the vaccine producers needed to wait until after a pandemic emerged before beginning vaccine manufacture. As Figure 14.2 highlights, the flu vaccine industry cannot supply sufficient vaccine quickly enough before a pandemic peaks in mortality; not even for the population of the nation or nations in which the vaccine is produced.[1480,1481]

The second problem in 2009 was that there was insufficient global vaccine production capacity, and the vaccine industry faced unpredictable technical delays in optimizing the manufacturing process and ramping up that production to full capacity. This full-scale production was not quick enough to equitably supply and immunize at-risk people before the pandemic peaked.[1482,1483] The world was extremely lucky that the 2009 pandemic flu outbreak was mild, and did not replicate the 1918 Spanish flu mortality rate of between one and three percent.

The above issues made obvious the inequity that exists when it comes to supplying vaccine to developing nations and nations that do not possess their own vaccine manufacturing capacity. Pandemic flu vaccine supply was prioritized for large advanced purchase agreements by rich nations.[1484] This inequity in vaccine supply happened in both 2005 (H5N1) and 2009 (H1N1). In 2009, this meant a WHO-donor consortium was unable to deliver any meaningful quantity of pandemic H1N1 flu vaccine to developing nations.[1485]

Under international law, the principle of sovereignty rules in the allocation of scarce resources, and this applies also to pandemic flu vaccine supply.[1486] This means that in times of crisis nations control their valuable national resources, placing their citizens and strategic priorities first. This applies to the supply of pandemic flu vaccine[1487,1488] as well as food[1489,1490,1491] and energy.

Can we manufacture and distribute equitably to all nations an adequate supply of pandemic flu vaccine before the peak of a pandemic's mortality? No. Not under the existing immunization and vaccine supply paradigm, irrespective of how much manufacturing capacity exists at the time. It will still take five or six months to begin to supply a vaccine, and much longer to supply billions of doses.[1492]

A pandemic like that of 1918–1919 will not only ravage the human population, but also severely damage the global economy for years afterwards, which would in turn undermine the switch to renewable energy. A pandemic will also likely create a global food crisis, because most people (including farmers) in the main food exporting and food producing nations[1493] will not be immunized in time.

In such a situation, finding a tenable strategic solution dictates *we must change the vaccine supply and immunization paradigm,* and not the behavior of governments (i.e., vaccine trade restrictions) or merely hope that the vaccine industry will cope better next time.

Generating Improved Antibody Protection Using Vaccine Adjuvants

One of the breakthrough findings from pandemic flu vaccine research in the last decade has been the use of oil-in-water emulsion *adjuvants, or immune response booster substances,* to improve vaccines. Oil-in-water emulsion adjuvants result in the generation of a special type of antibody response able to protect against a wider array of influenza-A virus infections (*broadly cross-reactive antibodies* is the technical term), compared with old-tech 1950s seasonal flu vaccines without an adjuvant that are still used today.

Approved H5N1 pre-pandemic flu vaccines stockpiled by the governments of wealthy nations, as well as H7N9 and H7N1 prototype vaccines (currently being stockpiled) containing oil-in-water adjuvants, will generate these broadly cross-reactive antibody responses. These adjuvants include the MF59C.1 adjuvant developed by Sequirus (formerly Novartis),[1494] GlaxoSmithKline's AS03 adjuvant,[1495] and other proprietary oil-in-water emulsion adjuvants.[1496]

Adjuvanted pandemic flu vaccine prototypes have consistently been shown (clinical data) to better protect people against pandemic flu. As a general summary, after an initial immunization with a vaccine containing viral antigens of, for example, one H7N9 viral subtype (H7N9a-prime), the primed antibody memory response generated can then be boosted at flexible intervals of up to six years with a second vaccine composed of a different virus subtype (H7N9b-booster). The booster immunization administered years later

(H7N9b-booster) will then generate an antibody response, which will protect against H7N9a-prime virus plus the second slightly different H7N9b virus (i.e., the mutant), as well as viral subtypes not immunized for (i.e., H7N9-unknown mutants). The same principle applies to H5N1, H7N1, as well as other pandemic Influenza-A strain threats.[1497,1498,1499,1500,1501,1502,1503,1504,1505]

The above cited pre-pandemic vaccine prototypes containing an oil-in-water adjuvant also generate higher magnitudes of antibody response, and more rapidly, in a higher percentage of immunized people, and with lower quantities of vaccine protein (i.e., antigens), compared with influenza-A vaccines without an adjuvant.

This type of adjuvanted influenza-A vaccine, plus the immunization flexibility it affords using different influenza-A subtype vaccines, means we can protect people ahead of a pandemic outbreak and before the peak of mortality, even if we don't know what pandemic strain will attack. If deployed, pre-pandemic immunization would blunt the first wave of a pandemic or stop it in its tracks, while manufacturers supply a strain-specific vaccine to booster the first immunization.[1506,1507,1508,1509]

The *critically important implication* of the above is that rather than stockpiling a pre-pandemic flu vaccine with a limited shelf life in cold-storage before a pandemic occurs, it would be *much better to "store" the vaccine as immunological memory* in the human population—as is normal with diseases for which we have vaccines. This immunological memory could then be flexibly boosted many years later, with the duration of this memory *well exceeding* the storage shelf life of a stockpiled vaccine (under the existing strategy of rich governments).

Leaders in vaccine R&D have suggested pre-pandemic immunization, because without it vaccine manufacturers will be *unable to provide two vaccine doses* (a prime and booster) per person before a pandemic peaks.[1510,1511] Pre-pandemic immunization would permit governments to eliminate the time associated with a priming immunization and a three-week wait before giving a booster immunization. Pre-pandemic immunization will save both valuable time and lives before the peak of a pandemic.[1512,1513,1514]

Why haven't we done this, after the experience of 2009? I think the most important reason is the impact of *vaccine and adjuvant activism.*[1515]

Activists have had a profound influence on politicians, drug regulators, and industry. In light of developments in vaccine technology, decisions must have been made not to immunize the human population until the risks are more significant, i.e., after a pandemic emerges and the perceived risk-to-benefit ratio changes. I say, "decisions must have been made" because industry (i.e., senior scientists, R&D leaders) has known and been vocal about changing the flu vaccine and immunization paradigm since 2009, without success.

The pharmaceutical industry and drug regulators have been "scarred and scared" by the history of drug safety issues impacting human health, and the resulting multi-billion dollar litigation. More recent activism concerning vaccine safety issues[1516] reminds industry of the possible risks associated with pre-pandemic immunization.

Our *perception of the risk of a pandemic flu outbreak must change*, as opposed to hoping that H5N1 or H7N9, etc. will not cause a pandemic. The association of pandemics with cold climates and extremes of solar activity (peaks and troughs) offers an *objective* reason for changing our perception of risk.

Pre-Pandemic Immunization with an Adjuvanted Flu Vaccine Will Solve the Problem

The human population has reasonably good immunity to H1N1 and H3N2 seasonal and pandemic influenza-A strains (i.e., broadly cross-reactive antibodies and cellular immunity) but has no preexisting immunity to H5N1, H7N9, H9N2, and H2N2 pandemic flu strains. This is why 2009's swine flu H1N1 was relatively benign (i.e., people had preexisting immunity), and why H5N1 and H7N9 infections will likely cause high mortality rates. This means we are *immunologically defenseless on two-thirds of our front line* (four of six influenza-A strains and all of their mutants), while major pandemic threats are on the horizon, preparing to infect us. So long as we have no preexisting immunity to these flu strains, a pandemic is likely to have a high rate of mortality.

All we've done since the emergence of highly pathogenic avian H5N1 virus in 2005 is effectively *tackle the vaccine problem from one end*. Industry has greatly expanded global flu vaccine manufacturing capacity, and is transitioning to larger-scale production methods (i.e., cell culture and recombinant methods). It has also reduced the time-line for vaccine batch release.[1517,1518,1519,1520] However, despite this major progress, it still takes too much time (about four to six months) to generate a vaccine prototype and have the production process optimized for large-scale production.

Importantly, the existing seasonal flu vaccines (without an adjuvant) are only 60 percent effective at preventing mild forms of influenza.[1521] And they are rendered largely ineffective by viral mutation.[1522,1523] And yet, despite all these previously cited shortcomings, *the only substantive change* made to vaccine technology since 2009 has been moving from a three-component (trivalent) to a four-component (quadrivalent) flu vaccine, by adding a second influenza-B strain.[1524] While this solved the B-strain vaccine efficacy issue, and expanded global manufacturing capacity (from manufacturing three to four vaccine components i.e., a 33 percent increase), *it did nothing* to provide immunological defenses (pre-immunity) against pandemic-potential H5, 7, 9, and H2 influenza-A strains. Nor did it do anything to enable pre-pandemic immunization.

This industry and government response (partially curtailed by anti-vaccine activism) *still leaves us with a reactive vaccination strategy*, in that we must wait until after a pandemic emerges before immunizing people. Pre-pandemic immunization will permit governments, industry, and WHO to *achieve herd immunity before a pandemic emerges, and before the peak of mortality*.

Drug Regulatory Pathways Already Exist for Pre-Pandemic Immunization

It should not be said that drug regulators would refuse to permit pre-pandemic immunization, if that was the government plan. Everything is technically and legally possible from a regulatory perspective in making a pre-pandemic, pandemic, or a seasonal flu vaccine, and with or without an adjuvant. Whatever permutation of vaccine strains and adjuvant is required, the development pathways and regulatory processes already exist to approve those vaccines. Where there is will on the part of government, there is a way on the part of industry and drug regulators. So perhaps voters need to demand that their governments make it a priority to take action that will ensure we are pre-immunized for pandemic flu (see Chapter 13's Bucket list).

The European Medicines Agency (EMA) and US Food and Drug Administration (FDA) both use a centralized and expedited approval procedure for pandemic flu vaccines. This consists of a preemptive *mock-up core dossier* (EMA)[1525,1526] or biologics license application (FDA) for a vaccine prototype,[1527] which is then supplemented during a pandemic, with final vaccine approval then being fast-tracked.

These regulatory applications for marketing pre-approval of a vaccine prototype contain preliminary manufacturing, safety, and human clinical data for a virus that has caused human infection but not a pandemic. Both the EMA and FDA anticipate the use of an adjuvant for a pandemic or pre-pandemic vaccine.

A number of pandemic and seasonal flu vaccines containing oil-in-water adjuvants are already on the market or sit in stockpiles ready for use. Fluad (by Seqirus) is the only seasonal flu vaccine containing an adjuvant (MF59C.1).[1528] It was approved for use in the US in 2015,[1529] and in Europe (1997).[1530] Two H5N1 influenza-A pre-pandemic vaccines containing an oil-in-water adjuvant (i.e., Prepandrix with AS03[1531] and Aflunov[1532] with MF59C.1) have also been approved for use. The US government has approved multiple manufacturers' vaccine components which are to be stockpiled in bulk for mix-and-match final vaccine formulations after a pandemic emerges. The USA maintains H5N1, H7N9, H1N1 vaccine protein components, and AS03 and MF59C.1 adjuvants in secure storage.[1533,1534,1535]

Collectively, the regulatory procedures mentioned above permit a seasonal flu vaccine to be changed annually (i.e., changing the influenza-A and B strain components). These procedures also permitted the introduction (in 2012-2013) of a second influenza-B viral antigen creating a quadrivalent vaccine.[1536] Regulatory procedures will permit a pre-pandemic vaccine prototype to be pre-approved and then updated for the pandemic-specific vaccine required once a pandemic emerges. This updated pandemic-specific vaccine would then be rapidly approved for market use within days of filing a new drug application.

Basically, government and its pharmaceutical regulators, in partnership with vaccine suppliers, *can do just about anything they want.* However, anti-vaccine activism does complicate things.

How Everyone's Needs Can Be Met

Ultimately, pre-pandemic immunization presupposes that Seqirus (MF59C.1) and GlaxoSmithKline (AS03) would sponsor a new vaccine product in the global marketplace, and also permit their adjuvant to be licensed to Sanofi and other vaccine companies. For this to happen there needs to be a sufficiently attractive government-supported market. However, risks must also be managed in order for GlaxoSmithKline, or Seqirus, or other companies to support a government decision to pre-immunize the population. This means someone has to own the responsibility for the risks and accept the liabilities. Put bluntly, who will pay the plaintiffs' compensation in the face of lawsuits if a vaccine causes side effects or death?

Activists and others who don't wish to be immunized don't need to be immunized, but this minority should not be allowed to put the rest of us in jeopardy.

Seasonal flu vaccines are administered annually to hundreds of millions of at-risk people, and these vaccines offer a ready-made formulation capable of being reconfigured to include pandemic potential H7N9 or H5N1 influenza-A strains. Novartis vaccine R&D leaders and lead scientists even suggested this vaccine reconfiguration in 2012 and 2013 for pre-pandemic immunization.[1537],[1538]

To make a pre-pandemic flu vaccine, Seqirus (for example) could theoretically add H7N9 and H5N1 antigens to the Fluad seasonal flu vaccine, with or without substituting one or more of the influenza-B strains. This would permit herd immunity to be preinstalled for H7N9 and H5N1 flu strains in at-risk populations under the guise of a seasonal flu immunization program. This proposed addendum to the seasonal flu vaccine could be used to improve overall immunization rates against influenza-A; that is, it would provide the public something of perceived value and get them into doctors' clinics to be immunized.

A new *pre-pandemic flu vaccine* could also be created. This vaccine would be composed of (for example) eight influenza-A virus strains. This new prepandemic flu vaccine would contain two different subtypes of the H5, H7, H2, and H9 influenza A strains, plus the MF59C.1 or AS03 adjuvant. This new vaccine would be distinct from the seasonal flu vaccine, and would complete our front line of antibody defense for the high-risk period we are about to enter (climate switch and grand solar minimum). Precedents exist for multi-component vaccines, e.g., Prevnar 13[1539] and Gardasil 9.[1540] The existence of these multi-component vaccines demonstrates that such vaccines are feasible from a technical and regulatory perspective.

God Help Us if Governments Think That Way

What staggers me as I discuss our pandemic flu and vaccine predicaments with all sorts of people from all walks of life is their almost unanimous response, namely:

"Oh well, we could probably do with a few less people on the planet."

They say that *until I remind them* how they and those dearest to them could die a horrific *yet preventable* death, struggling to breathe for one or two days as they bleed inside their lungs and body, gasping in agony while awaiting death.

I also remind them that the United Nation's Brundtland Report did not advocate allowing humans to be culled by pandemic flu as the means of "Ensuring a Sustainable Level of Population."[1541] Our failure to pre-immunize against pandemic flu will inadvertently achieve the Brundtland Report's objectives. *But do we really want that?*

CHAPTER 15

Conclusion

The Crux of It All

There is an urgent need to prepare the world for a 21st century climate switch. The current trough-to-peak phase of this centennial-scale climate oscillation is already late in switching back to a cooling phase, but that cooling will happen. We will face climate risks during this grand solar minimum—risks that unfortunately have been dismissed by the IPCC in the climate advice it has given to governments.

Decentralized sustainable development and switching the world's energy system to renewable energy are pivotal to mitigating the yet unperceived climate-related risks, and to living sustainably. Immunizing the world against pandemic flu is an essential insurance policy for preventing global food insecurity, massive economic disruption, and the derailment of the proposed energy system switch, all three of which would result from a high mortality pandemic influenza-A outbreak.

Ignoring Natural Climate Change before 1880 CE When Forecasting the Climate is Akin to Driving with Your Eyes Closed

We have been led to believe by the IPCC that human activity and greenhouse gas emissions have been the dominant cause of recent global warming. This blaming of humans (while ignoring nature) has occurred despite the fact that the science being promoted to governments in the IPCC's assessment reports since 1990 has never been justified by any form of correlation analysis. A detailed correlation analysis is a basic minimum requirement if you wish to justify a scientific theory. Such a climate correlation would need to be valid over all timescales (not just since 1880 or 1970), and would need to be compelling when compared with correlations for other climate sub-systems.

Without that basic correlation analysis, you can't justifiably blame humans for climate change, while *ignoring nature*. Even with that correlation analysis, you still can't prove cause and effect. However, you can eliminate carbon dioxide as the primary climate-controller when carbon dioxide does not move up and down with the multi-annual or decade-scale temperature oscillations (see Figures 4.3, 4.4, and 6.2), and when your forecasting is highly inaccurate over a 30-year period. Ultimately, nature will prove this argument.

This book has shown that the IPCC's science does not represent a consensus of climate scientists, but rather a United Nations *diktat* justified by the IPCC's founding Articles 1 and 2. In Chapter 1 you were shown the IPCC processes that enabled its scientific bias, and the harsh criticism provided by external scientific authorities. Articles 1 and 2 and the IPCC processes have enabled the IPCC's scientific bias for three decades. The IPCC selected carefully vetted government scientists to do its work, and then dressed that up as a scientific consensus—one that the media incessantly tries to brainwash us with today.

The proof of this scientific sham is in the IPCC's forecasts. If normal science were operating in the climate science field, then the IPCC's 30-year track record of generating highly inaccurate climate forecasts (1985–2018), should have been exposed by the peer review process. Likewise, the dismissal of the ice age by a statistically flawed 30,000 years should have been torn apart by the peer review process—but strangely, it wasn't.

The IPCC is the organization advising our governments on climate risks and risk mitigation strategies. Yet, strangely, you will learn virtually nothing about the natural, climate change-related risks which are detailed in this book. You have been shown exactly where to look in the various IPCC assessment reports to see how natural climate-related catastrophic risks were dismissed, downplayed, or ignored by the IPCC.

This means we are being inadvertently led to a climate switching "cliff edge." Why? Because the IPCC borrowed a redacted version of climate science decades ago to promote its political agenda—the switching of the world's energy system and moving humans to living sustainably before *we run out of oil and gas.*

The part of the climate science field not wedded to anthropogenic global warming has identified several alternative risk factors beyond human activity, which the IPCC has chosen to ignore, presumably to prevent the undermining of its political agenda. Solar scientists expert in climate change have warned us that this current grand solar minimum will lead to Little Ice Age-like conditions in the decades ahead. We really must listen to these experts if we wish to best prepare and mitigate the risks.

The IPCC's risk assessment effectively dismissed, downplayed, or ignored the plethora of climate-related human catastrophes that occurred throughout the Little Ice Age, as well as the event that triggered the Little Ice Age, namely the climate-forcing volcanic eruption at Rinjani in 1257. Climate scientists expert in Arctic glacier ice accumulation processes and in volcanism have pointed to the primary role of climate-forcing volcanism and grand solar minima in triggering and propagating centennial-scale ice accumulation and global cooling processes. Their science delineates the big risks that we will soon face.

Climate scientists expert in volcanism and agricultural food supply have advised us that if a large magnitude (VEI 7) or climate-forcing volcanic eruption were to occur today, it would have a devastating impact on global agriculture. *One year's food production for one-third of the world's population* would be lost. We don't have a plan for that, but we do have options if we are given sufficient time to prepare.

The climate optimum timings indicated in the ice core data and their relative phasings tell us earth entered an ice age 8 and 10.5 millennia ago in the Arctic and Antarctic, respectively. This climate data also highlights our slow descent into this ice age—slow relative to all other glacial cycles in the past 800,000 years (Antarctica) and 2 million years (globally). Northern Hemisphere summer temperatures have also declined over 8–10 millennia, and this has closely tracked the precession-driven decline in Northern Hemisphere solar irradiance.

I remind you of the above and what follows because this is fundamentally important to help you reorient yourself as to what stage of the glacial cycle we are in today. There was less ice at the poles at the Holocene Climate Optimum than there is today. Ice began to accumulate from about 5,000 years ago, and northeast Greenland was ice-locked by 3,000 years ago. Glacier ice rapidly accumulated during the Little Ice Age, and reached its peak in the mid-19th century. Much of this glacier ice then melted after the mid-19th century, as the sun entered its 20th century grand solar maximum phase, and human activity came into play.

Superimposed on the 8,000-year year decline in Greenland's temperature and glacier ice accumulation are dozens of centennial-scale climate oscillations of ±1°C. These oscillations reduced the temperature recorded in the ice core by 4.9°C at its lowest temperature trough in 1700. The subsequent trough-to-peak temperature rise between 1700 and 2016 tells us that today's climate peak is the largest statistical outlier in the past 8,000 years. That statistical analysis also tells us that *the higher the peak, the more abrupt the fall* in temperature when the climate switch happens. The Greenland ice core data also tells us that we are two centuries overdue for that fall, and it tells us very clearly that carbon dioxide did not cause the 1700-2016 trough-to-peak global warming phase (see Figure 4.3). Therefore, I conclude a fall is going to happen, and *this grand solar minimum is prime time* for it.

Some of the climate switches to a cooling phase since the Holocene Climate Optimum were termed (by the climate science field) rapid climate change events. These events were associated with the collapse of numerous ancient civilizations. Yet such climate risks were dismissed by the IPCC.

A Reminder of Our Vulnerability
to Climate-Switching Risks

We need to anticipate a climate switch to a global cooling phase and a return to ice age conditions. Risks and vulnerabilities are linked to colder climates, prolonged extremes of precipitation (drought, rainfall, and snow), the potential for climate-forcing volcanism, and the impact of all these on agriculture, water supply, and energy needs.

The big complication for any plan other than the global warming plan is that there is so much politicized and vested interest in anthropogenic global warming on the part of the IPCC, our governments, and the media. The global warming story, and blaming human activity for natural climate change, has an "oil supertanker-like momentum" to it. This momentum makes a change of direction almost impossible. This book will place its flag in the sand, and must wait for extreme weather to start raising questions in people's minds. Keep an eye on the global and Northern Hemisphere temperature if you want to know the truth.[1542] Don't fall prey to media manipulation.

The complication for any plan to mitigate climate change relates to non-climate factors that increase our vulnerability to climate-related risks. Such non-climate factors include uncontrolled population and economic growth, increasing urbanization, and our depletion of earth's non-renewable resources. Human activity has depleted two-thirds of the world's largest aquifers beyond sustainable limits, and increased water basin and transboundary river stress. This creates a major vulnerability, because we have reduced the buffering capacity of our water basins, which is where people live and, even more importantly, where most of our food is grown.

Compounding our vulnerability to climate risks is the fact that we only have 50 years of proven oil and gas reserves left. Making matters worse, more than half of the world's oil and gas reserves are *unproven guesstimates*, (this includes shale resources). This means the optimistic prospect for new oil and gas discoveries is more fiction than fact.

The high degree of urbanization constitutes another major risk and vulnerability the world will confront in a climate switch crisis. Urbanites have largely forgotten how to grow food at home, or they have lost the capacity for growing food. This has created a reliance on fewer food producers overall. This urban vulnerability is compounded by the high degree of centralized supply for essential resources (i.e., food, energy, and water) that are provided largely by big corporations and/or government.

The Plan: Decentralized Sustainable Development and Switching the World Energy System to Renewable Energy

While I may be critical of the IPCC, I do admire the master plan and spirit of the United Nations and its intent to switch the world's energy system and move us to a sustainable lifestyle.

Rapidly switching the world's energy system to renewable energy and implementing decentralized sustainable development are *pivotal strategies for mitigating a climate switch, catastrophic climate-related risks, and the coming crisis in the supply of fossil fuels.*

The focus on decentralization means that individuals, communities, and cities must embrace partial self-sufficiency for securing energy, water, and food. Decentralization anticipates that government, corporate, and commodity market actions will restrict resource supply in a crisis event, and so aims to reduce our vulnerability to that systematic risk.

In order to remove impediments to change, there is an urgent need for a scientific revolution that recognizes there is a high probability of a climate switch with its associated risks (i.e., cold, extremes of precipitation, climate-forcing volcanism, and pandemic flu). Articles 1 and 2 need to be amended or eliminated. A great sense of urgency is required. The correct climate change message and the right economic incentives are imperative if the world is to act.

By implementing a carbon tax, we can incentivize the energy system switch and drive energy efficiency innovations. By revealing that peak oil and gas discovery and production are behind us, and that we only have 50 years of proven oil and gas reserves left, we can convey a message of resource scarcity. In such a manner, a higher market price for oil and gas, necessary for achieving the energy system switch, will be achieved. The impact of higher energy costs and the fear of losing corporate profits will be a more effective strategy for switching the world's energy system than blaming human beings for destructive climate interference.

Massive renewable energy capacity will be required for this energy transition. To expand renewable energy supply will require regional super-grids and local smart grids, and plenty of high voltage direct current transmission interconnections to transport bulk electricity. Gigawatt-scale battery storage capacity will also be required, to ensure energy system resilience. In place of oil and gas pipelines and using shipping to transport fuels, we need to build the infrastructure for bringing renewable energy from resource-rich regions to regions with demand.

We need to anticipate the need for water supply solutions in regions that were impacted by extreme drought during the Little Ice

Age. Proactively improving the drought-buffering capacity of water basins through natural and artificial groundwater recharge is essential. The decentralized supply of both desalinated water and energy using coastal renewable energy systems offers sustainable solutions for coastal, near-coastal, and urban environments. Building pipelines for bulk water transportation from water-rich areas to cities, industries, and agriculture must be made a top priority, in preparation for extreme and prolonged drought.

Decentralizing food supply for cities and communities, and promoting food production at home, must become key priorities for central and municipal governments. Urban agriculture, high-tech indoor farming, urban soil-less food production, aquaculture, and home food production systems are all important means for improving city food supply and reducing urban vulnerability in a food crisis. Cold- and drought- adapted crops and farming practices already exist, but preparations for their deployment must be prioritized.

Climate-forcing volcanism that blocks out sunlight will decimate global food production. This means we need new food supply solutions that are independent of the sun and climate. Industrialized greenhouses surrounding cities, high-tech indoor farming, and bulk-scale single cell protein manufacture all offer climate- or sunlight-independent food production systems. These can be used to provide just-in-time food stockpiles (i.e., produced when required).

At the heart of decentralized, sustainable development are you and your family at home. The message is clear—we need to live sustainably, and become self-sufficient (or almost) in producing food and energy, and in managing our water supply during climate extremes. We need a backup plan to reduce our vulnerability to global-scale risks. Chapter 12 gives you best practice methods for living sustainably at home, and preparing your family for climate risks. Being prepared will make the difference between survival and extreme vulnerability or even death.

Chapter 13 provides a bucket list of things we need to do to prepare our homes, communities, municipalities, work places, and countries. This bucket list is geared toward voting leaders into power who will get the big jobs done on our behalf. Rather than be dependent on government, we must remind government leaders that they are in power because our votes put them there. There's much work to be done, and we need our leaders to act on our behalf before it's too late.

I hope you find this book provocative, and that it shatters the illusion that humans control the climate while denying Mother Nature—including the sun and earth—her primary role. I hope you find this book useful in helping you move toward living sustainably in your day-to-day life and in helping you prepare this world for the youth of today, who will face the cold and climate extremes of the future. My best wishes to us all in this endeavor.

About the Author and My Motivations
for Writing this Book

Prior to becoming an author and private researcher, I had a diverse career and education, and purposefully pursued an eclectic series of interests. I qualified as a veterinary surgeon in 1986 (Massey University, New Zealand) and worked in New Zealand and England as a veterinarian, predominantly in small animal practice. In 1997 I completed my Master of Business Administration at the London Business School, and pursued a career in biotechnology and global healthcare investment banking in Europe.

Between 2002 and 2012, as a vaccine innovator and CEO, I raised £23m from European life science investors and the UK government, and built a vaccine company in the UK that was sold in 2015. During this time a synthetic universal influenza-A vaccine, able to immunologically counter all potential pandemic influenza-A strains, was progressed into clinical testing (see LinkedIn). The underlying vaccine technology was developed to counter the threat of mutating viruses transmitted from animals to people (i.e., zoonosis). This background in innovating vaccines for zoonotic mutating viruses provided a doorway to my research on the Arctic climate and solar activity's influence on viral mutation and pandemic flu outbreaks.

My new career as an author began in 2012 and was enabled by two decades of hobby research in the field of pyramid archeology, resulting in the publication of *Discovering Ritual Meditation: Transcendental Healing and Self-Realization* (hyperlink) in 2014. I spent three years being a test pilot for the ancient priesthood ritual methods that I discovered during my archeology research. My interests in earth as a complex living system (Mother Earth), enabling the evolution of all life and consciousness, while living "in presence" (in the moment, in silence, aware of breath and all arisings) in the intense magnetic fields of Lake Atitlan's volcanoes (Guatemala), "inspired" this book.

Some important questions also helped shape my discoveries. How did climate change cause the demise of ancient Egypt's Old Kingdom and other ancient civilizations? Did we lose civilizations under the oceans, before the Holocene Climate Optimum, and as the ice of the last ice age began to melt? What was the significance of the word sun in sungod? What is really happening with the climate today? Are we witnessing more volcanism today?

I am interested in the sun's magnetic and electromagnetic control of earth systems and earth system risks (i.e., climate change, volcanism, earthquakes, disease)—after all, ancient civilizations worshipped the sungods and goddesses to protect against such risks.

End Notes

A number of the following endnotes point to places in copyrighted publications where important information can be found. Exposé summaries of relevant copyrighted information are provided and are accompanied by critiques and commentaries. You are urged to review the publicly available documents and see the information for yourself. This exposé is done without copying copyrighted text, and is done in the spirit of "fair use." You have a right to know what science has been discovered, and what science has been dismissed, ignored, or not detailed, relating to potentially catastrophic climate-related risks.

Tabulated Data Associated with Endnotes

Each table legend refers to the tabulated data above it.

KEY WORD TOTALS FOR EACH CLIMATE MECHANISM OF ACTION	IPCC 1990-92	IPCC 2014	Google Scholar (1977-1992)	Google Scholar (1999-2014)
Radiative Forcing (Greenhouse gases)	3,586	14,882	221,030	3,883,700
Solar Activity	515	1,458	144,247	1,677,320
Galactic Cosmic Rays-Albedo	678	3,122	107,637	1,336,600
Solar & Geo Magnetism	14	44	50,452	617,609
Atmospheric & Ocean Circulations	348	2,694	51,416	591,790
Earth Orbital Parameters	55	249	72,140	324,400
	5,196	22,449	646,922	8,431,419

	IPCC 1990-92	IPCC 2014	Google Scholar (1977-1992)	Google Scholar (1999-2014)
Radiative Forcing (Greenhouse gases)	69.0%	66.3%	34.2%	46.1%
Solar Activity	9.9%	6.5%	22.3%	19.9%
Galactic Cosmic Rays-Albedo	13.0%	13.9%	16.6%	15.9%
Solar & Geo Magnetism	0.3%	0.2%	7.8%	7.3%
Atmospheric & Ocean Circulations	6.7%	12.0%	7.9%	7.0%
Earth Orbital Parameters	1.1%	1.1%	11.2%	3.8%
	100%	100%	100%	100%

MECHANISMS OF ACTION

Radiative Forcing (Greenhouse gases): Carbon dioxide (CO2), Greenhouse gases(GHG), Aerosol(s), Water vapour (vapor), Volcanic(ism), Radiative

Solar Activity: Solar, Ultraviolet (UV), Carbon 14 (C14, 14 Carbon), sunspot(s), Beryllium 10 (10Be), solar flare(s)

Galactic Cosmic Rays-Albedo: Cloud (s), Albedo, Cosmic (ray/radiation), Ionization, Condensation nuclei, Cosmogenic.

Solar & Geo Magnetism: Magnetic (ism), Ionospher(eric), Geomagnet (ic, ism), Magnetosphere(ic), Heliosphere(ic), Solar magnet(ic, ism)

Atmospheric & Ocean Circulations: Monsoon(s), Atmospheric circulation(s), ENSO (El Nino, La Nina), Ocean circulation(s), Intertropical convergence zone (ITCZ), North Atlantic Oscillation

Earth Orbital Parameters: Orbit(s, al), Eccentric(ity), Tilt, Insolation, Milankovitch, Precession

1) Data referenced in and supporting Figure 1.1 and endnote 153. This pertains to document and search engine key word use associated with climate control mechanisms.

Climate Optimum Peak Interglacial duration (Kiloyears)	Today	124.8 Kyr	241.3 Kyr	329.4 Kyr	403.5 Kyr	522.7 Kyr	612.4 Kyr	697.9 Kyr	783 Kyr	859.2 Kyr	957.1 Kyr
	17.5	16.6	11.2	15.2	30.9	16.8	18.9	22.6	20.3	21.8	8.2
Climate Optimum Peak Interglacial duration (Kiloyears)	Delay	124.8 Kyr	241.3 Kyr	329.4 Kyr	403.5 Kyr	522.7 Kyr	612.4 Kyr	697.9 Kyr	783 Kyr	859.2 Kyr	957.1 Kyr
	49.6	16.6	11.2	15.2	30.9	16.8	18.9	22.6	20.3	21.8	8.2

© 2014 Carlton B. Brown, http://iceageearth.com, http://grandsolarminimum.com

2) Data referenced in and supporting endnote 242. Table of interglacial durations (Global data).

Climate Optimum Peak (Kyr)	2.1	124.8	241.3	329.4	403.5	492.2	612.4	697.9	783.0	859.2	957.1	
Climate Optima Interval (Kyr)	122.7	116.5	88.1	74.1	88.7	120.2	85.5	85.1	76.2	97.9	66.5	
Climate Optimum Peak (Kyr)	1023.6	1073.6	1167.9	1241.3	1281.7	1317.2	1355.3	1399.0	1438.9	1477.6	1521.7	
Climate Optima Interval (Kyr)	50.0	94.3	73.4	40.4	35.5	38.1	43.7	39.9	38.7	44.1	43.9	
Climate Optimum Peak (Kyr)	1565.6	1604.4	1632.0	1685.1	1738.0	1774.3	1811.1	1856.9	1889.0	1934.9	1981.8	2026.8
Climate Optima Interval (Kyr)	38.8	27.6	53.1	52.9	36.3	36.8	45.8	32.1	45.9	46.9	45.0	

© 2014 Carlton B. Brown, http://iceageearth.com, http://grandsolarminimum.com

3) Data referenced in and supporting endnote 243. Table of climate optimum intervals (Global data).

Antarctic Climate Optima (Kiloyears ago)	10.5	128.7	242.1	333.3	406.9	488.6	615.1	693.5	787.3
Global Climate Climate (Kiloyears ago)	2.1	124.8	241.3	329.4	403.5	492.2	612.4	697.9	783.0
Phasing gap (Antactic -v- Global)(Kiloyears)	8.4	3.9	0.8	3.9	3.4	-3.6	2.7	-4.4	4.3
Revised phasing gap (Antactic -v- Global)(Kiloyears)	40.5	3.9	0.8	3.9	3.4	-3.6	2.7	-4.4	4.3

© 2014 Carlton B. Brown, http://iceageearth.com, http://grandsolarminimum.com

4) Data referenced in and supporting endnote 244. Table of the phasing gaps between the climate optima, with and without a 30,000-year delay to the start of the ice age (Antarctica Dome-C versus global data).

Climate Reconstruction Location (Polar, Global)	GLACIAL CYCLE CLIMATE OPTIMA TIMINGS & PHASING (Kiloyear)											Mean
Antarctica climate optima (Dome-Fuji)(Uemura, R et al.)	10.1	131.9	243.3	336.1								
Climate optima interval (Dome Fuji)	121.8	111.4	92.8									108.7
Antarctica Dome EPICA-C climate optima (Jouzel et al.)	10.5	128.7	242.1	333.3	406.9	488.6	615.1	693.5	787.3			
Climate optima interval (EPICA Dome-C)	118.1	113.4	91.2	73.8	81.7		78.5	93.8				92.9
Arctic-Greenland climate optima (Sigfus J. Johnsen et al.	9.4	127.0										
Greenland climate optima interval	117.6											
Arctic-Greenland climate optima (B. M. Vinther, et al.)	8.0											
Global climate optima (Bintanja)	2.1	124.8	241.3	329.4	403.5	492.2	612.4	697.9	783.0	859.2	957.1	
Global climate optima interval	122.7	116.5	88.1	74.1	88.7	120.2	85.5	85.1	76.2	97.9	66.5	92.9
Dome EPICA-C climate optima B4 Global climate optima	8.43	3.86	0.77	3.88	3.36	-3.64	2.66	-4.39	4.30			2.1
Dome-Fuji climate optima B4 Global climate optima	8.00	7.10	2.00	6.70								6.0

© 2014 Carlton B. Brown, http://iceageearth.com, http://grandsolarminimum.com

5) Data referenced in and supporting endnote 279. Table of Polar and global climate optima, providing climate optima intervals and phasing gaps (bottom two lines in the table)

Time (Kyr before Y2K)	Temp (Deg.C)	Global Sea Level (m)
Y2K 0 Kyr	-0.05	1.46
Holocene Climate Optimum 2.1 Kyr	0.56	2.87
End of Younger Dryas 11.7 Kyr	-3.4	46.6
Glacial Maximum 19.6 Kyr	-15.9	123.4
	Deg.C	**Meters**
Total Holocene Interglacial Increase	16.5	120.5
Increase from Glacial Maximum to 11.7Kyr	12.5	76.8
% of Holocene Interglacial Increase by 11.7 Kyr	75.7%	63.7%

6) Data referenced in and supporting endnote 288. Table of climate and sea level parameters at the time of the Younger Dryas, relative to the glacial maximum and climate optimum.

Trough (YBP, Y2K)	Deg.C at Trough -1	Peak (YBP Y2K)	Deg.C at Peak (0)	Trough (YBP Y2K)	Deg.C at Trough +1	Temp Rise (deg.C)	Time to rise (trough-to-peak) (Years)	Temp Fall @1st Trough (deg.C)	Time to fall to 1st Trough (Years)	Temp Fall @ Max. Trough (deg.C)	Time to fall to Max. (Years)
300	-1.31	60	1.56			2.87	240				
4340	0.37	4200	2.81	4140	1.34	2.44	140	-1.47	60	-1.47	60
2200	0.05	2040	2.18	1980	-0.37	2.13	160	-2.55	60	-2.55	60
2640	-0.37	2540	1.61	2500	0.06	1.98	100	-1.55	40	-1.55	40
3000	0.98	2980	2.75	2960	0.63	1.77	20	-2.12	20	-2.12	20
Mean						2.24	132	-1.92	45	-1.92	45

7) Data referenced in and supporting endnote 343. Group 1 trough-to-peak temperature rises and falls (Greenland ice core).

Large Magnitude Explosive Volcanic Eruption Events (VEI ≥6)(VOGRIPA) (10yr resolution)	TOTALS	AVERAGE 10yr- SSN	STDEV
Number of VEI ≥6 Events	67	24.7	13.6
Peaks exact	10	30.1	12.7
Peak +/-10years	13	32.1	
Peak +/-20 years	2	41.9	
Peak +/->20 years	0	0.0	
Troughs exact	8	13.5	8.0
Trough +/-10 years	17	20.5	13.8
Trough +/-20 years	4	14.9	
Trough +/->20 years	1	14.9	
Midways +/-10 years	7	33.6	
Midways +/-20 years	1	42.9	
Midways >+/-20 years	4	10.8	
SSN: Sunspot numbers			
Peaks, Troughs, Midways +/-1 decade	82.1%		
Peaks, Troughs, Midways +/-2 decades	91.0%		

8) Data referenced in and supporting endnote 458. Vogripa/LaMEVE VEI >6 eruption event data and their association with sunspot number peaks and troughs.

8KYRS ERUPTION EVENTS	1	2	3	4	1&2	1-3 events	All Events
Eruption Impact W/M2	[≤-10W/m2]	[-5 to <-10W/m2]	[-2 to <-5W/m2]	[0 to <2 W/m2]			
Number of eruptions	10	33	63	166	43	106	272
Pearson	-0.35	-0.63	-0.30	-0.12	-0.72	-0.63	-0.57
P-Value 2-tailed	0.185	0.009	0.252	0.666	0.002	0.009	0.021
Spearman rank		-0.65			-0.72	-0.56	-0.54
P-Value 2-tailed		0.006			0.002	0.023	0.032
Shapiro-Wilks test		P = 0.011			P = 0.031	P = 0.499	P = 0.863
Outliers		No			No	No	No

9) Data referenced in and supporting endnote 459 and Figure 5.1.A. Volcanic eruption and statistical data supporting the volcanic forcing threshold used to identify climate forcing volcanic eruptions.

Periods (Decade)	0	±1	±2	±3	±4	±5	Total	
Big Trough	6	10	2	3	3		24	
Big Peak	6	8	2	2	1		19	
Big SSN Peaks & Troughs	12	18	4	5	4	0	43	59%
Percent of Total	16%	25%	5%	7%	5%	0%	59%	
Little Trough	7	9	2				18	
Little Peak	4	6	1	1			12	
Little SSN Peaks & Troughs	11	15	3	1	0	0	30	41%
Percent of Total	15%	21%	4%	1%	0%	0%	41%	
Total Peaks & Troughs	23	33	7	6	4	0	73	
Percent of Total	32%	45%	10%	8%	5%	0%	100%	

10) Data referenced in and supporting endnote 459 and Figure 5.1.B. This data pertains to the distribution of climate forcing volcanic eruptions relative to sunspot number peaks and troughs.

Endnotes

1   IPCC, 2014: Climate Change 2014: Synthesis Report. Contribution of Working Groups I, II and III to the Fifth Assessment Report of the Intergovernmental Panel on Climate Change [Core Writing Team, R.K. Pachauri and L.A. Meyer (eds.)]. IPCC, Geneva, Switzerland, 151 pages [Exposé: See the *Forward, page V.* The IPCC tells us they are 95 percent certain that human activity is the dominant cause of today's global warming.]

2   IPCC, 2014: Climate Change 2014: Synthesis Report. Contribution of Working Groups I, II and III to the Fifth Assessment Report of the Intergovernmental Panel on Climate Change [Core Writing Team, R.K. Pachauri and L.A. Meyer (eds.)]. IPCC, Geneva, Switzerland, 151 pages [Exposé: See page 8, section SPM 2.1. This section details the climate drivers according to the IPCC's version of climate science.]

3   IPCC, 2013: Climate Change 2013: The Physical Science Basis. Contribution of Working Group I to the Fifth Assessment Report of the Intergovernmental Panel on Climate Change [Stocker, T.F., D. Qin, G.-K. Plattner, M. Tignor, S.K. Allen, J. Boschung, A. Nauels, Y. Xia, V. Bex and P.M. Midgley (eds.)]. Cambridge University Press, Cambridge, United Kingdom and New York, NY, USA, 1535 pages [Exposé: See page 79, Box TS.6. This blue, boxed text details the Representative Concentration Pathway global warming scenarios and the Coupled Model Intercomparison Project Phase 5 Models used to provide its global warming climate forecasts.].

4   Definitions of Articles 1 and 2: United Nations Framework Convention on Climate Change. United Nations 1992. FCCC/INFORMAL/84, GE.05-62220 (E), 200705. [Exposé: See page 3 and 4 of the UNFCCC document for Article 1 and 2 definition of climate change and the mitigation objectives for climate change (respectively). Importantly, climate change is attributed to human activity, which changes the atmosphere, and which is in addition to natural climate change over the same period of time. This definition for climate change was present at the IPCC's 1988 founding. In other words, the science of climate change was predetermined (human activity, greenhouse gases) and had nothing to do with academic science. Article 2's objective is focused on stabilizing atmospheric greenhouse gases at levels that would prevent dangerous human interference with the climate system, while ensuring food production and sustainable economic development. In other words, in 1988 Article 2 had already determined that human activity was dangerous and that it needed to be mitigated.]. Last accessed 08/10/2018.

5   Influence of Articles 1 and 2. Climate Change: The IPCC Scientific Assessment (1990). Report prepared for Intergovernmental Panel on Climate Change by Working Group I. J.T. Houghton, G.J. Jenkins and J.J. Ephraums (eds.). Cambridge University Press, Cambridge, Great Britain, New York, NY, USA and Melbourne, Australia 410 pages [Exposé: See the Preface. This preface tells us the IPCC was set up by the United Nations Environment Programme and the World Meteorological Organization and was given responsibility for assessing the scientific information relating to aspects of climate change, such as greenhouse gas emissions and the modification of Earth's radiation balance caused by these emissions. The IPCC was also charged with evaluating the environmental, social, and economic consequences of climate change and proposing strategies for managing climate change. Critique: The IPCC's radiative forcing theory was predetermined from the outset. There is no mention of assessing the scientific information relating to natural climate change or the other mechanisms of climate control (see Figure 1.1). Articles 1 and 2 caused this scientific bias.].

6   Dismissing volcanism. IPCC, 2013: Climate Change 2013: The Physical Science Basis. Contribution of Working Group I to the Fifth Assessment Report of the Intergovernmental Panel on Climate Change [Stocker, T.F., D. Qin, G.-K. Plattner, M. Tignor, S.K. Allen, J. Boschung, A. Nauels, Y. Xia, V. Bex and P.M. Midgley (eds.)]. Cambridge University Press, Cambridge, United Kingdom and New York, NY, USA, 1535 pages [Exposé: (1) See page 1008-1009, FAQ 11.2. This section details how volcanic eruptions impact the climate. We are told that IPCC-promoted climate predictions do not include or reflect potential future volcanic eruptions (for various reasons, which is tantamount to confirmation bias). (2) See page 1009 sub-section 11.3.6.3 point 4. This point confirms that IPCC promoted climate forecasts do not include any impact of volcanism on the future climate, because of volcanism's unpredictability. Critique: By

not including the potential impact of volcanism in climate forecasts, natural planetary cooling factors (i.e., volcanic aerosols) known to counter carbon dioxide's global warming effect are thereby eliminated from IPCC promoted forecasts. Ignoring volcanism, solar activity, and natural climate change constitutes scientific bias, and fails to represent normal science and the full climate system. By dismissing nature the IPCC's climate forecasts are rendered invalid.].

7   Dismissing solar activity. IPCC, 2013: Climate Change 2013: The Physical Science Basis. Contribution of Working Group I to the Fifth Assessment Report of the Intergovernmental Panel on Climate Change [Stocker, T.F., D. Qin, G.-K. Plattner, M. Tignor, S.K. Allen, J. Boschung, A. Nauels, Y. Xia, V. Bex and P.M. Midgley (eds.)]. Cambridge University Press, Cambridge, United Kingdom and New York, NY, USA, 1535 pages [Exposé: (1) See page 1009 sub-section 11.3.6.3 point 4. This point tells us that the IPCC climate projections assume no changes in solar irradiance in their climate forecasts, because there is a low confidence in the solar activity projections. (2) See page 1007 sub-section 11.3.6.2.2. The IPCC dismissed the impact of solar forcing on the climate during this grand solar minimum, despite solar activity being at its lowest level for the longest period since the early 20th century. This scientific dismissal of solar activity's impact on the climate resulted from the low confidence the IPCC had in the solar activity projections (solar irradiance). But they had "high confidence" that greenhouse gases would offset any diminution in total solar irradiance's impact on the climate. The IPCC do not ascribe any climate modifying role to solar magnetism (magnetized solar wind) i.e., air circulations, geomagnetism-cosmic rays-low clouds (see Figures 4.3.A, 4.4, 6.2, 5.1-5.3, and Chapter 6). Critique: The IPCC's low confidence in these projections is counter to NASA's open acceptance of this grand solar minimum and its impact on our atmosphere and other earth systems linked to climate change (listen to the video, https://science.nasa.gov/science-news/news-articles/solar-minimum-is-coming). This low confidence is also counter to scientific experts advising the Russian Space program (i.e., Habibullo Abdussamatov), and leading solar activity scientists expert in climate change advising of a planetary cooling in the decades ahead. What legitimacy does the IPCC have to dismiss other scientific disciplines and space program (USA, Russia) sciences, which are known to impact earth systems? (i.e., magnetosphere, ionosphere, atmosphere, tectonic-volcanic processes)].

8   Not detailing rapid climate change: IPCC, 2012: Managing the Risks of Extreme Events and Disasters to Advance Climate Change Adaptation. A Special Report of Working Groups I and II of the Intergovernmental Panel on Climate Change [Field, C.B., V. Barros, T.F. Stocker, D. Qin, D.J. Dokken, K.L. Ebi, M.D. Mastrandrea, K.J. Mach, G.-K. Plattner, S.K. Allen, M. Tignor, and P.M. Midgley (eds.)]. Cambridge University Press, Cambridge, UK, and New York, NY, USA, 582 pages [Exposé and Critique: See page 122. The IPCC told us that it did not review the risk of abrupt or rapid climate change. This was despite this report being a Special Report on "Managing the Risks of Extreme Events and Disasters to Advance Climate Change Adaptation." As you will see from the citations in this book, there is much published climate science and climate history relating to the Little Ice Age, climate-forcing volcanism, and the specific rapid climate events known to have taken place at or since the Holocene Climate Optimum i.e., the 8.2, 5.9 and 4.2 kiloyear rapid climate events (and other events). None of this risk-relevant information was detailed in this special risk report.].

9   Not detailing abrupt or rapid climate change: IPCC, Climate Change 2013: The Physical Science Basis. Contribution of Working Group I to the Fifth Assessment Report of the Intergovernmental Panel on Climate Change [Stocker, T.F., D. Qin, G.-K. Plattner, M. Tignor, S.K. Allen, J. Boschung, A. Nauels, Y. Xia, V. Bex and P.M. Midgley (eds.)]. Cambridge University Press, Cambridge, United Kingdom and New York, NY, USA, 1535 pages [Exposé and Critique: (1) See pages 70, Section TFE.5. This section is supposed to detail abrupt climate change and so-called climate tipping points. We are told there is some information (when there is much information) on abrupt climate change but there is low confidence and little consensus that such an event would take place during this century. (2) See page 115, TS.6.4, and page 1033. These sections reiterate the message contained in section TFE.5 on page 70. Critique: This dismissal of abrupt or rapid climate change is despite the history of catastrophic climate change associated with the Little Ice Age (and glacier expansions), climate-forcing volcanism, and the specific rapid climate events known to have taken place at or since the

Holocene Climate Optimum i.e., the 8.2, 5.9 and 4.2 kiloyear rapid climate events (plus other events). Detecting confirmation bias: Once again this exemplifies the IPCC's confirmation bias. This confirmation bias is easy to spot in the assessment reports because when the IPCC wishes to disprove science that would be counter to, or undermine, its theory or forecasts, it has a low confidence and certainty in the predictions or the underlying science, or we are told there is little consensus on the issue. Conduct a PDF document search on the IPCC scientific assessment reports using the following key words contained, "confidence," "certainty," "uncertainty," and "consensus," and see how this bias is communicated relative to the arguments being created or supported.].

10   N. Scafetta, "Multi-scale harmonic model for solar and climate cyclical variation throughout the Holocene based on Jupiter-Saturn tidal frequencies plus the 11-year solar dynamo cycle." Journal of Atmospheric and Solar-Terrestrial Physics (2012). doi:10.1016/j.jastp.2012.02.016.

11   Theodor Landscheidt, "New Little Ice Age Instead of Global Warming? Energy & Environment. 2003." Volume 14, Issue 2, 327–350. https://doi.org/10.1260/095830503765184646.

12   R.J. Salvador, "A mathematical model of the sunspot cycle for the past 1000 years," Pattern Recognition Physics, 1, 117-122, doi:10.5194/prp-1-117-2013, 2013.

13   Habibullo Abdussamatov, "Current Long-Term Negative Average Annual Energy Balance of the Earth Leads to the New Little Ice age." Thermal Science. 2015 Supplement, Volume 19, S279-S288.

14   Jan-Erik Solheim, https://www.mwenb.nl/wp-content/uploads/2014/10/Blog-Jan-Erik-Solheim-def.pdf. Referred from http://www.climatedialogue.org/what-will-happen-during-a-new-maunder-minimum/. Citing blog for 4-5 solar-climate experts.

15   Boncho P. Bonev et al., "Long-Term Solar Variability and the Solar Cycle in the 21st Century." The Astrophysical Journal, 605:L81–L84, April 10, 2004.

16   Nils-Axel Mörner, "Solar Minima, Earth's rotation and Little Ice Ages in the past and in the future. The North Atlantic–European case." Global and Planetary Change 72 (2010) 282–293. doi:10.1016/j.gloplacha.2010.01.004.

17   A. Mazzarella, "The 60-year solar modulation of global air temperature: the Earth's rotation and atmospheric circulation connection." Theoretical and Applied Climatology. 88, 193–199 (2007). DOI 10.1007/s00704-005-0219-z.

18   David D. Zhang et al., "Global climate change, war, and population decline in recent human history." Proceedings of the National Academy of Sciences December, 2007, 104 (49) 19214-19219; DOI: 10.1073/pnas.0703073104.

19   Dian Zhang et al., "Climate change, social unrest and dynastic transition in ancient China." China Science Bulletin January, 2005, Volume 50, Issue 2, 137–144. https://doi.org/10.1007/BF02897517

20   D. Collet and M. Schuh (eds.), "Famines During the 'Little Ice Age'" (1300–1800), DOI 10.1007/978-3-319-54337-6_2. [See page 21].

21   Anthony J. McMichael, "Insights from past millennia into climatic impacts on human health and survival." Proceedings of the National Academy of Sciences March, 2012, 109 (13) 4730-4737; DOI: 10.1073/pnas.1120177109. [See page 4734, column 2, paragraph 2].

22   Geoffrey Parker, "Crisis and Catastrophe: The Global Crisis of the Seventeenth Century Reconsidered." The American Historical Review, Volume 113, No. 4 (October, 2008), 1053-1079. http://www.jstor.org/stable/30223245.

23   Michael J. Puma et al., "Exploring the potential impacts of historic volcanic eruptions on the contemporary global food system." Pages Magazine. Science Highlights. Volcanoes and Climate. Volume 23, No 2, December 2015.

24   Clive Oppenheimer, "Climatic, environmental and human consequences of the largest known historic eruption: Tambora volcano (Indonesia) 1815." Progress in Physical Geography: Earth and Environment (2003). Volume 27, Issue 2, 230 – 259. https://doi.org/10.1191/0309133303pp379ra.

25   Anthony J. McMichael, "Insights from past millennia into climatic impacts on human health and survival." Proceedings of the National Academy of Sciences March 2012, 109 (13) 4730-4737; DOI: 10.1073/pnas.1120177109. [See page 4735, column 2, paragraph 2].

26   R.B. Stothers, "Climatic and Demographic Consequences of the Massive Volcanic Eruption of 1258." Climatic Change (2000) 45: 361. https://doi.org/10.1023/A:1005523330643.

27   C. Oppenheimer, (2003). "Ice core and paleoclimate evidence for the timing and nature of the great mid-13th century volcanic eruption." International Journal of Climatology, 23: 417-426. doi:10.1002/joc.891.

28   David D. Zhang et al., "Global climate change, war, and population decline in recent human history." Proceedings of the National Academy of Sciences Dec 2007, 104 (49) 19214-19219; DOI: 10.1073/pnas.0703073104. [See page 19216].

29   P. Mayewski et al., (2004). "Holocene climate variability." Quaternary Research, 62(3), 243-255. doi:10.1016/j.yqres.2004.07.001.

30   Bernhard Weninger et al., "The Impact of Rapid Climate Change on prehistoric societies during the Holocene in the Eastern Mediterranean." Documenta Praehistorica XXXVI (2009). UDK 902(4-5)"631\637">551.583.

31   M. Staubwasser, M. and H. Weiss, (2006). "Holocene Climate and Cultural Evolution in Late Prehistoric–Early Historic West Asia." Quaternary Research, 66(3), 372-387. doi:10.1016/j.yqres.2006.09.001.

32   Robert K. Booth et al., "A severe centennial-scale drought in midcontinental North America 4200 years ago and apparent global linkages." The Holocene. Volume 15, Issue 3, 321 – 328. 2005. https://doi.org/10.1191/0959683605hl825ft.

33   Stanley J. Krom et al., (2003), Short contribution: "Nile flow failure at the end of the Old Kingdom, Egypt: Strontium isotopic and petrologic evidence." Geoarchaeology, 18: 395-402. doi:10.1002/gea.10065.

34   Ann Gibbons, "How the Akkadian Empire Was Hung Out to Dry." Science August 20, 1993: Volume 261, Issue 5124, 985. DOI: 10.1126/science.261.5124.985.

35   Jianjun Wang, "The abrupt climate change near 4,400 year BP on the cultural transition in Yuchisi, China and its global linkage." Scientific Reports | 6:27723 | DOI: 10.1038/srep27723. https://www.nature.com/articles/srep27723.pdf.

36   J. Wang et al., "The abrupt climate change near 4,400 year BP on the cultural transition in Yuchisi, China and its global linkage." Scientific Reports 2016 Jun 10;6:27723. doi: 10.1038/srep27723.

37   Fenggui Liu and Zhaodong Feng, "A dramatic climatic transition at ~4000 cal. year BP and its cultural responses in Chinese cultural domains." The Holocene. Volume 22, Issue 10, 1181–1197. April 12, 2012. https://doi.org/10.1177/0959683612441839.

38   A. Parker et al., (2006). "A Record of Holocene Climate Change from Lake Geochemical Analyses in Southeastern Arabia." Quaternary Research, 66(3), 465-476. doi:10.1016/j.yqres.2006.07.001.

39   J. Slawinska and A. Robock, 2018. "Impact of Volcanic Eruptions on Decadal to Centennial Fluctuations of Arctic Sea Ice Extent during the Last Millennium and on Initiation of the Little Ice Age." J. Climate, 31, 2145–2167, https://doi.org/10.1175/JCLI-D-16-0498.1.

40   F. Lehner et al., 2013. "Amplified inception of European Little Ice Age by sea ice–ocean–atmosphere feedbacks." J. Climate, 26, 7586–7602. https://doi.org/10.1175/JCLI-D-12-00690.1.

41   C. Newhall et al., 2018. "Anticipating future Volcanic Explosivity Index (VEI) 7 eruptions and their chilling impacts." Geosphere, v. 14, no. 2, p. 1–32, doi:10.1130/GES01513.1.

42   Odd Helge Otterå et al., "External forcing as a metronome for Atlantic multidecadal variability." Nature Geoscience Volume 3, 688–694 (2010).

43   Y. Zhong et al., "Centennial-scale climate change from decadally-paced explosive volcanism: a coupled sea ice-ocean mechanism." Climate Dynamics (2011) 37: 2373. https://doi.org/10.1007/s00382-010-0967-z.

44   B.M. Vinther et al., 2009. "Holocene thinning of the Greenland ice sheet." Nature, Vol. 461, pp. 385-388, September 17, 2009. National Centers for Environmental Information, NESDIS, NOAA, U.S. Department of Commerce. Greenland Ice Sheet Holocene d18O, Temperature, and Surface Elevation. doi:10.1038/nature08355. https://www.ncdc.noaa.gov/paleo-search/study/11148. Downloaded 05/05/2018. [See the data point at 7,980 years, +3.550C = peak temperature anomaly at the end of the Holocene interglacial or at the Holocene Climate Optimum compared with the reference year 2000 CE].

45   J.V. Jouzel et al., 2007. "Orbital and Millennial Antarctic Climate Variability over the Past 800,000 Years." Science, Volume 317, No. 5839, 793-797, 10 August 2007. National Centers for Environmental Information, NESDIS, NOAA, U.S. Department of Commerce. EPICA Dome C - 800KYr Deuterium Data and Temperature Estimates. https://www.ncdc.noaa.gov/ paleo/study/6080. Download data: Downloaded 08/02/2016. [See the data point at 10,527 years, +2.12°C = peak temperature anomaly at the end of the Holocene interglacial compared with 1950.].

46   R. Uemura, et al., 2012. "Ranges of moisture-source temperature estimated from Antarctic ice cores stable isotope records over glacial-interglacial cycles." Climate of the Past, 8, 1109-1125. doi: 10.5194/cp-8-1109-2012. National Centers for Environmental Information, NESDIS, NOAA, U.S. Department of Commerce. Dome Fuji 360KYr Stable Isotope Data and Temperature Reconstruction. https://www.ncdc.noaa.gov/paleo-search/study/13121. Downloaded 5/5/2018. [See the data point at Tsite(°C) at 10.1Kyr +1.35°C = peak temperature anomaly at the end of the Holocene interglacial compared with 2000 CE (B2K)].

47   Global mean surface temperature data, commonly referred to as HadCRUT4. https://www. metoffice.gov.uk/hadobs/hadcrut4/data/current/download.html. [Exposé: Look at the bottom left hand or first column for the current year-to-date temperature. Subtract that from the 2016 total to see the magnitude of the fall. Global Data: https://bit.ly/2nCgctz. Northern Hemisphere Data: https://bit.ly/2MRt75G, Southern Hemisphere Data: https://bit.ly/2nBfYTA. Tropics Data: https://bit.ly/2nFXJMM. [last downloaded 25/07/2018].

48   B.M.Vinther et al., 2009. "Holocene thinning of the Greenland ice sheet." Nature, Vol. 461, pp. 385-388, 17 September 2009. National Centers for Environmental Information, NESDIS, NOAA, U.S. Department of Commerce. Greenland Ice Sheet Holocene d18O, Temperature, and Surface Elevation. doi:10.1038/nature08355. https://www.ncdc.noaa.gov/paleo-search/ study/11148. Downloaded 05/05/2018. [Comment: This calculation was based on the difference between peak temperature 7,980 years ago at the Holocene Climate Optimum, and the temperature recorded in 1880.].

49   IPCC, 2007: Climate Change 2007: The Physical Science Basis. Contribution of Working Group I to the Fourth Assessment Report of the Intergovernmental Panel on Climate Change [Solomon, S., D. Qin, M. Manning, Z. Chen, M. Marquis, K.B. Averyt, M. Tignor and H.L. Miller (eds.)]. Cambridge University Press, Cambridge, United Kingdom and New York, NY, USA, 996 [See page 288 where we are told that sea surface temperature and the global mean temperature anomalies (1998) were the highest on record until 2005 (global mean temperature).].

50   Nicolaj K. Larsen et al., "The response of the southern Greenland ice sheet to the Holocene thermal maximum." Geology ; 43 (4): 291–294. doi: https://doi.org/10.1130/G36476.1.

51   D.S. Kaufman et al., Holocene thermal maximum in the western Arctic (0–1800W). Quaternary Science Reviews 23 (2004) 529–560.

52   J.P. Briner et al., "Holocene climate change in Arctic Canada and Greenland," Quaternary Science Reviews (2016), http://dx.doi.org/10.1016/j.quascirev.2016.02.010.

53   Data: (1) B.M. Vinther et al., 2009. "Holocene thinning of the Greenland ice sheet." Nature, Vol. 461, pp. 385-388, September 17, 2009. National Centers for Environmental Information, NESDIS, NOAA, U.S. Department of Commerce. Greenland Ice Sheet Holocene d18O, Temperature, and Surface Elevation. doi:10.1038/nature08355. https://www.ncdc.noaa.gov/ paleo-search/study/11148. Downloaded 05/05/2018. (2) HadCRUT4 near surface temperature data set for the Northern Hemisphere. http://www.metoffice.gov.uk/hadobs/hadcrut4/data/ current/download.html. Downloaded July 25, 2018. Personal Research: Between the Holocene Climate Optimum (HCO) 5980 BCE and 1700 CE the Greenland temperature anomaly (derived from the ice core) declined 4.860C. Between 1700-1940 the temperature rose 2.870C. From this data, it was calculated that the temperature in 2016 was 1.910C lower than at the Holocene Climate Optimum. 1700-1960 rise = 2.320C rise plus a "graft-on" of a 1960-2016 temperature anomaly rise of 0.630C. To determine the 1960-2016 temperature anomaly rise the HADCRUT4 Northern Hemisphere near surface temperature data was utilized to create a 20-year moving average, to graft on to the 1700-1960 (+0.630C). HCO-2016 approximation = -4.860C + 2.320C +0.630C = -1.910C.

54   R.B. Alley, 2004, "GISP2 Ice Core Temperature and Accumulation Data." National Centers for
      Environmental Information, NESDIS, NOAA, U.S. Department of Commerce. https://www.
      ncdc.noaa.gov/paleo/study/2475. Downloaded 5/5/2018. [Last Glacial Maximum's deepest
      temperature trough was 24,098 years ago (-530C) and the Holocene Climate Optimum was
      7,800 years ago (-28.860C). The difference between these time points is 16,297 years and
      24.560C.]

55   M. Walker et al., 2009, "Formal definition and dating of the GSSP (Global Stratotype Section
      and Point) for the base of the Holocene using the Greenland NGRIP ice core, and selected
      auxiliary records." J. Quaternary Sci., Volume 24 3–17. ISSN 0267-8179.

56   A.E. Carlson, 2013. "The Younger Dryas Climate Event." In: Elias S.A. (ed.) The Encyclopedia
      of Quaternary Science, Volume 3, 126-134. Amsterdam: Elsevier. http://people.oregonstate.
      edu/~carlsand/carlson_encyclopedia_Quat_2013_YD.pdf

57   R. Uemura, et al., 2012. "Ranges of moisture-source temperature estimated from Antarctic
      ice cores stable isotope records over glacial-interglacial cycles." Climate of the Past, 8,
      1109-1125. doi: 10.5194/cp-8-1109-2012.. National Centers for Environmental Information,
      NESDIS, NOAA, U.S. Department of Commerce. Dome Fuji 360KYr Stable Isotope Data
      and Temperature Reconstruction. https://www.ncdc.noaa.gov/paleo-search/study/13121.
      Downloaded 05/05/2018.

58   R.B. Alley, 2004. "GISP2 Ice Core Temperature and Accumulation Data." National Centers for
      Environmental Information, NESDIS, NOAA, U.S. Department of Commerce. https://www.
      ncdc.noaa.gov/paleo/study/2475. Downloaded 05/05/2018.

59   K. Nicolaj et al., "The response of the southern Greenland ice sheet to the Holocene thermal
      maximum." Geology; 43 (4): 291–294. doi: https://doi.org/10.1130/G36476.1.

60   J.P. Briner et al., "Holocene climate change in Arctic Canada and Greenland." Quaternary
      Science Reviews (2016), http://dx.doi.org/10.1016/j.quascirev.2016.02.010.

61   Leonid Polyak et al., "History of sea ice in the Arctic." Quaternary Science Reviews 29 (2010)
      1757–1778, https://doi.org/10.1016/j.quascirev.2010.02.010.

62   Jason P. Briner et al. "Holocene climate change in Arctic Canada and Greenland." Quaternary
      Science Reviews, Volume 147, 2016, 340-364, ISSN 0277-3791. https://doi.org/10.1016/j.
      quascirev.2016.02.010.

63   Ó Ingólfsson et al., 1998, "Antarctic glacial history since the Last Glacial Maximum:
      An overview of the record on land." Antarctic Science, 10(3), 326-344. doi:10.1017/
      S095410209800039X.

64   Leonid Polyak et al. "History of sea ice in the Arctic." Quaternary Science Reviews 29 (2010)
      1757–1778, https://doi.org/10.1016/j.quascirev.2010.02.010

65   N.L. Balascio et al. "Glacier response to North Atlantic climate variability during the
      Holocene." Climate of the Past, 11, 1587-1598, https://doi.org/10.5194/cp-11-1587-2015, 2015.

66   The RAISED Consortium1, Michael J. Bentley et al., "A community-based geological
      reconstruction of Antarctic Ice Sheet deglaciation since the Last Glacial Maximum."
      Quaternary Science Reviews. Volume 100, 15 September 2014, 1-9.

67   M. Frezzotti1 et al. "A synthesis of the Antarctic surface mass balance during the last 800
      years." The Cryosphere, 7, 303–319, 2013. www.the-cryosphere.net/7/303/2013/doi:10.5194/tc-7-
      303-2013 © Author(s) 2013. CC Attribution 3.0 License.

68   O.N. Solomina et al., 2016, "Glacier fluctuations during the past 2000 years." Quaternary
      Science Reviews, 149, 61-90. DOI: 10.1016/j.quascirev.2016.04.008. [See Figure 5, page 276.
      This figure collates a stacked time series of the number of glacier advances and recessions in
      each region into a global total.].

69   Michael E Mann. "Little Ice Age." Volume 1, The Earth system: physical and chemical
      dimensions of global environmental change, 504–509. In Encyclopedia of Global
      Environmental Change (ISBN 0-471-97796-9).

70   Leonid Polyak et al., "History of sea ice in the Arctic." Quaternary Science Reviews 29 (2010)
      1757–1778, https://doi.org/10.1016/j.quascirev.2010.02.010

71   Christophe Kinnard et al., "A changing Arctic seasonal ice zone: Observations from 1870–2003
      and possible oceanographic consequences." Geophysical Research Letters, Volume 35, L02507,
      doi:10.1029/2007GL032507, 2008.

72   O.N. Solomina et al., (2016). "Glacier fluctuations during the past 2000 years." Quaternary
      Science Reviews, 149, 61-90. DOI: 10.1016/j.quascirev.2016.04.008. [See Figure 5, page 276.
      This figure collates a stacked time series of the number of glacier advances and recessions in
      each region into a global total.].
73   A.S. Richey et al., (2015), "Quantifying renewable groundwater stress with GRACE." Water
      Resources Research, 51, 5217–5238, doi:10.1002/2015WR017349, and NASA via https://www.
      nasa.gov/jpl/grace/study-third-of-big-groundwater-basins-in-distress.
74   H Munia et al., "Water stress in global transboundary river basins: significance
      of upstream water use on downstream stress." Environment Research Letter
      11 (2016) 014002. doi:10.1088/1748-9326/11/1/014002 http://iopscience.iop.org/
      article/10.1088/1748-9326/11/1/014002/pdf.
75   Limited oil and gas reserves: IPCC, Climate Change 2014: Mitigation of Climate Change.
      Contribution of Working Group III to the Fifth Assessment Report of the Intergovernmental
      Panel on Climate Change [Edenhofer, O., R. Pichs-Madruga, Y. Sokona, E. Farahani, S.
      Kadner, K. Seyboth, A. Adler, I. Baum, S. Brunner, P. Eickemeier, B. Kriemann, J. Savolainen,
      S. Schlömer, C. von Stechow, T. Zwickel and J.C. Minx (eds.)]. Cambridge University Press,
      Cambridge, United Kingdom and New York, NY, USA. [Exposé: See page 379. The IPCC
      tells us that we have about 50 years of oil reserves and 70 years of natural gas reserves. They
      also confirm that energy discovery is more challenging and more costly, with a limited scope
      for profitable extraction of those reserves (i..e, extraction will require greater technology
      deployment). Question: How will humans generate the carbon dioxide required to produce the
      global warming predicted by the IPCC for the 21st century with only about 50 or so years of oil
      and gas left?].
76   Limited oil and gas reserves: IPCC, Climate Change 2007: Mitigation. Contribution of Working
      Group III to the Fourth Assessment Report of the Intergovernmental Panel on Climate Change
      [B. Metz, O.R. Davidson, P.R. Bosch, R. Dave, L.A. Meyer (eds)], Cambridge University Press,
      Cambridge, United Kingdom and New York, NY, USA. [Exposé: See page 265, section 4.3.1.
      This section details fossil fuels reserves, telling us the proven and probable oil and gas reserves
      will only last for decades, while coal will last for centuries. Question: How will humans
      generate the carbon dioxide required to produce the global warming predicted by the IPCC for
      the 21st century with only about 50 or so years of oil and gas left?].
77   Data. 50 years of proven oil and gas reserves (see Chapter 8): Energy Information
      Administration data was obtained from: International Energy Statistics. These calculations
      utilized the following data files. Natural gas https://bit.ly/2LC6GBo, Crude Oil https://bit.
      ly/2IWeEaP, Coal data https://bit.ly/2L6pk3w. [Comment: These reserve timeline estimates are
      calculated by dividing the 2013 Energy Information Agency's proven global oil, natural gas,
      and coal reserves by 2013 levels of production. This calculation tells us there are 50 years of
      proven oil and gas, and 130 years of coal reserves left. These reserve timeline estimates do not
      assume any population or economic growth, or a switch to a cold climate regime, which would
      accelerate energy demand and reduce the timelines.].
78   U.S. Energy Information Administration report. Technically Recoverable Shale Oil and Shale
      Gas Resources. An Assessment of 137 Shale Formations in 41 Countries Outside the United
      States. June 2013. [Critique: See Table 2, page 3 for proven and unproven conventional and
      non-conventional energy reserves. The US Energy Information Agency reserve revisions
      mean that one-third of world gas and one-tenth of world oil resources are projections for shale
      resources. These revisions also mean that 50 percent and 70 percent of total conventional and
      unconventional oil and gas projections respectively are classified as unproven reserves (i.e.,
      guesstimates). See pages 15-19, Methodology: These 2013 reserve revisions were based on
      predictions involving the application of historic US shale oil and gas recovery rates to foreign
      petroliferous basins with similar geophysical characteristics. These revisions assumed the same
      optimum operating context internationally as in the USA. See Chapter 8 for a more detailed
      critique on this tenuous assumption.].
79   T.R. Klett et al., 2015, U.S. Geological Survey assessment of reserve growth outside of the
      United States: U.S. Geological Survey Scientific Investigations Report 2015–5091. http://dx.doi.
      org/10.3133/sir20155091. [Exposé: (1) See page 1; "The U.S. Geological Survey estimated
      volumes of potential additions to oil and gas reserves for the United States by reserve growth in

discovered accumulations. These volumes were derived by using a new methodology developed by the U.S. Geological Survey." (2) See page 4; Assessment of Reserve Growth Outside of the United States "Because recoverable volumes for individual reservoirs were not reported for many fields outside of the United States, the individual accumulation analysis was not used. Data acquired from individually analyzed U.S. accumulations were used as analogs in this study." Critique: Significant increases in US fossil fuel reserves resulted from the deployment of new, non-validated forecasting methodology. Internationally, the reserve revisions were guesstimates, based on transferring historical precedents for the USA to overseas. None of these methods involved physically verifying the new reserves in the oil and gas wells or fields. That means these are unproven reserves].

80   All-time low for discovered resources in 2017: Around 7 billion barrels of oil equivalent was discovered. December 21, 2017. https://www.rystadenergy.com/newsevents/news/press-releases/all-time-low-discovered-resources-2017/.

81   Declining Reserve Replacement Ratios Deceiving In Resource Play Environment. November. 28, 2017. View Issue. Maurice Smith. JWN Energy. Daily Oil Bulletin. https://www.sproule. com/application/files/2415/1188/2978/Sproule-Declining-Reserve-Replacement-Ratios-Nora-Stewart-Steve-Golko.pdf.

82   Tom Whipple, Online article. "Peak Oil Review." December 26, 2017. Originally published by ASPO-US. December 26, 2017. https://www.resilience.org/stories/2017-12-26/peak-oil-review-dec-26-2017/

83   Kjell Aleklett and Colin J. Campbell, "The peak and decline of world oil and gas production." Minerals and Energy-Raw Materials Report 18.1 (2003): 5-20.

84   Ian Chapman, 2014, "The end of Peak Oil? Why this topic is still relevant despite recent denials." Energy Policy, 64 . 93-101. http://insight.cumbria.ac.uk/id/eprint/1708/.

85   Eric Beinhocker et al., Report. "The carbon productivity challenge: Curbing climate change and sustaining economic growth." McKinsey Global Institute. June 2008.

86   Martin L. Weitzman, "A Review of The Stern Review on the Economics of Climate Change." Journal of Economic Literature. Volume XLV (September 2007), 703–724.

87   S.J. Richard et al., "A Review of the Stern Review." World Economics. Volume 7, No. 4, October–December 2006.

88   S. Yamayoshi et al., "Enhanced Replication of Highly Pathogenic Influenza A(H7N9) Virus in Humans." Emerging Infectious Diseases Journal 2018;24(4):746-750. https://dx.doi.org/10.3201/eid2404.171509.

89   Qi Tang et al., "China is closely monitoring an increase in infection with avian influenza A (H7N9) virus." BioScience Trends. 2017; 11(1):122-124. DOI: 10.5582/bst.2017.01041.

90   J. Artois et al., "Changing Geographic Patterns and Risk Factors for Avian Influenza A(H7N9) Infections in Humans, China." Emerging Infectious Diseases Journal 2018;24(1):87-94. https://dx.doi.org/10.3201/eid2401.171393.

91   Centers for Disease Control and Prevention. Highly Pathogenic Asian Avian Influenza A (H5N1) in People. https://www.cdc.gov/flu/avianflu/h5n1-people.htm.

92   L.O. Durand et al., "Timing of Influenza A(H5N1) in Poultry and Humans and Seasonal Influenza Activity Worldwide, 2004–2013." Emerging Infectious Diseases Journal 2015;21(2):202-208. https://dx.doi.org/10.3201/eid2102.140877.

93   Rino Rappuoli and Philip R. Dormitzer, "Influenza: Options to Improve Pandemic Preparation." Science 22 Jun 2012: Volume 336, Issue 6088, 1531-1533. DOI: 10.1126/science.1221466.

94   Ali H. Ellebedy et al., "H5N1 immunization broadens immunity to influenza." Proceedings of the National Academy of Sciences Sep 2014, 111 (36) 13133-13138; DOI: 10.1073/pnas.1414070111.

95   Paul Gillard et al., "Long-term booster schedules with AS03A-adjuvanted heterologous H5N1 vaccines induces rapid and broad immune responses in Asian adults." BMC Infectious Diseases201414:142. https://doi.org/10.1186/1471-2334-14-142.

96   Lopez P et al., "Combined Administration of MF59-Adjuvanted A/H5N1 Prepandemic and Seasonal Influenza Vaccines: Long-Term Antibody Persistence and Robust Booster Responses 1 Year after a One-Dose Priming Schedule." Clinical and Vaccine Immunology: CVI. 2013;20(5):753-758. doi:10.1128/CVI.00626-12.

97 Anuradha Madan et al., "Immunogenicity and Safety of an AS03-Adjuvanted H7N9 Pandemic Influenza Vaccine in a Randomized Trial in Healthy Adults." The Journal of Infectious Diseases, Volume 214, Issue 11, 1 December 2016, 1717–1727, https://doi.org/10.1093/infdis/jiw414.

98 Thomas S. Kuhn, The Structure of Scientific Revolutions. Second Edition, Enlarged. International Encyclopedia of Unified Science. © 1962, 1970 by The University of Chicago. All rights reserved. Published 1962. Second Edition, enlarged, 1970.

99 See endnote 1.

100 See endnote 2.

101 See endnote 3.

102 See endnote 77.

103 See endnote 76.

104 See endnote 75.

105 See endnote 4.

106 See endnote 5.

107 Impact of Articles 1 and 2: Conduct a Google search for the IPCC document, "Principles Governing IPCC Work" to witness firsthand its operational constraints. [See pages 1 and 2. Look specifically at the IPCC's role. The IPCC was tasked only with assessing the scientific basis of the risk of human activity on climate change, to assess its impact, and to provide adaptation and mitigation options. There is no mention of assessing or mitigating natural climate change within its role (i.e., the biggest cause of climate change and natural disasters).].

108 Slawinska, J. and A. Robock, 2018, "Impact of Volcanic Eruptions on Decadal to Centennial Fluctuations of Arctic Sea Ice Extent during the Last Millennium and on Initiation of the Little Ice Age." J. Climate, 31, 2145–2167, https://doi.org/10.1175/JCLI-D-16-0498.1.

109 G.H. Miller et al., 2012, "Abrupt onset of the Little Ice Age triggered by volcanism and sustained by sea-ice/ocean feedbacks." Geophysical Research Letters, 39, L02708, doi:10.1029/2011GL050168.

110 F. Lehner et al., 2013, "Amplified inception of European Little Ice Age by sea ice–ocean–atmosphere feedbacks." J. Climate, 26, 7586–7602. https://doi.org/10.1175/JCLI-D-12-00690.1.

111 C. Gao et al., 2008, "Volcanic forcing of climate over the past 1500 years: An improved ice core-based index for climate models." Journal of Geophysical Research, 113, D23111, doi: 10.1029/2008JD010239. [See Figure 2, page 5].

112 Y. Zhong et al., "Centennial-scale climate change from decadally-paced explosive volcanism: a coupled sea ice-ocean mechanism." Climate Dynamics (2011) 37: 2373. https://doi.org/10.1007/s00382-010-0967-z.

113 C. Newhall et al., 2018, "Anticipating future Volcanic Explosivity Index (VEI) 7 eruptions and their chilling impacts." Geosphere, v. 14, no. 2, p. 1–32, doi:10.1130/GES01513.1.

114 Narrow scope of climate risk assessment: IPCC, Climate Change 2014: Impacts, Adaptation, and Vulnerability. Part A: Global and Sectoral Aspects. Contribution of Working Group II to the Fifth Assessment Report of the Intergovernmental Panel on Climate Change [Field, C.B., V.R. Barros, D.J. Dokken, K.J. Mach, M.D. Mastrandrea, T.E. Bilir, M. Chatterjee, K.L. Ebi, Y.O. Estrada, R.C. Genova, B. Girma, E.S. Kissel, A.N. Levy, S. MacCracken, P.R. Mastrandrea, and L.L. White (eds.)]. Cambridge University Press, Cambridge, United Kingdom and New York, NY, USA, 1132 pages [Exposé: See pages 59-65, (1) Section B-1. The IPCC specifically tells us that key climate risks assessed relate to severe impacts relative to Article 2, which refers to dangerous human interference with the climate system. (2) See Table TS.3 and TS.3. Critique: There is no mention that the IPCC assessed severe impacts related to natural climate change, just risks related to human activity. In other words, historical climate catastrophes (i.e., Little Ice Age, grand solar minimum cooling events, rapid climate events, climate-forcing volcanism) were either not assessed or were not reported in the IPCC assessment reports, because they fall outside Article 2's scope.].

115 See endnote 6.

116 See endnote 7.

117 See endnote 8.

118 See endnote 9.

119 Climate change assessments. Review of the processes and procedures of the IPCC. October 2010. InterAcademy Council. Committee to Review the Intergovernmental Panel on Climate Change. Report available at http://reviewipcc.interacademycouncil.net/.

120 Craig D. Idso et al., "Why Scientists Disagree About Global Warming." The NIPCC Report on Scientific Consensus. https://www.heartland.org/_template-assets/documents/Books/Why%20Scientists%20Disagree%20Second%20Edition%20with%20covers.pdf.

121 The Logic of Scientific Discovery (Routledge Classics) (Volume 56) 2nd Edition by Karl Popper. Available at https://www.amazon.com/Logic-Scientific-Discovery-Routledge-Classics/dp/0415278449.

122 Thomas S. Kuhn, The Structure of Scientific Revolutions. Second Edition, Enlarged. International Encyclopedia of Unified Science. © 1962, 1970 by The University of Chicago. All rights reserved. Published 1962. Second Edition, enlarged, 1970.

123 Forecast inaccuracy: IPCC, Climate Change 2013: The Physical Science Basis. Contribution of Working Group I to the Fifth Assessment Report of the Intergovernmental Panel on Climate Change [Stocker, T.F., D. Qin, G.-K. Plattner, M. Tignor, S.K. Allen, J. Boschung, A. Nauels, Y. Xia, V. Bex and P.M. Midgley (eds.)]. Cambridge University Press, Cambridge, United Kingdom and New York, NY, USA, 1535 pages [Exposé: (1) See page 61, Box TS.3. This section reviews the high degree of inaccuracy associated with IPCC-promoted climate forecasts. The IPCC tells us that 111 of its 114 forecasts (paragraph 2) were higher (i.e., over-forecasted) than the actual global mean surface temperature between 1998 and 2012 (HadCRUT4), and that nearly all their forecasts missed the 15-year hiatus in global warming during this time. (2) See page 61, Box TS.3. In the penultimate paragraph, we are told that 93 of 114 forecasts were below the actual global mean surface temperature (HadCRUT4) in the 15-year period prior to 1998. Despite 30 years of climate forecasting inaccuracies and missing the 15-year climate hiatus, the IPCC tells us it still has a "very high confidence" that its long-term forecasts are aligned with the real world data. (3) See page 62, Box TS.3. In the penultimate paragraph, a weak explanation is provided for this forecasting inaccuracy. The IPCC then suggests they should scale back their near-term forecasts by 10 percent (rather than reject their theory). However, they admit this proposed down-scaling is still insufficient to account for the 15-year climate hiatus. In Box TS.3 the IPCC indicate the difference between forecasts and actual temperature could be due to internal climate variability, incorrect theory and model assumptions about radiative forcing, or model response errors (page 61 paragraph 2). The IPCC also tell us this inaccuracy could be linked to when the forecasting models were "initialized" relative to the real world phase of temperature oscillation or natural climate change. This latter point is tantamount to admitting that natural climate change is the cause of the promoted climate change, rather than carbon dioxide. Critique: The explanations provided by the IPCC for their inaccurate forecasts hide the fact that natural climate change, in operation for billions of years, was ignored or dismissed by the IPCC in its theory and climate forecasting. This natural climate change (and the changes of phase from warming to cooling to warming) is the reason for the 15-year periods of under-forecasting then over-forecasting, while missing the 15-year climate hiatus. The IPCC also failed to explain that during the 15-year hiatus, carbon dioxide levels continued to increase by 7.4 percent. This continued rise in atmospheric carbon dioxide levels during the climate hiatus highlights a complete disconnect between the global mean air temperature and atmospheric carbon dioxide levels, and in fact disproves the IPCC's radiative forcing theory.].

124 Thomas S. Kuhn, The Structure of Scientific Revolutions. Second Edition, Enlarged. International Encyclopedia of Unified Science. © 1962, 1970 by The University of Chicago.

125 Ole Humlum et al., "The phase relation between atmospheric carbon dioxide and global temperature." Global and Planetary Change. Volume 100, January 2013, 51-69.

126 Manfred Mudelsee, "The phase relations among atmospheric CO2 content, temperature and global ice volume over the past 420 ka." Quaternary Science Reviews 20 (2001) 583-58.

127 Eric Monnin et al., "Atmospheric CO2 Concentrations over the Last Glacial Termination." By Science 05 Jan 2001: 112-114.

128 N. Caillon et al., 2003. "Timing of atmospheric CO2 and Antarctic temperature changes across Termination III." Science 299: 1728-1731.

129 H. Fischer et al., 1999, "Ice core records of atmospheric CO2 around the last three glacial terminations." Science, 283 , 1712-1714.

130 George C. Reid, "Solar Variability and the Earth's Climate: Introduction and Overview." Space Science Reviews (2000) 94: 1. https://doi.org/10.1023/A:1026797127105. [See 2-4, Figure 1].

131 E. Friis-Christensen, K. Lassen, "Length of the Solar Cycle: An Indicator of Solar Activity Closely Associated with Climate." Science, New Series, Volume 254, No. 5032. (Nov. 1, 1991), 698-700. [See Figures 1-3, 699-700].

132 Adriano Mazzarella, "Solar Forcing of Changes in Atmospheric Circulation, Earth's Rotation and Climate." The Open Atmospheric Science Journal, 2008, 2, 181-184.

133 A. Mazzarella, "The 60-year solar modulation of global air temperature: the Earth's rotation and atmospheric circulation connection." Theoretical and Applied Climatology. 88, 193–199 (2007). DOI 10.1007/s00704-005-0219-z.

134 Boian Kirov, Katya Georgieva, "Long-term variations and interrelations of ENSO, NAO and solar activity." Physics and Chemistry of the Earth Parts. 2002. A/B/C 27(6-8):441-448. DOI10.1016/S1474-7065(02)00024-4.

135 V. Bucha, "Geomagnetic activity and the North Atlantic Oscillation." Studia Geophysica et Geodaetica. July 2014, Volume 58, Issue 3, 461–472. https://doi.org/10.1007/s11200-014-0508-z.

136 Mazzarella A. and N. Scafetta, 2012, "Evidences for a quasi 60-year North Atlantic Oscillation since 1700 and its meaning for global climate change." Theoretical Applied Climatology 107, 599-609. DOI: 10.1007/s00704-011-0499-4.

137 Vincent Courtillot et al., "Are there connections between the Earth's magnetic field and climate?" Earth and Planetary Science Letters 253 (2007) 328–339.

138 Dimitar Todorov Valev, "Statistical relationships between the surface air temperature anomalies and the solar and geomagnetic activity indices." 2006, Physics and Chemistry of the Earth Parts A/B/C 31(1):109-112. DOI10.1016/j.pce.2005.03.005.

139 Thomas S. Kuhn, The Structure of Scientific Revolutions. Second Edition, Enlarged. International Encyclopedia of Unified Science. © 1962, 1970 by The University of Chicago.

140 Thomas S. Kuhn, The Structure of Scientific Revolutions. Second Edition, Enlarged. International Encyclopedia of Unified Science. © 1962, 1970 by The University of Chicago.

141 Conduct a Google search for the IPCC document, "Principles Governing IPCC Work" to witness firsthand its operational constraints (pages 1 and 2).

142 See endnote 4.

143 Climate Change: The IPCC Scientific Assessment (1990). Report prepared for Intergovernmental Panel on Climate Change by Working Group I. J.T. Houghton, G.J. Jenkins and J.J. Ephraums (eds.). Cambridge University Press, Cambridge, Great Britain, New York, NY, USA and Melbourne, Australia 410 pages [See the Preface's introduction].

144 The non-International Panel on Climate Change (NIPCC). http://climatechangereconsidered. org/about-the-ipcc/. (last accessed 17/05/2018).

145 The Nongovernmental International Panel on Climate Change (NIPCC). http:// climatechangereconsidered.org/about-the-nipcc/.

146 On Maurice Strong. The Heartland Institute. Nongovernmental International Panel on Climate Change. http://climatechangereconsidered.org/about-the-ipcc/. [Exposé: This page also cited John Izzard's blog. "Maurice Strong, Climate Crook." Quadrant Online December 02nd 2015. Available at http://quadrant.org.au/opinion/doomed-planet/2010/01/discovering-maurice-strong/.Maurice Strong. (last accessed 21/03/2018)].

147 Climate Change: The IPCC Scientific Assessment (1990). Report prepared for Intergovernmental Panel on Climate Change by Working Group I. J.T. Houghton, G.J. Jenkins and J.J. Ephraums (eds.). Cambridge University Press, Cambridge, Great Britain, New York, NY, USA and Melbourne, Australia 410 pages. [Exposé: See the front matter immediately after the cover and title page. We are told that the scientific review was conducted by the international scientific community and represents the most authoritative statement about climate change. Government nominated scientists conducted this scientific review, which was focused on anthropogenic global warming.].

148 See endnote 5.

149 Climate change assessments. Review of the processes and procedures of the IPCC. October 2010. InterAcademy Council. Committee to Review the Intergovernmental Panel on Climate

Change. Report available at http://reviewipcc.interacademycouncil.net/. [Exposé: See page 18, "Confirmation bias"].

150 IPCC, Climate Change 2007: The Physical Science Basis. Contribution of Working Group I to the Fourth Assessment Report of the Intergovernmental Panel on Climate Change [Solomon, S., D. Qin, M. Manning, Z. Chen, M. Marquis, K.B. Averyt, M. Tignor and H.L. Miller (eds.)]. Cambridge University Press, Cambridge, United Kingdom and New York, NY, USA, 996 pages [Exposé: See the Preface, page vii].

151 ., Climate Change 2013: The Physical Science Basis. Contribution of Working Group I to the Fifth Assessment Report of the Intergovernmental Panel on Climate Change [Stocker, T.F., D. Qin, G.-K. Plattner, M. Tignor, S.K. Allen, J. Boschung, A. Nauels, Y. Xia, V. Bex and P.M. Midgley (eds.)]. Cambridge University Press, Cambridge, United Kingdom and New York, NY, USA, 1535 pages. [Exposé: See Preface page viii] (Accessed on 21/03/2018).

152 See endnote 4.

153 Personal Research: This utilized the above nine cited references pertaining to IPCC documents from 1990-1992 and 2014. Climate control mechanisms were identified from the scientific literature by the existence of significant climate correlations detailed in the peer review science (see Chapter 6). A group of six key words were identified for each climate control mechanism by reviewing scientific publications. Provisionally, key words were identified from Google Scholar and scientific publications available online using a sighting study. These key words were finally selected because they represented the highest percentage of key words within each climate control mechanism. Results were tabulated (at the start of the endnotes, referencing this endnote.), summarized, and graphed. All 1990 First Assessment Reports and 1992 Supplementary Assessment Reports were pooled (1990-1992), and Google Scholar was searched for the 15 years prior to and including the year of the IPCC publication dates (i.e., 1977-1992, 1999-2014).

154 Climate Change: The IPCC Scientific Assessment (1990). Report prepared for Intergovernmental Panel on Climate Change by Working Group I. J.T. Houghton, G.J. Jenkins and J.J. Ephraums (eds.). Cambridge University Press, Cambridge, Great Britain, New York, NY, USA and Melbourne, Australia 410 pages.

155 Climate Change: The IPCC Impacts Assessment (1990). Report prepared for Intergovernmental Panel on Climate Change by Working Group II. W.J. McG. Tegart, G.W. Sheldon and D.C. Griffiths (eds.). Australian Government Publishing Service, Camberra, Australia 294 pages.

156 Climate Change: The IPCC Response Strategies (1990). Report prepared for Intergovernmental Panel on Climate Change by Working Group III 330 pages.

157 Climate Change 1992: The Supplementary Report to the IPCC Scientific Assessment. Report prepared for Intergovernmental Panel on Climate Change by Working Group I combined with Supporting Scientific Material. J.T. Houghton, B.A. Callander and S.K. Varney (eds.). Cambridge University Press, Cambridge, Great Britain, New York, NY, USA, and Victoria, Australia. 218 pages.

158 Climate Change 1992: The Supplementary Report to the IPCC Impacts Assessment. Report prepared for Intergovernmental Panel on Climate Change by Working Group II combined with Supporting Scientific Material W.J. McG Tegart and G.W. Sheldon (eds.). Australian Government Publishing Service, Canberra, Australia. 128 pages.

159 IPCC, Climate Change 2013: The Physical Science Basis. Contribution of Working Group I to the Fifth Assessment Report of the Intergovernmental Panel on Climate Change [Stocker, T.F., D. Qin, G.-K. Plattner, M. Tignor, S.K. Allen, J. Boschung, A. Nauels, Y. Xia, V. Bex and P.M. Midgley (eds.)]. Cambridge University Press, Cambridge, United Kingdom and New York, NY, USA, 1535 pages.

160 IPCC, Climate Change 2014: Impacts, Adaptation, and Vulnerability. Part A: Global and Sectoral Aspects. Contribution of Working Group II to the Fifth Assessment Report of the Intergovernmental Panel on Climate Change [Field, C.B., V.R. Barros, D.J. Dokken, K.J. Mach, M.D. Mastrandrea, T.E. Bilir, M. Chatterjee, K.L. Ebi, Y.O. Estrada, R.C. Genova, B. Girma, E.S. Kissel, A.N. Levy, S. MacCracken, P.R. Mastrandrea, and L.L. White (eds.)]. Cambridge University Press, Cambridge, United Kingdom and New York, NY, USA, 1132 pages.

161 IPCC, Climate Change 2014: Impacts, Adaptation, and Vulnerability. Part B: Regional Aspects. Contribution of Working Group II to the Fifth Assessment Report of the Intergovernmental

Panel on Climate Change [Barros, V.R., C.B. Field, D.J. Dokken, M.D. Mastrandrea, K.J. Mach, T.E. Bilir, M. Chatterjee, K.L. Ebi, Y.O. Estrada, R.C. Genova, B. Girma, E.S. Kissel, A.N. Levy, S. MacCracken, P.R. Mastrandrea, and L.L. White (eds.)]. Cambridge University Press, Cambridge, United Kingdom and New York, NY, USA, pages 688.

162   IPCC, Climate Change 2014: Mitigation of Climate Change. Contribution of Working Group III to the Fifth Assessment Report of the Intergovernmental Panel on Climate Change [Edenhofer, O., R. Pichs-Madruga, Y. Sokona, E. Farahani, S. Kadner, K. Seyboth, A. Adler, I. Baum, S. Brunner, P. Eickemeier, B. Kriemann, J. Savolainen, S. Schlömer, C. von Stechow, T. Zwickel and J.C. Minx (eds.)]. Cambridge University Press, Cambridge, United Kingdom and New York, NY, USA.

163   Climate change assessments. Review of the processes and procedures of the IPCC. October 2010. InterAcademy Council. Committee to Review the Intergovernmental Panel on Climate Change. Report available at http://reviewipcc.interacademycouncil.net/.

164   Craig D. Idso et al., "Why Scientists Disagree About Global Warming." The NIPCC Report on Scientific Consensus (non-International Panel on Climate Change). https://www.heartland.org/_template-assets/documents/Books/Why%20Scientists%20Disagree%20Second%20Edition%20with%20covers.pdf.

165   Conduct a Google Search for the following IPCC document. "Procedures for the Preparation, Review, Acceptance, Adoption, Approval and Publication of IPCC Reports" or Appendix A to the Principles Governing IPCC Work Procedures.

166   Climate change assessments. Review of the processes and procedures of the IPCC. October 2010. InterAcademy Council. Committee to Review the Intergovernmental Panel on Climate Change. Report available at http://reviewipcc.interacademycouncil.net/.

167   Craig D. Idso et al., "Why Scientists Disagree About Global Warming." The NIPCC Report on Scientific Consensus. https://www.heartland.org/_template-assets/documents/Books/Why%20Scientists%20Disagree%20Second%20Edition%20with%20covers.pdf.

168   Conduct a Google Search for the following IPCC document. "Procedures for the Preparation, Review, Acceptance, Adoption, Approval and Publication of IPCC Reports" or Appendix A to the Principles Governing IPCC Work Procedures. [Exposé: See pages 5, 16].

169   Climate change assessments. Review of the processes and procedures of the IPCC. October 2010. InterAcademy Council. Committee to Review the Intergovernmental Panel on Climate Change. Report available at http://reviewipcc.interacademycouncil.net/. [Exposé: See page 14. We are told that governments do not always put forward the names of the best climate scientist volunteers for the IPCC work. They also inform us that political considerations are prioritized over scientific expertise and qualifications in the IPCC scientist selection process.].

170   Conduct a Google Search for the following IPCC document. "Procedures for the Preparation, Review, Acceptance, Adoption, Approval and Publication of IPCC Reports" or Appendix A to the Principles Governing IPCC Work Procedures.

171   Climate change assessments. Review of the processes and procedures of the IPCC. October 2010. InterAcademy Council. Committee to Review the Intergovernmental Panel on Climate Change. Report available at http://reviewipcc.interacademycouncil.net/. (Exposé: see page 18, "confirmation bias").

172   Climate change assessments. Review of the processes and procedures of the IPCC. October 2010. InterAcademy Council. Committee to Review the Intergovernmental Panel on Climate Change. Report available at http://reviewipcc.interacademycouncil.net/. [Exposé: See page 14, section "Author selection"].

173   Climate change assessments. Review of the processes and procedures of the IPCC. October 2010. InterAcademy Council. Committee to Review the Intergovernmental Panel on Climate Change. Report available at http://reviewipcc.interacademycouncil.net/. [Exposé: See page 48].

174   Climate change assessments. Review of the processes and procedures of the IPCC. October 2010. InterAcademy Council. Committee to Review the Intergovernmental Panel on Climate Change. Report available at http://reviewipcc.interacademycouncil.net/. [Exposé: See page 21].

175   Conduct a Google Search for the following IPCC document. "Procedures for the Preparation, Review, Acceptance, Adoption, Approval and Publication of IPCC Reports" or Appendix A to the Principles Governing IPCC Work Procedures. [Exposé: See page 9].

176 Climate change assessments. Review of the processes and procedures of the IPCC. October 2010. InterAcademy Council. Committee to Review the Intergovernmental Panel on Climate Change. Report available at http://reviewipcc.interacademycouncil.net/. [Exposé: See page 23].

177 Conduct a Google Search for the following IPCC document. "Procedures for the Preparation, Review, Acceptance, Adoption, Approval and Publication of IPCC Reports" or Appendix A to the Principles Governing IPCC Work Procedures. [Exposé: See pages 7 and 8].

178 Climate change assessments. Review of the processes and procedures of the IPCC. October 2010. InterAcademy Council. Committee to Review the Intergovernmental Panel on Climate Change. Report available at http://reviewipcc.interacademycouncil.net/. [See page 24].

179 Karl Popper, The Logic of Scientific Discovery (Routledge Classics) (Volume 56), 2nd Edition. Available at https://www.amazon.com/Logic-Scientific-Discovery-Routledge-Classics/dp/0415278449

180 Karl Popper, The Logic of Scientific Discovery (Routledge Classics) (Volume 56), 2nd Edition. Available at https://www.amazon.com/Logic-Scientific-Discovery-Routledge-Classics/dp/0415278449

181 Dismissing volcanism and solar activity. IPCC, 2013: Climate Change 2013: The Physical Science Basis. Contribution of Working Group I to the Fifth Assessment Report of the Intergovernmental Panel on Climate Change [Stocker, T.F., D. Qin, G.-K. Plattner, M. Tignor, S.K. Allen, J. Boschung, A. Nauels, Y. Xia, V. Bex and P.M. Midgley (eds.)]. Cambridge University Press, Cambridge, United Kingdom and New York, NY, USA, 1535 pages [Exposé: (1) See page 1008-1009, FAQ 11.2. This section details how volcanic eruptions impact the climate. We are told that IPCC promoted climate predictions do not include or reflect potential future volcanic eruptions (for various reasons, which are tantamount to confirmation bias). (2) See page 1009 sub-section 11.3.6.3 point 4. This point confirms that IPCC promoted climate forecasts do not include any impact of volcanism on the future climate, for reasons of volcanism's unpredictability. Critique: By not including the potential impact of volcanism in climate forecasts natural planetary cooling factors (i.e., volcanic aerosols) are eliminated from forecasts, which would naturally counter carbon dioxide's global warming effect. Ignoring volcanism, solar activity, and natural climate change constitutes scientific bias, and invalidates the IPCC's climate forecasts.]. (2) See page 1009 sub-section 11.3.6.3 point 4. This point tells us that the IPCC climate projections assume no changes in solar irradiance in their climate forecasts, because there is a low confidence in the solar activity projections. (3) See page 1007 sub-section 11.3.6.2.2. The IPCC dismissed the impact of solar forcing on the climate during this grand solar minimum, despite solar activity being at its lowest level for the longest period since the early 20th century. This scientific dismissal of solar activity's impact on the climate was because they had low confidence in the solar activity projections (solar irradiance), but they had "high confidence" that greenhouse gases would offset any diminution in total solar irradiance's impact on the climate. The IPCC do not ascribe any climate modifying role to solar magnetism (magnetized solar wind) i.e., air circulations, geomagnetism-cosmic rays-low clouds (see Figures 4.3.A, 4.4, 6.2, 5.1-5.3, and Chapter 6). Critique: The IPCC's low confidence in these projections is counter to NASA's open acceptance of this grand solar minimum and its impact on our atmosphere and other earth systems linked to climate change (listen to the video, https://science.nasa.gov/science-news/news-articles/solar-minimum-is-coming). This low confidence is also counter to scientific experts advising the Russian Space program (i.e., Habibullo Abdussamatov), and leading solar activity scientists expert in climate change advising of a planetary cooling in the decades ahead. What legitimacy does the IPCC have to dismiss other scientific disciplines and space program (USA, Russia) sciences, which are known to impact earth systems? (i.e., magnetosphere, ionosphere, atmosphere, tectonic-volcanic processes)].

182 Deferring the ice age 30,000 years. IPCC, Climate Change 2007: The Physical Science Basis. Contribution of Working Group I to the Fourth Assessment Report of the Intergovernmental Panel on Climate Change [Solomon, S., D. Qin, M. Manning, Z. Chen, M. Marquis, K.B. Averyt, M. Tignor and H.L. Miller (eds.)]. Cambridge University Press, Cambridge, United Kingdom and New York, NY, USA, 996 pages [Exposé: See page 56, Box TS.6. This section details the influence of earth's orbit of the sun on the onset of an ice age. We are told how the Milankovitch theory, associated with controlling ice ages, is a well-developed theory. It

posits that minima of solar irradiance at high latitudes in the Northern Hemisphere enable snow to persist through the summer, which causes the ice sheets to expand. The IPCC tells us that global warming won't be mitigated by changes in earth's orbit (i.e., the orbits impact on decreasing solar irradiance), and that earth will not enter the next ice age for 30,000 years. More worrying, on page 85, section TS.6.2.4, it is re-emphasized that earth will not enter another ice age for 30,000 years or more, and that this was a "robust finding." Critique: I am unable to find any peer-reviewed scientific publication for this IPCC ice age delay hypothesis. There is a statistical consequence in delaying the ice age by 30,000-years because this delay impacts the interglacial period duration, the inter-climate optimum interval, and global-to-Antarctica climate optimum phasing gaps relative to all other glacial cycles for the global, Antarctic and Arctic climate data.). This ice age delay hypothesis is being passed off as though it is a scientific fact, when in reality it is an unproven, non-peer reviewed, and readily falsifiable hypothesis.] (last accessed 21/03/2018).

183 Ole Humlum et al., "The phase relation between atmospheric carbon dioxide and global temperature." Global and Planetary Change. Volume 100, January 2013, 51-69.
184 Geophysical Fluid Dynamics Laboratory discussion on: "Transient and Equilibrium Climate Sensitivity." https://www.gfdl.noaa.gov/transient-and-equilibrium-climate-sensitivity/.
185 Craig D. Idso et al., "Why Scientists Disagree About Global Warming." The NIPCC Report on Scientific Consensus. https://www.heartland.org/_template-assets/documents/Books/Why%20Scientists%20Disagree%20Second%20Edition%20with%20covers.pdf.
186 Scaling back forecasts by 10 percent. IPCC, Climate Change 2013: The Physical Science Basis. Contribution of Working Group I to the Fifth Assessment Report of the Intergovernmental Panel on Climate Change [Stocker, T.F., D. Qin, G.-K. Plattner, M. Tignor, S.K. Allen, J. Boschung, A. Nauels, Y. Xia, V. Bex and P.M. Midgley (eds.)]. Cambridge University Press, Cambridge, United Kingdom and New York, NY, USA, 1535 pages [Exposé: (1) See page 62, Box TS.3. In the penultimate paragraph, a weak explanation is provided for this forecasting inaccuracy. The IPCC then suggests they should scale back their near-term forecasts by 10 percent (rather than reject their theory). However, they admit this proposed down-scaling is still not enough to account for the 15-year global climate hiatus. Critique: The explanations provided by the IPCC for their forecast inaccuracy hide the fact that natural climate change, in operation for billions of years, was ignored or dismissed by the IPCC in its theory and climate forecasting. This natural climate change (and changing phase from warming to cooling to warming) is the reason for the 15 year periods of under-forecasting then over-forecasting, while missing the 15 year climate hiatus. The IPCC also failed to explain that during the 15-year hiatus, carbon dioxide continued its unabated increase by 7.4 percent. This continued rise in atmospheric carbon dioxide during the climate hiatus highlights a complete disconnect between the global mean temperature and atmospheric carbon dioxide levels, and disproves the IPCC's radiative forcing theory.].
187 Manfred Mudelsee, "The phase relations among atmospheric CO2 content, temperature and global ice volume over the past 420 ka." Quaternary Science Reviews 20 (2001) 583-58.
188 Eric Monnin et al., "Atmospheric CO2 Concentrations over the Last Glacial Termination." By Science 05 Jan 2001: 112-114.
189 N. Caillon et al., 2003, "Timing of atmospheric CO2 and Antarctic temperature changes across Termination III." Science 299: 1728-1731.
190 H. Fischer et al., 1999, "Ice core records of atmospheric CO2 around the last three glacial terminations." Science, 283, 1712-1714.
191 Ole Humlum et al., "The phase relation between atmospheric carbon dioxide and global temperature." Global and Planetary Change. Volume 100, January 2013, 51-69.
192 Ole Humlum et al., "The phase relation between atmospheric carbon dioxide and global temperature." Global and Planetary Change. Volume 100, January 2013, 51-69.
193 IPCC, Climate Change 2014: Synthesis Report. Contribution of Working Groups I, II and III to the Fifth Assessment Report of the Intergovernmental Panel on Climate Change [Core Writing Team, R.K. Pachauri and L.A. Meyer (eds.)]. IPCC, Geneva, Switzerland, 151 pages [Exposé: See page 57, Box 2.2. This section provides an explanation of the Coupled Model Intercomparison Project Phase 5 forecasting models, and their four scenarios of human-induced global warming.].

194   The carbon dioxide data (expressed as a mole fraction in dry air, micromol/mol, abbreviated as ppm) used to support this statement was provided by NASA (see link), which cited the National Oceanic and Atmospheric Administration (NOAA) and Earth System Research Laboratory (ESRL). https://data.giss.nasa.gov/modelforce/ghgases//CMIP5/CO2_OBS_1850-2005.lpl.

195   See endnote 123.

196   Ole Humlum et al., "The phase relation between atmospheric carbon dioxide and global temperature." Global and Planetary Change. Volume 100, January 2013, 51-69.

197   IPCC, Climate Change 2013: The Physical Science Basis. Contribution of Working Group I to the Fifth Assessment Report of the Intergovernmental Panel on Climate Change [Stocker, T.F., D. Qin, G.-K. Plattner, M. Tignor, S.K. Allen, J. Boschung, A. Nauels, Y. Xia, V. Bex and P.M. Midgley (eds.)]. Cambridge University Press, Cambridge, United Kingdom and New York, NY, USA, 1535 pages [Exposé: See page 20, section E.1. This section summarizes the IPCC's most recent climate forecasts.].

198   Global mean surface temperature data, commonly referred to as HadCRUT4. https://www.metoffice.gov.uk/hadobs/hadcrut4/data/current/download.html. [Exposé: Look at the bottom left hand or first column for the current year-to-date temperature. Subtract that from the 2016 total to see the magnitude of the fall. Global Data: https://bit.ly/2nCgctz. Northern Hemisphere Data: https://bit.ly/2MRt75G, Southern Hemisphere Data: https://bit.ly/2nBfYTA. Tropics Data: https://bit.ly/2nFXJMM. [last downloaded 25/07/2018].

199   IPCC, Climate Change 2013: The Physical Science Basis. Contribution of Working Group I to the Fifth Assessment Report of the Intergovernmental Panel on Climate Change [Stocker, T.F., D. Qin, G.-K. Plattner, M. Tignor, S.K. Allen, J. Boschung, A. Nauels, Y. Xia, V. Bex and P.M. Midgley (eds.)]. Cambridge University Press, Cambridge, United Kingdom and New York, NY, USA, 1535 pages [Exposé: See page 14, Figure SPM.5. This figure provides the IPCC's radiative forcing contribution estimates and uncertainties (2011 versus 1750) for an array of factors involved in the IPCC's version of climate change. Critique: According to the IPCC theory nearly all radiative forcing factors are attributable to anthropogenic casues while ignoring natural climate change. Factors involved in natural climate change include secular changes to solar activity (electromagnetism, magnetism), geomagnetism, cosmic rays and low clouds, volcanic aerosols, cloud formation at higher altitudes and different latitudes, climate and ocean circulation systems, water vapor (irrigation, fossil fuel combustion, global warming) and ocean degassing (carbon dioxide) consequent to global warming.]

200   See endnote 186.

201   The International Institute of Forecasters (IIF). https://forecasters.org/.

202   The International Journal of Forecasting (IJF) is the leading journal in its field and is an official publication of the International Institute of Forecasters (IIF). https://ijf.forecasters.org/.

203   International Journal of Forecasting Official Publication of the International Institute of Forecasters. https://www.journals.elsevier.com/international-journal-of-forecasting/.

204   Principles of Forecasting. A Handbook for Researchers and Practitioners, edited by J. Scott Armstrong, Kluwer Academic Publishers, 2001. Available at https://www.amazon.com/Principles-Forecasting-Researchers-Practitioners-International/dp/0792374010.

205   To: Senator James M. Inhofe From: Drs. J. Scott Armstrong and Kesten C. Green Re: "Your Request for an Analysis of the U.S. Environmental Protection Agency's Advanced Notice of Proposed Rulemaking for Greenhouse Gases." https://www.parliament.nz/resource/0000077829.

206   K.C. Green et al., 2008, "Benchmark Forecasts for Climate Change." Retrieved from http://repository.upenn.edu/marketing_papers/138.

207   J.S. Armstrong et al., 2011, "Research on Forecasting for the Manmade Global Warming Alarm: Testimony to Committee on Science, Space and Technology Subcommittee on Energy and Environment on "Climate Change: Examining the processes used to create science and policy". Retrieved from http://repository.upenn.edu/marketing_papers/139.

208   To: Senator James M. Inhofe From: Drs. J. Scott Armstrong and Kesten C. Green Re: Your Request for an Analysis of the U.S. Environmental Protection Agency's Advanced Notice of Proposed Rulemaking for Greenhouse Gases. https://www.parliament.nz/resource/0000077829. [Exposé].

209 K.C. Green et al., 2008, "Benchmark Forecasts for Climate Change." Retrieved from http://repository.upenn.edu/marketing_papers/138. [Exposé].

210 J.S. Armstrong et al., 2011, "Research on Forecasting for the Manmade Global Warming Alarm: Testimony to Committee on Science, Space and Technology Subcommittee on Energy and Environment on "Climate Change: Examining the processes used to create science and policy". Retrieved from http://repository.upenn.edu/marketing_papers/139. [Exposé: Expert witness statement.].

211 See endnote 193.

212 B.M. Vinther et al., 2009, "Holocene thinning of the Greenland ice sheet." Nature, Vol. 461, pp. 385-388, 17 September 2009. National Centers for Environmental Information, NESDIS, NOAA, U.S. Department of Commerce. Greenland Ice Sheet Holocene d18O, Temperature, and Surface Elevation. doi:10.1038/nature08355. https://www.ncdc.noaa.gov/paleo-search/study/11148. Downloaded 05/05/2018.

213 R.B. Alley, 2004, "GISP2 Ice Core Temperature and Accumulation Data." National Centers for Environmental Information, NESDIS, NOAA, U.S. Department of Commerce. https://www.ncdc.noaa.gov/paleo/study/2475. Downloaded 05/05/2018. [Last Glacial Maximum's deepest temperature trough was 24,098 years ago (-530C) and the Holocene Climate Optimum was 7,800 years ago (-28.860C). The difference between these time points is 16,297 years and 24.560C.]

214 J. V. Jouzel et al., 2007, "Orbital and Millennial Antarctic Climate Variability over the Past 800,000 Years." Science, Volume 317, No. 5839, 793-797, 10 August 2007. National Centers for Environmental Information, NESDIS, NOAA, U.S. Department of Commerce. EPICA Dome C - 800KYr Deuterium Data and Temperature Estimates. https://www.ncdc.noaa.gov/paleo/study/6080. Download data: Downloaded 08/02/2016.

215 R.V. Uemura et al., 2012, "Ranges of moisture-source temperature estimated from Antarctic ice cores stable isotope records over glacial-interglacial cycles." Climate of the Past, 8, 1109-1125. doi: 10.5194/cp-8-1109-2012. National Centers for Environmental Information, NESDIS, NOAA, U.S. Department of Commerce. Dome Fuji 360KYr Stable Isotope Data and Temperature Reconstruction. https://www.ncdc.noaa.gov/paleo-search/study/13121. Downloaded 05/05/2018.

216 T. Kobashi et al., 2013, "Causes of Greenland temperature variability over the past 4000 year: implications for northern hemispheric temperature changes." Climate of the Past, 9(5), 2299-2317. doi: 10.5194/cp-9-2299-2013. National Centers for Environmental Information, NESDIS, NOAA, U.S. Department of Commerce. Northern Hemisphere 4000 Year Temperature Reconstructions. https://www.ncdc.noaa.gov/paleo/study/15535. Downloaded 05/05/2018.

217 B.M. Vinther et al., 2009, "Holocene thinning of the Greenland ice sheet." Nature, Vol. 461, pp. 385-388, 17 September 2009. National Centers for Environmental Information, NESDIS, NOAA, U.S. Department of Commerce. Greenland Ice Sheet Holocene d18O, Temperature, and Surface Elevation. doi:10.1038/nature08355. https://www.ncdc.noaa.gov/paleo-search/study/11148. Downloaded 05/05/2018.

218 N. Scafetta, "Multi-scale harmonic model for solar and climate cyclical variation throughout the Holocene based on Jupiter-Saturn tidal frequencies plus the 11-year solar dynamo cycle." Journal of Atmospheric and Solar-Terrestrial Physics (2012). doi:10.1016/j.jastp.2012.02.016.

219 Theodor Landscheidt, "New Little Ice Age Instead of Global Warming?" Energy & Environment. 2003. Volume 14, Issue 2, 327–350. https://doi.org/10.1260/095830503765184646.

220 R. J. Salvador, "A mathematical model of the sunspot cycle for the past 1000 years." Pattern Recognition Physics, 1, 117-122, doi:10.5194/prp-1-117-2013, 2013.

221 Habibullo Abdussamatov, "Current Long-Term Negative Average Annual Energy Balance of the Earth Leads to the New Little Ice age." Thermal Science. 2015 Supplement, Volume 19, S279-S288.

222 Jan-Erik Solheim, https://www.mwenb.nl/wp-content/uploads/2014/10/Blog-Jan-Erik-Solheim-def.pdf. Referred from http://www.climatedialogue.org/what-will-happen-during-a-new-maunder-minimum/. Citing blog for 4-5 solar-climate experts.

223 Boncho P. Bonev et al., "Long-Term Solar Variability and the Solar Cycle in the 21st Century." The Astrophysical Journal, 605:L81–L84, 2004 April 10.

224 Nils-Axel Mörner. "Solar Minima, Earth's rotation and Little Ice Ages in the past and in the future. The North Atlantic–European case." Global and Planetary Change 72 (2010) 282–293. doi:10.1016/j.gloplacha.2010.01.004.

225 A. Mazzarella, "The 60-year solar modulation of global air temperature: the Earth's rotation and atmospheric circulation connection." Theoretical and Applied Climatology. 88, 193–199 (2007). DOI 10.1007/s00704-005-0219-z.

226 See endnote 182.

227 D. H. Tarling, "Milankvitch Cycles in Climate Change." Geology and Geophysics. Proc. 6th International Symposium on Geophysics, Tanta, Egypt (2010), 1- 8.

228 Mark A. Maslin and Andy J. Ridgwell, "Mid-Pleistocene revolution and the 'eccentricity myth'." Geological Society, London, Special Publications, 247, 19-34, 1 January 2005, https://doi.org/10.1144/GSL.SP.2005.247.01.02. Available online at http://sp.lyellcollection.org/content/247/1/19.

229 Richard A. Muller and Gordon J. MacDonald, "Spectrum of 100-kyr glacial cycle: Orbital inclination, not eccentricity." Proc. Natl. Acad. Sci. USA Volume 94, 8329–8334, August 1997 Colloquium Paper. Available online at http://www.pnas.org/content/pnas/94/16/8329.full.pdf

230 J. A. Rial, 2004, "Earth's orbital eccentricity and the rhythm of the Pleistocene ice ages: The concealed pacemaker." Global and Planetary Change, 41(2), 81-93. DOI:10.1016/j.gloplacha.2003.10.003.

231 J. Kirkby et al., "The glacial cycles and cosmic rays." CERN-PH-EP/2004-027. https://arxiv.org/abs/physics/0407005.

232 Gerald E. Marsh, "Interglacials, Milankovitch Cycles, and Carbon Dioxide." DOI: 10.1155/2014/345482. Available at https://arxiv.org/abs/1002.0597.

233 H. Wanner et al., "Structure and origin of Holocene cold events." Quaternary Science Reviews (2011), doi:10.1016/j.quascirev.2011.07.010. [Comment: See Figure 5a, page 9, depicting the steady decline in Northern Hemisphere summer solar insolation at north 15 and 65 degree latitudes, and indicating that insolation has declined by a whopping 40 W/m2. This is based on the landmark research by Berger, 1978 (André Berger, Long-Term Variations of Daily Insolation and Quaternary Climatic Changes. 1978. Journal of the Atmospheric Sciences 35(12):2362-2367. DOI: 10.1175/1520-0469(1978)035<2362:LTVODI>2.0.CO;2).].

234 D.S. Kaufman et al., "Holocene thermal maximum in the western Arctic (0–180°W)." Quaternary Science Reviews, Volume 23, Issues 5–6, 2004, 529-560. https://doi.org/10.1016/j.quascirev.2003.09.007. [Comment: See the abstract. We are told that the precession-driven summer insolation anomaly peaked 12,000-10,000 years ago. See also Figure 9a which depicts the 65°N insolation anomaly at different times of the year, indicating a whopping 50 Wm-2 decline in summer solstice insolation from its peak 12,000-10,000 years ago.].

235 G.H. Miller et al., 2012, "Abrupt onset of the Little Ice Age triggered by volcanism and sustained by sea-ice/ocean feedbacks." Geophysical Research Letters, 39, L02708, doi:10.1029/2011GL050168. [Comment: We are told in the opening sentence that the Northern Hemisphere summer temperatures track a precession-driven decline in summer insolation for 8,000 years, and that the summer temperature changes are the greatest in the Arctic. This article cites CAPE Project Members, 2001; Kaufman et al., 2004; Vinther et al., 2009.].

236 Y. Zhong et al., "Centennial-scale climate change from decadally-paced explosive volcanism: a coupled sea ice-ocean mechanism." Climate Dynamics (2011) 37: 2373. https://doi.org/10.1007/s00382-010-0967-z. [Comment: The abstract tells us that the Northern Hemisphere Holocene summer cooling was driven predominantly by the decline in precession- modulated summer insolation. Page 2, top left, tells us this summer decline in insolation from 8,000 years ago in the Northern Hemisphere led to glacier ice expansion, especially from 5,000 years ago.].

237 Darrell Kaufman et al., "Recent Warming Reverses Long-Term Arctic Cooling." September 2009. Science 325(5945):1236-1239. DOI: 10.1126/science.1173983. [Comment: This publication details the Arctic cooling that has been in progress for the last 2,000 years until this recent global warming phase. This millennial-scale cooling trend correlates (r = +0.87 with a R-squared 0.76, see Figure 4.) with a reduction in precession of the solstice driven summer insolation (6 W m−2 insolation at 65°N) for the last 2,000 years. See Figure 3F. The publication indicates a temperature decline of 0.22° ± 0.06°C per 1000 years, which tracks the slow decline in orbitally driven summer insolation at high northern latitudes.].

238 US Environmental Protection Agency. Climate Change Indicators: Climate Forcing. https://
www.epa.gov/climate-indicators/climate-change-indicators-climate-forcing#ref1.
239 See endnote 182.
240 Bintanja, R. and R.S.W. van de Wal, "North American ice-sheet dynamics and the onset of
100,000-year glacial cycles." Nature, Volume 454, 869-872, 14 August 2008. doi:10.1038/
nature07158. National Centers for Environmental Information, NESDIS, NOAA,
U.S. Department of Commerce. Global 3Ma Temperature, Sea Level, and Ice Volume
Reconstructions. https://www.ncdc.noaa.gov/paleo-search/study/11933. Downloaded
10/27/2015.
241 J. V. Jouzel et al., 2007, "Orbital and Millennial Antarctic Climate Variability over the Past
800,000 Years." Science, Volume 317, No. 5839, 793-797, 10 August 2007. National Centers for
Environmental Information, NESDIS, NOAA, U.S. Department of Commerce. EPICA Dome
C - 800KYr Deuterium Data and Temperature Estimates. https://www.ncdc.noaa.gov/paleo/
study/6080. Download data: Downloaded 08/02/2016.
242 <u>Data</u>: Bintanja, R. and R.S.W. van de Wal, "North American ice-sheet dynamics and the
onset of 100,000-year glacial cycles." Nature, Volume 454, 869-872, 14 August 2008.
doi:10.1038/nature07158. National Centers for Environmental Information, NESDIS, NOAA,
U.S. Department of Commerce. Global 3Ma Temperature, Sea Level, and Ice Volume
Reconstructions. https://www.ncdc.noaa.gov/paleo-search/study/11933. Downloaded
10/27/2015. <u>Personal Research</u>: The following table was created to summarize 965,300 years
of global interglacial periods (Kiloyears). A Grubb's test (extreme studentized deviate)
was performed to determine if the IPCC's proposed 30,000-year extension to the Holocene
interglacial period created a statistically significant outlier in this group of eleven glacial cycle
comparators. To test the statistical validity, the current interglacial duration of 17,500 years was
extended by 30,000, plus 2,100 years from its existing climate optimum peak to bring it up to
today, yielding a revised 49,600 year interglacial period. By extending this current interglacial
by 30,000 years the group interglacial duration mean was 21.1Kyr, standard deviation 11.2Kyr,
N=11, and an outlier was detected (two-sided P<0.05). The critical value of Z = 2.35. A
goodness of fit test was performed with the original data using the d'Agostino-Pearson test: P
= 0.386, indicating the original group was not different from a Gaussian or normal distribution.
On this basis the Grubbs test was selected to test for an outlier. By delaying the Holocene
30,000 years the interglacial period was extended to 49,600 years and the d'Agostino-Pearson
test: P = 0.011. This 30,000 year interglacial extension changed the data-population distribution
to a non-normal or non-Gaussian distribution. This change in data distribution adds further
support that the 30,000-year delay can not be statistically justified.
243 <u>Data</u>: Bintanja, R. and R.S.W. van de Wal, "North American ice-sheet dynamics and the
onset of 100,000-year glacial cycles." Nature, Volume 454, 869-872, 14 August 2008.
doi:10.1038/nature07158. National Centers for Environmental Information, NESDIS, NOAA,
U.S. Department of Commerce. Global 3Ma Temperature, Sea Level, and Ice Volume
Reconstructions. https://www.ncdc.noaa.gov/paleo-search/study/11933. Downloaded
10/27/2015. <u>Personal Research</u>: The following table data summary was created detailing
2,026,800 years of global climate optima timings and intervals (from peak-to-peak; Kiloyears).
Without any change to the 2.1Kyr climate optimum 2,100 years ago i.e., without a 30,000-year
extension to the Holocene interglacial period, the current largest climate optimum interval of
122.7 kiloyears (2.1-124.8Kyr peak) is not a statistically significant outlier (P>0.05) compared
with the comparator group (N=33). However, when this interglacial period is extended 30,000
years the revised climate optima interval (154.8 kiloyears) becomes a statistically significant
outlier (P<0.05). For the revised dataset (with 30,000-year extension): the Mean = 62.3,
standard deviation = 30.2, N = 33. Outlier detected? Yes. Significance level: 0.05 (two-sided).
Critical value of Z: 2.95. A goodness of fit was pre-assessed for the original unmodified data
(i.e., no 30Kyr delay) using the d'Agostino-Pearson test: P = 0.095. On this basis the Grubb's
outlier test was selected because the original data was normally distributed. When the 122.7Kyr
climate optimum interval was substituted with the 154.8Kyr data point the d'Agostino-Pearson
test: P = 0.011. This 30,000-year modification changed the data-population distribution to a
non-normal distribution. This change in data distribution adds further support that the 30,000-
year delay can not be statistically justified.

244 Data: (1) Jouzel, J., V. et al. 2007. "Orbital and Millennial Antarctic Climate Variability over the Past 800,000 Years." Science, Volume 317, No. 5839, 793-797, 10 August 2007. National Centers for Environmental Information, NESDIS, NOAA, U.S. Department of Commerce. EPICA Dome C - 800KYr Deuterium Data and Temperature Estimates. https://www.ncdc. noaa.gov/paleo/study/6080. Download data: Downloaded 08/02/2016. (2) R. Bintanja and R.S.W. van de Wal, North American ice-sheet dynamics and the onset of 100,000-year glacial cycles. Nature, Volume 454, 869-872, 14 August 2008. doi:10.1038/nature07158. National Centers for Environmental Information, NESDIS, NOAA, U.S. Department of Commerce. Global 3Ma Temperature, Sea Level, and Ice Volume Reconstructions. https://www.ncdc. noaa.gov/paleo-search/study/11933. Downloaded 10/27/2015. Personal Research: The data is tabulated above (at the start of the endnotes, referencing this endnote). detailing 787,300 years of Antarctic-to-global climate optima phasing gaps (Kiloyears). A Grubb's test (extreme studentized deviate) was performed on the Antarctic-to-global climate optima phasing gaps to determine if the IPCC's proposed 30,000 year ice age delay created a statistically significant outlier from the other 8 glacial cycle comparator phasing gaps (Antarctica versus global; glacial cycles 1-9). By delaying the Holocene Climate Optima 30,000 years, the phasing gap changes from the current 8,400 years to a statistically significant 40,500 years (P<0.05)(40,500 years =30,000+2,100+8,400 years). The 30,000-year phasing gap increase was only applied to the global climate data's climate optimum because Antarctica's climate optimum was set in the ice record 10,500 years ago and new ice has accumulated since, indicating its ice age has already started (i.e., the inner dome is 100 meters higher today than it was at the Holocene Climate Optimum, see Chapter 3). A goodness of fit test was performed with the original data using the d'Agostino-Pearson test: P = 0.678, indicating the original group was not different from a Gaussian or normal distribution. On this basis the Grubbs test was selected to test for an outlier. By delaying the Holocene 30,000 years the phasing gap was increased to 40,500 years and the d'Agostino-Pearson test: P< 0.001. This 30,000-year interglacial extension changed the data-population distribution from a normal to a non-normal distribution. This change in data distribution adds further support that the 30,000-year delay can not be statistically justified.

245 A.E. Carlson, 2013, "The Younger Dryas Climate Event." In: Elias S.A. (ed.) The Encyclopedia of Quaternary Science, Volume 3, 126-134. Amsterdam: Elsevier. http://people.oregonstate. edu/~carlsand/carlson_encyclopedia_Quat_2013_YD.pdf.

246 Nicolaj K. Larsen et al., "The response of the southern Greenland ice sheet to the Holocene thermal maximum." Geology ; 43 (4): 291–294. doi: https://doi.org/10.1130/G36476.1.

247 D.S. Kaufman et al., "Holocene thermal maximum in the western Arctic (0–1800W)." Quaternary Science Reviews 23 (2004) 529–560.

248 J.P. Briner et al., "Holocene climate change in Arctic Canada and Greenland." Quaternary Science Reviews (2016), http://dx.doi.org/10.1016/j.quascirev.2016.02.010.

249 Highest temperature on record. IPCC, Climate Change 2007: The Physical Science Basis. Contribution of Working Group I to the Fourth Assessment Report of the Intergovernmental Panel on Climate Change [Solomon, S., D. Qin, M. Manning, Z. Chen, M. Marquis, K.B. Averyt, M. Tignor and H.L. Miller (eds.)]. Cambridge University Press, Cambridge, United Kingdom and New York, NY, USA, 996 pages [Exposé: See page 288. The IPCC tells us the 1998 global mean air temperature was the highest on record until 2005.].

250 Bintanja, R. and R.S.W. van de Wal, "North American ice-sheet dynamics and the onset of 100,000-year glacial cycles." Nature, Volume 454, 869-872, 14 August 2008. doi:10.1038/ nature07158. National Centers for Environmental Information, NESDIS, NOAA, U.S. Department of Commerce. Global 3Ma Temperature, Sea Level, and Ice Volume Reconstructions. https://www.ncdc.noaa.gov/paleo-search/study/11933. Downloaded 10/27/2015.

251 H. Wanner et al., "Structure and origin of Holocene cold events." Quaternary Science Reviews (2011), doi:10.1016/j.quascirev.2011.07.010.

252 D.S. Kaufman et al., "Holocene thermal maximum in the western Arctic (0–180°W)." Quaternary Science Reviews, Volume 23, Issues 5–6, 2004, 529-560. https://doi.org/10.1016/j. quascirev.2003.09.007.

253 G.H. Miller et al., 2012, "Abrupt onset of the Little Ice Age triggered by volcanism and sustained by sea-ice/ocean feedbacks." Geophysical Research Letters, 39, L02708, doi:10.1029/2011GL050168.

254 Y. Zhong et al., "Centennial-scale climate change from decadally-paced explosive volcanism: a coupled sea ice-ocean mechanism." Climate Dynamics (2011) 37: 2373. https://doi.org/10.1007/s00382-010-0967-z.

255 Darrell Kaufman et al., "Recent Warming Reverses Long-Term Arctic Cooling." September 2009. Science 325(5945):1236-1239. DOI: 10.1126/science.1173983.

256 Nicolaj K. Larsen et al., "The response of the southern Greenland ice sheet to the Holocene thermal maximum." Geology ; 43 (4): 291–294. doi: https://doi.org/10.1130/G36476.1.

257 J.P. Briner et al., "Holocene climate change in Arctic Canada and Greenland." Quaternary Science Reviews (2016), http://dx.doi.org/10.1016/j.quascirev.2016.02.010.

258 Leonid Polyak et al., "History of sea ice in the Arctic." Quaternary Science Reviews 29 (2010) 1757–1778, https://doi.org/10.1016/j.quascirev.2010.02.010.

259 Ó. Ingólfsson et al., 1998, "Antarctic glacial history since the Last Glacial Maximum: An overview of the record on land." Antarctic Science, 10(3), 326-344. doi:10.1017/S095410209800039X.

260 The RAISED Consortium1, Michael J. Bentley et al. "A community-based geological reconstruction of Antarctic Ice Sheet deglaciation since the Last Glacial Maximum." Quaternary Science Reviews. Volume 100, 15 September 2014, 1-9.

261 Jason P. Briner et al., "Holocene climate change in Arctic Canada and Greenland." Quaternary Science Reviews, Volume 147, 2016, 340-364, ISSN 0277-3791. https://doi.org/10.1016/j.quascirev.2016.02.010.

262 O.N. Solomina et al., 2016, "Glacier fluctuations during the past 2000 years." Quaternary Science Reviews, 149, 61-90. DOI: 10.1016/j.quascirev.2016.04.008. [See Figure 5, page 276. This figure collates a stacked time series of the number of glacier advances and recessions in each region into a global total.].

263 M. Frezzotti1 et al., "A synthesis of the Antarctic surface mass balance during the last 800 years." The Cryosphere, 7, 303–319, 2013. www.the-cryosphere.net/7/303/2013/doi:10.5194/tc-7-303-2013.

264 Jan Veizer et al., "Evidence for decoupling of atmospheric CO2 and global climate during the Phanerozoic eon." January 2001. Nature 408(6813):698-701. DOI10.1038/35047044. Available at https://www.nature.com/articles/35047044. [See Figures 1 and 2].

265 James Zachos et al., "Trends, Rhythms, and Aberrations in Global Climate 65 Ma to Present." Science 27 Apr 2001. Volume 292, Issue 5517, 686-693. DOI: 10.1126/science.1059412. Available at http://science.sciencemag.org/content/292/5517/686.full.

266 H. C. Larsen et al., "Seven Million Years of Glaciation in Greenland." Science, New Series, Volume 264, No. 5161 (May 13, 1994), 952-955. Available at https://www.researchgate.net/publication/234051692_Seven_Million_Years_of_Glaciation_in_Greenland.

267 J.V. Jouzel et al., 2007, "Orbital and Millennial Antarctic Climate Variability over the Past 800,000 Years." Science, Volume 317, No. 5839, 793-797, 10 August 2007. National Centers for Environmental Information, NESDIS, NOAA, U.S. Department of Commerce. EPICA Dome C - 800KYr Deuterium Data and Temperature Estimates. https://www.ncdc.noaa.gov/paleo/study/6080. Download data: Downloaded 08/02/2016.

268 Walker et al., 2009, "Formal definition and dating of the GSSP (Global Stratotype Section and Point) for the base of the Holocene using the Greenland NGRIP ice core, and selected auxiliary records." J. Quaternary Sci., Volume 24 3–17. ISSN 0267-8179.

269 Data: R. Bintanja and R.S.W. van de Wal, "North American ice-sheet dynamics and the onset of 100,000-year glacial cycles." Nature, Volume 454, 869-872, 14 August 2008. doi:10.1038/nature07158. National Centers for Environmental Information, NESDIS, NOAA, U.S. Department of Commerce. Global 3Ma Temperature, Sea Level, and Ice Volume Reconstructions. https://www.ncdc.noaa.gov/paleo-search/study/11933. Downloaded 10/27/2015. [Comment: Based on summing up the 'Eurasian + North American ice volume' / 'total ice volume' for each 100-year time point (in meters of sea level equivalents), and averaging this over 19,600 years = 87%.].

270 Bintanja, R. and R.S.W. van de Wal, "North American ice-sheet dynamics and the onset of
     100,000-year glacial cycles." Nature, Volume 454, 869-872, 14 August 2008. doi:10.1038/
     nature07158. National Centers for Environmental Information, NESDIS, NOAA,
     U.S. Department of Commerce. Global 3Ma Temperature, Sea Level, and Ice Volume
     Reconstructions. https://www.ncdc.noaa.gov/paleo-search/study/11933. Downloaded
     10/27/2015.

271 Bintanja, R. and R.S.W. van de Wal, "North American ice-sheet dynamics and the onset of
     100,000-year glacial cycles." Nature, Volume 454, 869-872, 14 August 2008. doi:10.1038/
     nature07158. National Centers for Environmental Information, NESDIS, NOAA,
     U.S. Department of Commerce. Global 3Ma Temperature, Sea Level, and Ice Volume
     Reconstructions. https://www.ncdc.noaa.gov/paleo-search/study/11933. Downloaded
     10/27/2015.

272 J.V. Jouzel et al., 2007, "Orbital and Millennial Antarctic Climate Variability over the Past
     800,000 Years." Science, Volume 317, No. 5839, 793-797, 10 August 2007. National Centers for
     Environmental Information, NESDIS, NOAA, U.S. Department of Commerce. EPICA Dome
     C - 800KYr Deuterium Data and Temperature Estimates. https://www.ncdc.noaa.gov/paleo/
     study/6080. Download data: Downloaded 08/02/2016.

273 R.V. Uemura et al., 2012, "Ranges of moisture-source temperature estimated from Antarctic
     ice cores stable isotope records over glacial-interglacial cycles." Climate of the Past, 8,
     1109-1125. doi: 10.5194/cp-8-1109-2012. National Centers for Environmental Information,
     NESDIS, NOAA, U.S. Department of Commerce. Dome Fuji 360KYr Stable Isotope Data
     and Temperature Reconstruction. https://www.ncdc.noaa.gov/paleo-search/study/13121.
     Downloaded 05/05/2018.

274 Sigfus J. Johnsen et al., 1997, "The d18O record along the Greenland Ice Core Project deep ice
     core and the problem of possible Eemian climatic instability." Journal of Geophysical Research:
     Oceans, 102(C12), 26397-26410. doi: 10.1029/97JC00167. National Centers for Environmental
     Information, NESDIS, NOAA, U.S. Department of Commerce. GRIP Ice Core 248KYr Oxygen
     Isotope Data. https://www.ncdc.noaa.gov/paleo-search/study/17839.

275 B.M. Vinther et al., 2009, "Holocene thinning of the Greenland ice sheet." Nature, Vol. 461,
     pp. 385-388, 17 September 2009. National Centers for Environmental Information, NESDIS,
     NOAA, U.S. Department of Commerce. Greenland Ice Sheet Holocene d18O, Temperature,
     and Surface Elevation. doi:10.1038/nature08355. https://www.ncdc.noaa.gov/paleo-search/
     study/11148. Downloaded 05/05/2018.

276 R.B. Alley, 2004, "GISP2 Ice Core Temperature and Accumulation Data." National Centers for
     Environmental Information, NESDIS, NOAA, U.S. Department of Commerce. https://www.
     ncdc.noaa.gov/paleo/study/2475. Downloaded 05/05/2018.

277 Sigfus J. Johnsen et al., 1997, "The d18O record along the Greenland Ice Core Project deep ice
     core and the problem of possible Eemian climatic instability." Journal of Geophysical Research:
     Oceans, 102(C12), 26397-26410. doi: 10.1029/97JC00167. National Centers for Environmental
     Information, NESDIS, NOAA, U.S. Department of Commerce. GRIP Ice Core 248KYr Oxygen
     Isotope Data. https://www.ncdc.noaa.gov/paleo-search/study/17839.

278 B.M. Vinther et al., 2009, "Holocene thinning of the Greenland ice sheet." Nature, Vol. 461,
     pp. 385-388, 17 September 2009. National Centers for Environmental Information, NESDIS,
     NOAA, U.S. Department of Commerce. Greenland Ice Sheet Holocene d18O, Temperature,
     and Surface Elevation. doi:10.1038/nature08355. https://www.ncdc.noaa.gov/paleo-search/
     study/11148. Downloaded 05/05/2018.

279 Data: (1) J.V. Jouzel et al., 2007, "Orbital and Millennial Antarctic Climate Variability over
     the Past 800,000 Years." Science, Volume 317, No. 5839, 793-797, 10 August 2007. National
     Centers for Environmental Information, NESDIS, NOAA, U.S. Department of Commerce.
     EPICA Dome C - 800KYr Deuterium Data and Temperature Estimates. https://www.ncdc.noaa.
     gov/paleo/study/6080. Download data: Downloaded 08/02/2016. (2) Bintanja, R. and R.S.W.
     van de Wal, "North American ice-sheet dynamics and the onset of 100,000-year glacial cycles."
     Nature, Volume 454, 869-872, 14 August 2008. doi:10.1038/nature07158. National Centers for
     Environmental Information, NESDIS, NOAA, U.S. Department of Commerce. Global 3Ma
     Temperature, Sea Level, and Ice Volume Reconstructions. https://www.ncdc.noaa.gov/paleo-
     search/study/11933. Downloaded 10/27/2015. (3) R.V. Uemura et al., 2012, "Ranges of moisture-

source temperature estimated from Antarctic ice cores stable isotope records over glacial-interglacial cycles." Climate of the Past, 8, 1109-1125. doi: 10.5194/cp-8-1109-2012. National Centers for Environmental Information, NESDIS, NOAA, U.S. Department of Commerce. Dome Fuji 360KYr Stable Isotope Data and Temperature Reconstruction. https://www.ncdc.noaa.gov/paleo-search/study/13121. Downloaded 05/05/2018. Personal Research: Based on the above-cited climate data the climate optima timings and inter-climate optima intervals were tabulated using the Dome Fuji, EPICA Dome-C, and Greenland ice core data. This table (at the start of the endnotes, referencing this endnote) also details 787,300 years of Antarctic-to-global climate optima phasing gaps (Kiloyears) for both EPICA Dome-C and Dome-Fuji.

280   A.E. Carlson, 2013, "The Younger Dryas Climate Event." In: Elias S.A. (ed.) The Encyclopedia of Quaternary Science, Volume 3, 126-134. Amsterdam: Elsevier. http://people.oregonstate.edu/~carlsand/carlson_encyclopedia_Quat_2013_YD.pdf.

281   Anthony D. Barnosky et al., "Approaching a state shift in Earth's biosphere." Nature Volume 486, 52–58 (07 June 2012). doi:10.1038/nature11018.

282   R. B. Firestone et al., "Evidence for an extraterrestrial impact 12,900 years ago that contributed to the megafaunal extinctions and the Younger Dryas cooling." PNAS October 9, 2007. 104 (41) 16016-16021; https://doi.org/10.1073/pnas.0706977104.

283   R.B. Alley, 2004, "GISP2 Ice Core Temperature and Accumulation Data." National Centers for Environmental Information, NESDIS, NOAA, U.S. Department of Commerce. https://www.ncdc.noaa.gov/paleo/study/2475. Downloaded 05/05/2018.

284   R.V. Uemura et al., 2012, "Ranges of moisture-source temperature estimated from Antarctic ice cores stable isotope records over glacial-interglacial cycles." Climate of the Past, 8, 1109-1125. doi: 10.5194/cp-8-1109-2012. National Centers for Environmental Information, NESDIS, NOAA, U.S. Department of Commerce. Dome Fuji 360KYr Stable Isotope Data and Temperature Reconstruction. https://www.ncdc.noaa.gov/paleo-search/study/13121. Downloaded 05/05/2018.

285   Bintanja, R. and R.S.W. van de Wal, "North American ice-sheet dynamics and the onset of 100,000-year glacial cycles." Nature, Volume 454, 869-872, 14 August 2008. doi:10.1038/nature07158. National Centers for Environmental Information, NESDIS, NOAA, U.S. Department of Commerce. Global 3Ma Temperature, Sea Level, and Ice Volume Reconstructions. https://www.ncdc.noaa.gov/paleo-search/study/11933. Downloaded 10/27/2015.

286   J. V. Jouzel et al., 2007, "Orbital and Millennial Antarctic Climate Variability over the Past 800,000 Years." Science, Volume 317, No. 5839, 793-797, 10 August 2007. National Centers for Environmental Information, NESDIS, NOAA, U.S. Department of Commerce. EPICA Dome C - 800KYr Deuterium Data and Temperature Estimates. https://www.ncdc.noaa.gov/paleo/study/6080. Download data: Downloaded 08/02/2016.

287   Sigfus J. Johnsen et al., 1997, "The d18O record along the Greenland Ice Core Project deep ice core and the problem of possible Eemian climatic instability." Journal of Geophysical Research: Oceans, 102(C12), 26397-26410. doi: 10.1029/97JC00167. National Centers for Environmental Information, NESDIS, NOAA, U.S. Department of Commerce. GRIP Ice Core 248KYr Oxygen Isotope Data. https://www.ncdc.noaa.gov/paleo-search/study/17839.

288   Data: R. Bintanja and R.S.W. van de Wal, "North American ice-sheet dynamics and the onset of 100,000-year glacial cycles." Nature, Volume 454, 869-872, 14 August 2008. doi:10.1038/nature07158. National Centers for Environmental Information, NESDIS, NOAA, U.S. Department of Commerce. Global 3Ma Temperature, Sea Level, and Ice Volume Reconstructions. https://www.ncdc.noaa.gov/paleo-search/study/11933. Downloaded 10/27/2015. Personal Research: Using the above-cited file, the temperature and sea level data were extracted; from the Last glacial maximum 19,600 years ago to the Holocene Climate Optimum (2,100 years ago). The Younger Dryas 11,700 years ago was included to help us see that the last ice age did not end 11,700 years ago. Results: by 11,700 years ago the sea level had already risen 64% and the temperature 76% of their total Holocene interglacial rise (from glacial maximum to climate optimum). This confirms the last ice age did not end 11,700 years ago, but rather it was the Younger Dryas that ended. The data very clearly tells us the last ice age ended 19,600 years ago, after which the sea level began to rise and the ice mass decrease.].

289 R. Bintanja and R.S.W. van de Wal, "North American ice-sheet dynamics and the onset of 100,000-year glacial cycles." Nature, Volume 454, 869-872, 14 August 2008. doi:10.1038/nature07158. National Centers for Environmental Information, NESDIS, NOAA, U.S. Department of Commerce. Global 3Ma Temperature, Sea Level, and Ice Volume Reconstructions. https://www.ncdc.noaa.gov/paleo-search/study/11933. Downloaded 10/27/2015.

290 R.V. Uemura et al., 2012, "Ranges of moisture-source temperature estimated from Antarctic ice cores stable isotope records over glacial-interglacial cycles." Climate of the Past, 8, 1109-1125. doi: 10.5194/cp-8-1109-2012. National Centers for Environmental Information, NESDIS, NOAA, U.S. Department of Commerce. Dome Fuji 360KYr Stable Isotope Data and Temperature Reconstruction. https://www.ncdc.noaa.gov/paleo-search/study/13121. Downloaded 05/05/2018.

291 R.B. Alley., 2004, "GISP2 Ice Core Temperature and Accumulation Data." National Centers for Environmental Information, NESDIS, NOAA, U.S. Department of Commerce. https://www.ncdc.noaa.gov/paleo/study/2475. Downloaded 05/05/2018.

292 Nicolaj K. Larsen et al., "The response of the southern Greenland ice sheet to the Holocene thermal maximum." Geology ; 43 (4): 291–294. doi: https://doi.org/10.1130/G36476.1.

293 D.S. Kaufman et al., "Holocene thermal maximum in the western Arctic (0–1800W)." Quaternary Science Reviews 23 (2004) 529–560.

294 J.P. Briner et al., "Holocene climate change in Arctic Canada and Greenland." Quaternary Science Reviews (2016), http://dx.doi.org/10.1016/j.quascirev.2016.02.010.

295 Nicolaj K. Larsen et al., "The response of the southern Greenland ice sheet to the Holocene thermal maximum." Geology ; 43 (4): 291–294. doi: https://doi.org/10.1130/G36476.1.

296 J.P. Briner et al., "Holocene climate change in Arctic Canada and Greenland." Quaternary Science Reviews (2016), http://dx.doi.org/10.1016/j.quascirev.2016.02.010.

297 Leonid Polyak et al., "History of sea ice in the Arctic." Quaternary Science Reviews 29 (2010) 1757–1778, https://doi.org/10.1016/j.quascirev.2010.02.010.

298 Leonid Polyak et al., "History of sea ice in the Arctic." Quaternary Science Reviews 29 (2010) 1757–1778, https://doi.org/10.1016/j.quascirev.2010.02.010.

299 N. L. Balascio et al., "Glacier response to North Atlantic climate variability during the Holocene." Climate of the Past, 11, 1587-1598, https://doi.org/10.5194/cp-11-1587-2015, 2015.

300 O. N. Solomina et al., 2016, "Glacier fluctuations during the past 2000 years." Quaternary Science Reviews, 149, 61-90. DOI: 10.1016/j.quascirev.2016.04.008. [See Figure 5, page 276. This figure collates a stacked time series of the number of glacier advances and recessions in each region into a global total.].

301 Michael E Mann, "Little Ice Age." Volume 1, The Earth system: physical and chemical dimensions of global environmental change, 504–509. In Encyclopedia of Global Environmental Change (ISBN 0-471-97796-9).

302 Leonid Polyak et al., "History of sea ice in the Arctic." Quaternary Science Reviews 29 (2010) 1757–1778, https://doi.org/10.1016/j.quascirev.2010.02.010

303 Christophe Kinnard et al., "A changing Arctic seasonal ice zone: Observations from 1870–2003 and possible oceanographic consequences." Geophysical Research Letters, Volume 35, L02507, doi:10.1029/2007GL032507, 2008.

304 O.N. Solomina et al., 2016, "Glacier fluctuations during the past 2000 years." Quaternary Science Reviews, 149, 61-90. DOI: 10.1016/j.quascirev.2016.04.008. [See Figure 5, page 276. This figure collates a stacked time series of the number of glacier advances and recessions in each region into a global total.].

305 Jason P. Briner et al., "Holocene climate change in Arctic Canada and Greenland." Quaternary Science Reviews, Volume 147, 2016, 340-364, ISSN 0277-3791. https://doi.org/10.1016/j.quascirev.2016.02.010.

306 O.N. Solomina et al., 2016, "Glacier fluctuations during the past 2000 years." Quaternary Science Reviews, 149, 61-90. DOI: 10.1016/j.quascirev.2016.04.008. [See Figure 5, page 276. This figure collates a stacked time series of the number of glacier advances and recessions in each region into a global total.].

307 A.N. Mackintosh et al., 2014, "Retreat history of the East Antarctic Ice Sheet since the Last Glacial Maximum." Quaternary Science Reviews 100, 10e30. http://dx.doi.org/10.1016/j. quascirev.2013.07.024.

308 The RAISED Consortium1, Michael J. Bentley et al. "A community-based geological reconstruction of Antarctic Ice Sheet deglaciation since the Last Glacial Maximum." Quaternary Science Reviews. Volume 100, 15 September 2014, 1-9.

309 Ó. Ingólfsson et al., 1998, "Antarctic glacial history since the Last Glacial Maximum: An overview of the record on land." Antarctic Science, 10(3), 326-344. doi:10.1017/ S095410209800039X.

310 Ó. Ingólfsson et al., 1998, "Antarctic glacial history since the Last Glacial Maximum: An overview of the record on land. "Antarctic Science, 10(3), 326-344. doi:10.1017/ S095410209800039X.

311 The RAISED Consortium1, Michael J. Bentley et al. "A community-based geological reconstruction of Antarctic Ice Sheet deglaciation since the Last Glacial Maximum." Quaternary Science Reviews. Volume 100, 15 September 2014, 1-9.

312 M. Frezzotti1 et al., "A synthesis of the Antarctic surface mass balance during the last 800 years." The Cryosphere, 7, 303–319, 2013. www.the-cryosphere.net/7/303/2013/doi:10.5194/tc-7-303-2013. [See Figure 5.A, 312.].

313 H. Wanner et al., "Structure and origin of Holocene cold events." Quaternary Science Reviews (2011), doi:10.1016/j.quascirev.2011.07.010. [Comment: See Figure 5a, page 9, depicting the steady decline in Northern Hemisphere summer solar insolation at north 15 and 65 degree latitudes, and indicating that insolation has declined by a whopping 40 W/m2. This is based on the landmark research by Berger, 1978 (André Berger, Long-Term Variations of Daily Insolation and Quaternary Climatic Changes. 1978. Journal of the Atmospheric Sciences 35(12):2362-2367. DOI: 10.1175/1520-0469(1978)035<2362:LTVODI>2.0.CO;2).].

314 D.S. Kaufman et al., "Holocene thermal maximum in the western Arctic (0–180°W)." Quaternary Science Reviews, Volume 23, Issues 5–6, 2004, 529-560. https://doi.org/10.1016/j. quascirev.2003.09.007. [Comment: See the abstract. We are told that the precession-driven summer insolation anomaly peaked 12,000-10,000 years ago. See also Figure 9a which depicts the 65°N insolation anomaly at different times of the year, indicating a whopping 50 Wm-2 decline in summer solstice insolation from its peak 12,000-10,000 years ago.].

315 G.H. Miller et al., 2012, "Abrupt onset of the Little Ice Age triggered by volcanism and sustained by sea-ice/ocean feedbacks." Geophysical Research Letters, 39, L02708, doi:10.1029/2011GL050168. [Comment: We are told in the opening sentence that the Northern Hemisphere summer temperatures track a precession-driven decline in summer insolation for 8,000 years, and that the summer temperature changes are the greatest in the Arctic. This article cites CAPE Project Members, 2001; Kaufman et al., 2004; Vinther et al., 2009.].

316 Y. Zhong et al., "Centennial-scale climate change from decadally-paced explosive volcanism: a coupled sea ice-ocean mechanism." Climate Dynamics (2011) 37: 2373. https://doi.org/10.1007/ s00382-010-0967-z. [Comment: The abstract tells us that the Northern Hemisphere Holocene summer cooling was driven predominantly by the decline in precession- modulated summer insolation. Page 2, top left, tells us this summer decline in insolation from 8,000 years ago in the Northern Hemisphere led to glacier ice expansion, especially from 5,000 years ago.].

317 Darrell Kaufman et al., "Recent Warming Reverses Long-Term Arctic Cooling." September 2009. Science 325(5945):1236-1239. DOI: 10.1126/science.1173983. [Comment: This publication details the Arctic cooling that has been in progress for the last 2,000 years until this recent global warming phase. This millennial-scale cooling trend correlates (r = +0.87 with a R-squared 0.76, see Figure 4.) with a reduction in precession of the solstice driven summer insolation (6 W m−2 insolation at 65°N) for the last 2,000 years. See Figure 3F. The publication indicates a temperature decline of 0.22° ± 0.06°C per 1000 years, which tracks the slow decline in orbitally driven summer insolation at high northern latitudes.].

318 R. Bintanja and R.S.W. van de Wal, "North American ice-sheet dynamics and the onset of 100,000-year glacial cycles." Nature, Volume 454, 869-872, 14 August 2008. doi:10.1038/ nature07158. National Centers for Environmental Information, NESDIS, NOAA, U.S. Department of Commerce. Global 3Ma Temperature, Sea Level, and Ice Volume

Reconstructions. https://www.ncdc.noaa.gov/paleo-search/study/11933. Downloaded 10/27/2015.

319　R. Bintanja and R.S.W. van de Wal, "North American ice-sheet dynamics and the onset of 100,000-year glacial cycles." Nature, Volume 454, 869-872, 14 August 2008. doi:10.1038/ nature07158. National Centers for Environmental Information, NESDIS, NOAA, U.S. Department of Commerce. Global 3Ma Temperature, Sea Level, and Ice Volume Reconstructions. https://www.ncdc.noaa.gov/paleo-search/study/11933. Downloaded 10/27/2015.

320　　　J.V. Jouzel et al., 2007, "Orbital and Millennial Antarctic Climate Variability over the Past 800,000 Years." Science, Volume 317, No. 5839, 793-797, 10 August 2007. National Centers for Environmental Information, NESDIS, NOAA, U.S. Department of Commerce. EPICA Dome C - 800KYr Deuterium Data and Temperature Estimates. https://www.ncdc.noaa.gov/paleo/ study/6080. Download data: Downloaded 08/02/2016.

321　Data: R. Bintanja and R.S.W. van de Wal, "North American ice-sheet dynamics and the onset of 100,000-year glacial cycles." Nature, Volume 454, 869-872, 14 August 2008. doi:10.1038/nature07158. National Centers for Environmental Information, NESDIS, NOAA, U.S. Department of Commerce. Global 3Ma Temperature, Sea Level, and Ice Volume Reconstructions. https://www.ncdc.noaa.gov/paleo-search/study/11933. Downloaded 10/27/2015. Personal Research: The temperature data for the first 2,100 years from the climate optimum was extracted for the last 34 glacial cycles. This temperature time-series was rebased to zero degrees and zero time so all glacial cycles could be compared on the same basis, i.e., from their peaks. The temperature declined by 0.610C after the Holocene Climate Optimum, which was 1.260C above the average of all other glacial cycles in 2,026,800 years. The current glacial cycle's slow decline was not a significant outlier in the group.

322　Data: J.V. Jouzel et al., 2007, "Orbital and Millennial Antarctic Climate Variability over the Past 800,000 Years." Science, Volume 317, No. 5839, 793-797, 10 August 2007. National Centers for Environmental Information, NESDIS, NOAA, U.S. Department of Commerce. EPICA Dome C - 800KYr Deuterium Data and Temperature Estimates. https://www.ncdc. noaa.gov/paleo/study/6080. Download data: Downloaded 08/02/2016. Personal Research: The global temperature for this current ice age inception has declined 1.20C since the Holocene Climate Optimum 10,500 years ago. By contrast, the temperature had declined by an average of 4.30C 10,500 years after the climate optimum for all eight previous glacial cycles from the last 800,000 years. The current glacial cycle's slow decline was not a significant outlier in the group.].

323　R. Bintanja and R.S.W. van de Wal, "North American ice-sheet dynamics and the onset of 100,000-year glacial cycles." Nature, Volume 454, 869-872, 14 August 2008. doi:10.1038/ nature07158. National Centers for Environmental Information, NESDIS, NOAA, U.S. Department of Commerce. Global 3Ma Temperature, Sea Level, and Ice Volume Reconstructions. https://www.ncdc.noaa.gov/paleo-search/study/11933. Downloaded 10/27/2015.

324　N. Scafetta, "Multi-scale harmonic model for solar and climate cyclical variation throughout the Holocene based on Jupiter-Saturn tidal frequencies plus the 11-year solar dynamo cycle." Journal of Atmospheric and Solar-Terrestrial Physics (2012). doi:10.1016/j.jastp.2012.02.016.

325　Theodor Landscheidt, "New Little Ice Age Instead of Global Warming?" Energy & Environment. 2003. Volume 14, Issue 2, 327 – 350. https://doi. org/10.1260/095830503765184646.

326　R.J. Salvador, "A mathematical model of the sunspot cycle for the past 1000 years." Pattern Recognition Physics, 1, 117-122, doi:10.5194/prp-1-117-2013, 2013.

327　Habibullo Abdussamatov, "Current Long-Term Negative Average Annual Energy Balance of the Earth Leads to the New Little Ice age." Thermal Science. 2015 Supplement, Volume 19, S279-S288.

328　Jan-Erik Solheim, https://www.mwenb.nl/wp-content/uploads/2014/10/Blog-Jan-Erik-Solheim-def.pdf. Referred from http://www.climatedialogue.org/what-will-happen-during-a-new-maunder-minimum/. Citing blog for 4-5 solar-climate experts.

329　Boncho P. Bonev et al., "Long-Term Solar Variability and the Solar Cycle in the 21st Century." The Astrophysical Journal, 605:L81–L84, 2004 April 10.

330 Nils-Axel Mörner, "Solar Minima, Earth's rotation and Little Ice Ages in the past and in the future." The North Atlantic–European case. Global and Planetary Change 72 (2010) 282–293. doi:10.1016/j.gloplacha.2010.01.004.
331 Data: B.M. Vinther et al., 2009, "Holocene thinning of the Greenland ice sheet." Nature, Vol. 461, pp. 385-388, 17 September 2009. National Centers for Environmental Information, NESDIS, NOAA, U.S. Department of Commerce. Greenland Ice Sheet Holocene d18O, Temperature, and Surface Elevation. doi:10.1038/nature08355. https://www.ncdc.noaa.gov/paleo-search/study/11148. Downloaded 05/05/2018. Personal Research: Between the Holocene Climate Optimum 5980 BCE (+3.550C) and the deepest temperature trough in 1700 CE (-1.310C) the temperature declined 4.860C. Between 1700 and 1940 the temperature then raised 2.870C.].
332 Global mean surface temperature data, commonly referred to as HadCRUT4. https://www. metoffice.gov.uk/hadobs/hadcrut4/data/current/download.html. [Exposé: Look at the bottom left hand or first column for the current year-to-date temperature. Subtract that from the 2016 total to see the magnitude of the fall. Global Data: https://bit.ly/2nCgctz. Northern Hemisphere Data: https://bit.ly/2MRt75G, Southern Hemisphere Data: https://bit.ly/2nBfYTA. Tropics Data: https://bit.ly/2nFXJMM. [last downloaded 25/07/2018].
333 B.M. Vinther et al., 2009, "Holocene thinning of the Greenland ice sheet." Nature, Vol. 461, pp. 385-388, 17 September 2009. National Centers for Environmental Information, NESDIS, NOAA, U.S. Department of Commerce. Greenland Ice Sheet Holocene d18O, Temperature, and Surface Elevation. doi:10.1038/nature08355. https://www.ncdc.noaa.gov/paleo-search/study/11148. Downloaded 05/05/2018.
334 Global mean surface temperature data, commonly referred to as HadCRUT4. https://www. metoffice.gov.uk/hadobs/hadcrut4/data/current/download.html. [Exposé: Look at the bottom left hand or first column for the current year-to-date temperature. Subtract that from the 2016 total to see the magnitude of the fall. Global Data: https://bit.ly/2nCgctz. Northern Hemisphere Data: https://bit.ly/2MRt75G, Southern Hemisphere Data: https://bit.ly/2nBfYTA. Tropics Data: https://bit.ly/2nFXJMM. [last downloaded 25/07/2018].
335 See endnote 249.
336 Nicolaj K. Larsen et al., "The response of the southern Greenland ice sheet to the Holocene thermal maximum." Geology ; 43 (4): 291–294. doi: https://doi.org/10.1130/G36476.1.
337 D.S. Kaufman et al., "Holocene thermal maximum in the western Arctic (0–1800W)." Quaternary Science Reviews 23 (2004) 529–560.
338 J.P. Briner et al., "Holocene climate change in Arctic Canada and Greenland." Quaternary Science Reviews (2016), http://dx.doi.org/10.1016/j.quascirev.2016.02.010.
339 Data: B.M. Vinther et al., 2009, "Holocene thinning of the Greenland ice sheet." Nature, Vol. 461, pp. 385-388, 17 September 2009. National Centers for Environmental Information, NESDIS, NOAA, U.S. Department of Commerce. Greenland Ice Sheet Holocene d18O, Temperature, and Surface Elevation. doi:10.1038/nature08355. https://www.ncdc.noaa.gov/paleo-search/study/11148. Downloaded 05/05/2018. Personal Research: Between the Holocene Climate Optimum in 5980 BCE (temperature anomaly +3.550C) and the deepest temperature trough in 1700 CE (temperature anomaly -1.310C) the temperature declined by 4.860C (4.860C = +3.550C less -1.310C). From this deep temperature trough in 1700 up to 1940, the temperature rose 2.870C.].
340 B.M. Vinther et al., 2009, "Holocene thinning of the Greenland ice sheet." Nature, Vol. 461, pp. 385-388, 17 September 2009. National Centers for Environmental Information, NESDIS, NOAA, U.S. Department of Commerce. Greenland Ice Sheet Holocene d18O, Temperature, and Surface Elevation. doi:10.1038/nature08355. https://www.ncdc.noaa.gov/paleo-search/study/11148. Downloaded 05/05/2018.
341 Data: (1) B.M. Vinther et al., 2009, "Holocene thinning of the Greenland ice sheet." Nature, Vol. 461, pp. 385-388, 17 September 2009. National Centers for Environmental Information, NESDIS, NOAA, U.S. Department of Commerce. Greenland Ice Sheet Holocene d18O, Temperature, and Surface Elevation. doi:10.1038/nature08355. https://www.ncdc.noaa.gov/paleo-search/study/11148. Downloaded 05/05/2018. (2) HadCRUT4 near surface temperature data set for the Northern Hemisphere. http://www.metoffice.gov.uk/hadobs/hadcrut4/data/current/download.html. Downloaded 25 July 2018. Personal Research: Between the Holocene

Climate Optimum in 5980 BCE (temperature anomaly +3.550C) and the deepest temperature trough in 1700 CE (temperature anomaly -1.310C) the temperature declined by 4.860C (4.860C = +3.550C less -1.310C). Between 1700-1940 the temperature rose 2.870C. From this data, it was calculated that the temperature at 2016's peak was approximately 1.90C lower than it was at the Holocene Climate Optimum. To counter the devil's advocate argument that 1700 to 1960 actually represented two sequentially stacked peaks, a second trough-to-peak rise (+2.810C) was created. This utilized the existing 1840-1940 trough-to-1940 phase (2.070C) with an additional temperature anomaly (+0.740C) grafted on (+2.810C = 2.070C +0.740C). The HADCRUT4 Northern Hemisphere temperature data was utilized to create a 20-year moving average temperature anomaly graft from 1940-2016 (+0.740C). Conclusion: Both the 2.870C and +2.810C trough-to-peak rises are statistically significant outliers among the 39 trough-to-peak temperature rises exceeding 0.990C since the Holocene Climate Optimum.].

342 Data: (1) B.M. Vinther et al., 2009, "Holocene thinning of the Greenland ice sheet." Nature, Vol. 461, pp. 385-388, 17 September 2009. National Centers for Environmental Information, NESDIS, NOAA, U.S. Department of Commerce. Greenland Ice Sheet Holocene d18O, Temperature, and Surface Elevation. doi:10.1038/nature08355. https://www.ncdc.noaa.gov/paleo-search/study/11148. Downloaded 05/05/2018. (2) HadCRUT4 near surface temperature data set for the Northern Hemisphere. http://www.metoffice.gov.uk/hadobs/hadcrut4/data/current/download.html. Downloaded 25 July 2018. Personal Research: All 39 climate trough-to-peak temperature rises exceeding +0.990C, between 5980 BCE and 1940 CE were extracted from the temperature data, derived from the Greenland ice core, for group analysis (range, +0.990C to +2.870C, average 77.4 years trough-to-peak, n=39). These trough-to-peak temperature increases selected trough-to-peaks to start from the deepest time point in the maximum trough preceding the tallest peak. A goodness-of-fit test of all 39 trough-to-peak temperature rises showed that the data did not follow a normal distribution. This indicates the possibility that more than one global warming process may be involved with the bigger climate oscillation outliers (i.e., an extreme grand solar maxima). Results: Prior to stratifying the data an Iglewicz and Hoaglin's robust test (two-sided test) for multiple outliers was performed using a modified Z score of ≥ 1.5 and ≥ 5 as the outlier criteria. The modified Z score of ≥ 1.5 highlighted significant outliers above +1.770C. A higher modified Z score of ≥ 5 yielded the most extreme outlier the +2.870C trough-to-peak between 1700 and 1940. Given the outliers that were revealed, the data was stratified into two groups (0.990C - 1.770C or $\geq 1.770C$). This stratification yielded 2 normally distributed groups (Group-1, N=5, Group-2 N=34), that were, statistically, significantly different from one another (unpaired Welch T-Test, 2-tailed P-value = 0.007). Group 1's smallest temperature rise was 0.210C greater than Group 2's largest temperature rise, highlighting the gap between the two groups. On the basis of the above, the peak-to-trough temperature rise from 1700 to 1940 (+2.870C) was confirmed as the most significant outlier. This process was repeated for the grafted peak from 1840-2016 (+2.810C) as detailed in Figure 4.1. Group-1 swapped the +2.870C with the +2.810C, which was also statistically, significantly different from Group-2 (unpaired Welch T-Test, two-sided P-value = 0.0061). Conclusion: Group-1 (N=5) composed of trough-to-peak outliers $\geq 1.770C$ were significantly larger global warmings than Group-2 (N=34), and the +2.870C or +2.810C were the largest outliers.

343 Data: (1) B.M. Vinther et al., 2009, "Holocene thinning of the Greenland ice sheet." Nature, Vol. 461, pp. 385-388, 17 September 2009. National Centers for Environmental Information, NESDIS, NOAA, U.S. Department of Commerce. Greenland Ice Sheet Holocene d18O, Temperature, and Surface Elevation. doi:10.1038/nature08355. https://www.ncdc.noaa.gov/paleo-search/study/11148. Downloaded 05/05/2018. Personal Research: Groups 1 and 2 (previous citation) were compared by their magnitudes of decline from the temperature peak to see if there was a difference between them once the climate switched to a cooling phase. The time to reach the first post-peak trough, and to their maximum troughs was calculated. Each group had a normal distribution (d'Agostino-Pearson test and Shapiro-Wilks test: P>0.05) but a different variance. As such a Welch T-test (unpaired) was used to assess group differences. Results: Group 1 ($\geq 1.770C$ trough-to-peak) showed a mean temperature decline at the 1st trough after the peak of 1.920C versus Group 2's ($\leq 1.770C$ trough-to-peak) mean temperature decline of 1.030C, which represented a statistically significant difference in temperature

decline over Group 1 (2-tailed P-value = 0.0433). Group 1 showed a mean temperature decline at the maximum trough after the peak of 1.920C versus Group 2's mean temperature decline of 1.230C, but this difference was not significantly different (-0.690C, P-value 0.0784). Moreover, Group 1 rapidly declined such that its first post-peak trough was the same as its maximum trough i.e., Group 1 temperature fell abruptly. Group 2 showed a difference between its first and maximum trough of -0.200C, which was significantly different (P-value = 0.001928). Group 1 took two intervals (i.e., 45 years) to drop -1.920C with its first and maximum trough being the same (-1.920C). By contrast, Group 2 took on mean 1.82 intervals (i.e., 36 years) to reach its first trough and 3.15 intervals (i.e., 63 years) to reach its deepest trough. Conclusion: The higher the preceding trough-to-peak temperature rise (statistical outlier, or tall temperature peaks) the greater and more abrupt the temperature falls to near its maximum trough when the climate switches.

344 Olga N. Solomina et al., "Holocene glacier fluctuations." Quaternary Science Reviews. Volume 111, 2015, 9-34. https://doi.org/10.1016/j.quascirev.2014.11.018.

345 C. Andersen et al., "A highly unstable Holocene climate in the subpolar North Atlantic: evidence from diatoms." Quaternary Science Reviews, Volume 23, Issues 20–22, 2004, 2155-2166. https://doi.org/10.1016/j.quascirev.2004.08.004.

346 H. Wanner et al., "Structure and origin of Holocene cold events." Quaternary Science Reviews (2011), doi:10.1016/j.quascirev.2011.07.010.

347 T. Kobashi et al., 2013, "Causes of Greenland temperature variability over the past 4000 year: implications for northern hemispheric temperature changes." Climate of the Past, 9(5), 2299-2317. doi: 10.5194/cp-9-2299-2013. National Centers for Environmental Information, NESDIS, NOAA, U.S. Department of Commerce. Northern Hemisphere 4000 Year Temperature Reconstructions. https://www.ncdc.noaa.gov/paleo/study/15535. Downloaded 05/05/2018.

348 Manfred Mudelsee, "The phase relations among atmospheric CO2 content, temperature and global ice volume over the past 420 ka." Quaternary Science Reviews 20 (2001) 583-58.

349 Eric Monnin et al., "Atmospheric CO2 Concentrations over the Last Glacial Termination." By Science 05 Jan 2001: 112-114.

350 N. Caillon et al., 2003, "Timing of atmospheric CO2 and Antarctic temperature changes across Termination III." Science 299: 1728-1731.

351 H. Fischer et al., 1999, "Ice core records of atmospheric CO2 around the last three glacial terminations." Science, 283 , 1712-1714.

352 Ole Humlum et al., "The phase relation between atmospheric carbon dioxide and global temperature." Global and Planetary Change. Volume 100, January 2013, 51-69.

353 R.C. Finkel and K. Nishiizumi, 1997, "Beryllium 10 concentrations in the Greenland Ice Sheet Project 2 ice core from 3–40 ka." J. Geophys. Res., 102(C12), 26699–26706, doi: 10.1029/97JC01282.

354 Data: (1) A.M. Berggren et al., 2009, "A 600-year annual 10Be record from the NGRIP ice core, Greenland." Geophysical Research Letters, 36, L11801, doi:10.1029/2009GL038004. National Centers for Environmental Information, NESDIS, NOAA, U.S. Department of Commerce. North GRIP - 600 Year Annual 10Be Data. https://www.ncdc.noaa.gov/paleo-search/study/8618. Downloaded 05/05/2018. (2) T. Kobashi et al., 2013, "Causes of Greenland temperature variability over the past 4000 years: implications for northern hemispheric temperature changes." Climate of the Past, 9(5), 2299-2317. doi: 10.5194/cp-9-2299-2013. National Centers for Environmental Information, NESDIS, NOAA, U.S. Department of Commerce. Northern Hemisphere 4000 Year Temperature Reconstructions. https://www.ncdc.noaa.gov/paleo/study/15535. Downloaded 05/05/2018. (3) Etheridge, D.M., et al., 2001, "Law Dome Atmospheric CO2 Data," IGBP PAGES/World Data Center for Paleoclimatology. Data Contribution Series #2001-083. NOAA/NGDC Paleoclimatology Program, Boulder CO, USA. https://www1.ncdc.noaa.gov/pub/data/paleo/icecore/antarctica/law/law2006.xls. Downloaded 28 August 2018. Personal Research: A 18-year moving average of the Beryllium-10 concentration anomaly (relative to the 1960-1986 average) and carbon dioxide concentration anomaly (relative to the 1961-1990 average) were rendered from the raw data and plotted against the Northern Hemisphere temperature anomaly (relative to the 1961-1990 average) to create Figures 4.3.A and B.]

355   Michael E Mann, "Little Ice Age." Volume 1, The Earth system: physical and chemical dimensions of global environmental change, 504–509, citing see Bradley and Jones, 1993; Pfister, 1995

356   G.H. Miller et al., "Temperature and precipitation history of the Arctic." Quaternary Science Reviews, Volume 29, Issues 15–16, 2010. 1679-1715. https://doi.org/10.1016/j.quascirev.2010.03.001.

357   O.N. Solomina et al., 2016, "Glacier fluctuations during the past 2000 years." Quaternary Science Reviews, 149, 61-90. DOI: 10.1016/j.quascirev.2016.04.008. [See Figure 5, page 276. This figure collates a stacked time series of the number of glacier advances and recessions in each region into a global total.].

358   Data: (1) A.M. Berggren et al., 2009, "A 600-year annual 10Be record from the NGRIP ice core, Greenland." Geophysical Research Letters, 36, L11801, doi:10.1029/2009GL038004. National Centers for Environmental Information, NESDIS, NOAA, U.S. Department of Commerce. North GRIP - 600 Year Annual 10Be Data. https://www.ncdc.noaa.gov/paleo-search/study/8618. Downloaded 05/05/2018. (2) T. Kobashi et al., 2013, "Causes of Greenland temperature variability over the past 4000 year: implications for Northern Hemispheric temperature changes." Climate of the Past, 9(5), 2299-2317. doi: 10.5194/cp-9-2299-2013. National Centers for Environmental Information, NESDIS, NOAA, U.S. Department of Commerce. Northern Hemisphere 4000 Year Temperature Reconstructions. https://www.ncdc.noaa.gov/paleo/study/15535. Downloaded 05/05/2018. Statistics Software Utilized: Spearman rank calculator utilized: Wessa P., (2017), Spearman Rank Correlation (v1.0.3) in Free Statistics Software (v1.2.1), Office for Research Development and Education, URL https://www.wessa.net/rwasp_spearman.wasp/. Personal Research: Figure 4.4.A.) Spearman rank correlation r= -0.76, two-tailed P-value = <0.00001, N=484 annual pairings. Both the Northern Hemisphere temperature and the Beryllium-10 concentration anomaly (18-year moving average) anomalies were not normally distributed, though they did not contain outliers. A scatter plot of the data indicated a linear relationship. A Spearman rank correlation was utilized given the non-normal distributions. The correlation was optimized using an 18-year moving average Beryllium-10 concentration anomaly. This 18-year moving average was selected using the scatterplot trend line in Microsoft Excel to maximize the R-squared (versus an 11-year, 5-year, and no moving average). Figure 4.4.B) A Spearman rank correlation r= -0.876, two-tailed P-value = <0.00001, N=205 annual pairings. A Pearson correlation r= -0.91, two-tailed P-value = <0.00001, N=205. The grand solar minima temperature decline phases and their corresponding 18-year trailing average Beryllium-10 data were extracted from the full data set and compiled into a single time series (as linked sequential periods). Each grand solar minimum period was analyzed as a stand-alone grand solar minimum data set (Data not shown) and as fusion of four grand solar minima. The results and conclusion are the same. The temperature data is normally distributed. The 18-year moving average Beryllium-10 concentration anomaly is not normally distributed, indicated by a d'Agostino-Pearson test that yielded a p=0.019, indicating a non-normal distribution. However, the scatter plot demonstrates a linear relationship, and there were no outliers. The correlation was optimized using a 18-year moving average Beryllium-10 concentration anomaly, selected using the scatterplot trend line in Microsoft Excel to maximize the R-squared (versus an 11-year, 5-year, and no moving average). Note: A Pearson correlation was also calculated for both data sets supporting Figures 4.4.A and B, yielding a similar level of correlation, statistical significance, and the same conclusion (Data not shown).

359   N. Scafetta, "Multi-scale harmonic model for solar and climate cyclical variation throughout the Holocene based on Jupiter-Saturn tidal frequencies plus the 11-year solar dynamo cycle." Journal of Atmospheric and Solar-Terrestrial Physics (2012). doi:10.1016/j.jastp.2012.02.016.

360   Theodor Landscheidt, "New Little Ice Age Instead of Global Warming?" Energy & Environment. 2003. Volume 14, Issue 2, 327 – 350. https://doi.org/10.1260/095830503765184646.

361   R.J. Salvador, "A mathematical model of the sunspot cycle for the past 1000 years." Pattern Recognition Physics, 1, 117-122, doi:10.5194/prp-1-117-2013, 2013.

362   Habibullo Abdussamatov, "Current Long-Term Negative Average Annual Energy Balance of the Earth Leads to the New Little Ice age." Thermal Science. 2015 Supplement, Volume 19, S279-S288.

363 Jan-Erik Solheim, https://www.mwenb.nl/wp-content/uploads/2014/10/Blog-Jan-Erik-Solheim-def.pdf. Referred from http://www.climatedialogue.org/what-will-happen-during-a-new-maunder-minimum/. Citing blog for 4-5 solar-climate experts.

364 Boncho P. Bonev et al., "Long-Term Solar Variability and the Solar Cycle in the 21st Century." The Astrophysical Journal, 605:L81–L84, 2004 April 10.

365 Nils-Axel Mörner, "Solar Minima, Earth's rotation and Little Ice Ages in the past and in the future. The North Atlantic–European case." Global and Planetary Change 72 (2010) 282–293. doi:10.1016/j.gloplacha.2010.01.004.

366 N. Scafetta, "Multi-scale harmonic model for solar and climate cyclical variation throughout the Holocene based on Jupiter-Saturn tidal frequencies plus the 11-year solar dynamo cycle." Journal of Atmospheric and Solar-Terrestrial Physics (2012). doi:10.1016/j.jastp.2012.02.016.

367 Habibullo Abdussamatov, "Current Long-Term Negative Average Annual Energy Balance of the Earth Leads to the New Little Ice age." Thermal Science. 2015 Supplement, Volume 19, S279-S288.

368 N.A. Mörner et al., "General conclusions regarding the planetary–solar–terrestrial interaction." Pattern Recognition Physics, 1, 205–206, 2013. www.pattern-recogn-phys.net/1/205/2013/. doi:10.5194/prp-1-205-2013.

369 J.E. Solheim, "The sunspot cycle length – modulated by planets?" Pattern Recognition Physics, 1, 159–164, 2013. www.pattern-recogn-phys.net/1/159/2013/. doi:10.5194/prp-1-159-2013.

370 Wilson, I. R. G et al., "Does a Spin-Orbit Coupling Between the Sun and the Jovian Planets Govern the Solar Cycle?" Astronomical Society of Australia, Volume 25, Issue 2, 85-93. DOI:10.1071/AS06018.

371 Katya Georgieva, "Effects of interplanetary disturbances on the Earth's atmosphere and climate." http://www.issibern.ch/teams/interplanetarydisturb/wp-content/uploads/2014/01/proposal.pdf.

372 See endnote 181.

373 K.J. Anchukaitis et al., 2010, "The influence of volcanic eruptions on the climate of the Asian monsoon region." Geophysical Research Letters, 37, L22703, doi:10.1029/2010GL044843.

374 Feng Shi et al., "A tree-ring reconstruction of the South Asian summer monsoon index over the past millennium." Scientific Reports Volume 4, Article number: 6739 (2014). DOI: 10.1038/srep06739.

375 Z. Zhuo et al., 2014, "Proxy evidence for China's monsoon precipitation response to volcanic aerosols over the past seven centuries." Journal of Geophysical Research Atmos., 119, 6638–6652, doi:10.1002/2013JD021061.

376 Chaochao Gao, "Volcanic monsoon influence revealed from multi-proxy evidence." PAGES Magazine. Science Highlights. Volcanoes and Climate. Volume 23, No 2, December 2015.

377 C. Shen et al., "Exceptional drought events over eastern China during the last five centuries." Climatic Change (2007) 85: 453. https://doi.org/10.1007/s10584-007-9283-y.

378 Jim M. Haywood et al., "Asymmetric forcing from stratospheric aerosols impacts Sahelian rainfall." Nature Climate Change Volume 3, 660–665 (2013).

379 L. Oman et al., 2006, "High-latitude eruptions cast shadow over the African monsoon and the flow of the Nile." Geophysical Research Letters, 33, L18711, doi:10.1029/2006GL027665.

380 K.E. Trenberth and A. Dai, 2007, "Effects of Mount Pinatubo volcanic eruption on the hydrological cycle as an analog of geoengineering." Geophysical Research Letters, 34, L15702, doi:10.1029/2007GL030524.

381 Drew T. Shindell et al., "Dynamic winter climate response to large tropical volcanic eruptions since 1600." Journal of Geophysical Research. Volume 109, D05104, doi:10.1029/2003JD004151, 2004.

382 E.M. Fischer et al., 2007, "European climate response to tropical volcanic eruptions over the last half millennium." Geophysical Research Letters, 34, L05707, doi:10.1029/2006GL027992.

383 Ben B. B. Booth et al., "Aerosols implicated as a prime driver of twentieth-century North Atlantic climate variability." Nature Volume 484, 228–232 (2012). DOI:10.1038/nature10946.

384 Mignot, J., Khodri et al., "Volcanic impact on the Atlantic Ocean over the last millennium." Climate of the Past, 7, 1439-1455, https://doi.org/10.5194/cp-7-1439-2011, 2011.

385 M. Sigl et al., 2015, "Timing and climate forcing of volcanic eruptions for the past 2,500 years." Nature, 523(7562), 543–549. DOI:10.1038/nature14565.

386 Michael J. Puma et al., "Exploring the potential impacts of historic volcanic eruptions on the contemporary global food system." PAGES Magazine. Science Highlights. Volcanoes and Climate. Volume 23, No 2, December 2015.

387 C. Newhall et al., 2018, "Anticipating future Volcanic Explosivity Index (VEI) 7 eruptions and their chilling impacts." Geosphere, v. 14, no. 2, p. 1–32, doi:10.1130/GES01513.1.

388 J. Slawinska and A. Robock, 2018, "Impact of Volcanic Eruptions on Decadal to Centennial Fluctuations of Arctic Sea Ice Extent during the Last Millennium and on Initiation of the Little Ice Age." J. Climate, 31, 2145–2167, https://doi.org/10.1175/JCLI-D-16-0498.1.

389 G.H. Miller et al., 2012, "Abrupt onset of the Little Ice Age triggered by volcanism and sustained by sea-ice/ocean feedbacks." Geophysical Research Letters, 39, L02708, doi:10.1029/2011GL050168.

390 F. Lehner et al., 2013, "Amplified inception of European Little Ice Age by sea ice–ocean–atmosphere feedbacks." J. Climate, 26, 7586–7602. https://doi.org/10.1175/JCLI-D-12-00690.1.

391 C. Gao et al., 2008, "Volcanic forcing of climate over the past 1500 years: An improved ice core-based index for climate models." Journal of Geophysical Research, 113, D23111, doi: 10.1029/2008JD010239. [See Figure 2, page 5].

392 Y. Zhong et al., "Centennial-scale climate change from decadally-paced explosive volcanism: a coupled sea ice-ocean mechanism." Climate Dynamics (2011) 37: 2373. https://doi.org/10.1007/s00382-010-0967-z.

393 Odd Helge Otterå et al., "External forcing as a metronome for Atlantic multidecadal variability." Nature Geoscience Volume 3, 688–694 (2010).

394 A.V. Kurbatov et al., 2006, "A 12,000 year record of explosive volcanism in the Siple Dome Ice Core, West Antarctica." Journal of Geophysical Research, 111, D12307, doi:10.1029/2005JD006072.

395 C. Oppenheimer, 2003, "Ice core and paleoclimate evidence for the timing and nature of the great mid-13th century volcanic eruption." International Journal of Climatology, 23: 417-426. doi:10.1002/joc.891. [See Figure 1, 418, depicting the volcanic sulphate record in Greenland's GISP2 ice core].

396 O.N. Solomina et al., 2016, "Glacier fluctuations during the past 2000 years." Quaternary Science Reviews, 149, 61-90. DOI: 10.1016/j.quascirev.2016.04.008. [See Figure 5, 276. This figure collates a stacked time series of the number of glacier advances and recessions in each region into a global total.].

397 G.H. Miller et al., "Temperature and precipitation history of the Arctic." Quaternary Science Reviews, Volume 29, Issues 15–16, 2010. 1679-1715. https://doi.org/10.1016/j.quascirev.2010.03.001.

398 G.H. Miller et al., 2012, "Abrupt onset of the Little Ice Age triggered by volcanism and sustained by sea-ice/ocean feedbacks." Geophysical Research Letters, 39, L02708, doi:10.1029/2011GL050168.

399 E. Martin et al., 2014, "Volcanic sulfate aerosol formation in the troposphere." Journal of Geophysical Research Atmos., 119, 12,660–12,673, doi:10.1002/2014JD021915.

400 J.P. Vernier et al., 2011, "Major influence of tropical volcanic eruptions on the stratospheric aerosol layer during the last decade." Geophysical Research Letters, 38, L12807, doi: 10.1029/2011GL047563.

401 Alan Robock, "Volcanic Eruptions and Climate." Reviews of Geophysics, 38, 2 / May 2000. 191–219.

402 Bethan Harris, "The potential impact of super-volcanic eruptions on the Earth's atmosphere." Weather – August 2008, Volume 63, No. 8. DOI: 10.1002/wea.263. https://rmets.onlinelibrary.wiley.com/doi/pdf/10.1002/wea.263.

403 E.S. Martin et al., 2014, "Volcanic sulfate aerosol formation in the troposphere." Journal of Geophysical Research Atmos., 119, 12,660–12,673, doi:10.1002/2014JD021915.

404 C. Gao et al., 2008, "Volcanic forcing of climate over the past 1500 years: An improved ice core-based index for climate models." Journal of Geophysical Research, 113, D23111, doi: 10.1029/2008JD010239.

405 Didier Swingedouw et al., 2015, "Bidecadal North Atlantic ocean circulation variability controlled by timing of volcanic eruptions." Nature Communications. 6:6545 | DOI: 10.1038/ncomms7545.

406 D. Zanchettin et al., 2013, "Background conditions influence the decadal climate response to strong volcanic eruptions." Journal of Geophysical Research Atmos., 118, 4090–4106, doi:10.1002/jgrd.50229.

407 "Super-volcano; Global effects and future threats." Report of the Geological Society of London, a working group.

408 M. Sigl et al., 2015, "Timing and climate forcing of volcanic eruptions for the past 2,500 years." Nature, 523(7562), 543–549. DOI:10.1038/nature14565.

409 M. Sigl et al., 2015, "Timing and climate forcing of volcanic eruptions for the past 2,500 years." Nature, 523(7562), 543–549. DOI:10.1038/nature14565.

410 Alan Robock, "Volcanic Eruptions and Climate." Reviews of Geophysics, 38, 2 / May 2000. 191–219.

411 M. Sigl et al., 2015, "Timing and climate forcing of volcanic eruptions for the past 2,500 years." Nature, 523(7562), 543–549. DOI:10.1038/nature14565.

412 K.J. Anchukaitis et al., 2010, "The influence of volcanic eruptions on the climate of the Asian monsoon region." Geophysical Research Letters, 37, L22703, doi:10.1029/2010GL044843.

413 Feng Shi et al., "A tree-ring reconstruction of the South Asian summer monsoon index over the past millennium." Scientific Reports Volume 4, Article number: 6739 (2014). DOI: 10.1038/srep06739.

414 Z. Zhuo et al., 2014, "Proxy evidence for China's monsoon precipitation response to volcanic aerosols over the past seven centuries." Journal of Geophysical Research Atmos., 119, 6638–6652, doi:10.1002/2013JD021061.

415 Chaochao Gao, "Volcanic monsoon influence revealed from multi-proxy evidence." PAGES Magazine. Science Highlights. Volcanoes and Climate. Volume 23, No 2, December 2015.

416 C. Shen et al., "Exceptional drought events over eastern China during the last five centuries." Climatic Change (2007) 85: 453. https://doi.org/10.1007/s10584-007-9283-y.

417 Jim M. Haywood et al., "Asymmetric forcing from stratospheric aerosols impacts Sahelian rainfall." Nature Climate Change Volume 3, 660–665 (2013).

418 L. Oman et al., 2006, "High-latitude eruptions cast shadow over the African monsoon and the flow of the Nile." Geophysical Research Letters, 33, L18711, doi:10.1029/2006GL027665.

419 K.E. Trenberth and A. Dai, 2007, "Effects of Mount Pinatubo volcanic eruption on the hydrological cycle as an analog of geoengineering." Geophysical Research Letters, 34, L15702, doi:10.1029/2007GL030524.

420 Drew T. Shindell et al., "Dynamic winter climate response to large tropical volcanic eruptions since 1600." Journal of Geophysical Research. Volume 109, D05104, doi:10.1029/2003JD004151, 2004.

421 E.M. Fischer et al., 2007, "European climate response to tropical volcanic eruptions over the last half millennium." Geophysical Research Letters, 34, L05707, doi:10.1029/2006GL027992.

422 Ben B. B. Booth et al., "Aerosols implicated as a prime driver of twentieth-century North Atlantic climate variability." Nature Volume 484, 228–232 (2012). DOI:10.1038/nature10946.

423 J. Mignot et al, "Volcanic impact on the Atlantic Ocean over the last millennium." Climate of the Past, 7, 1439-1455, https://doi.org/10.5194/cp-7-1439-2011, 2011.

424 J. Slawinska and A. Robock, 2018, "Impact of Volcanic Eruptions on Decadal to Centennial Fluctuations of Arctic Sea Ice Extent during the Last Millennium and on Initiation of the Little Ice Age." J. Climate, 31, 2145–2167, https://doi.org/10.1175/JCLI-D-16-0498.1.

425 Didier Swingedouw et al., 2015, "Bidecadal North Atlantic ocean circulation variability controlled by timing of volcanic eruptions." Nature Communications. 6:6545 | DOI: 10.1038/ncomms7545.

426 D.O. Zanchettin et al., 2013, "Background conditions influence the decadal climate response to strong volcanic eruptions." Journal of Geophysical Research Atmos., 118, 4090–4106, doi:10.1002/jgrd.50229.

427 Y. Zhong et al., "Centennial-scale climate change from decadally-paced explosive volcanism: a coupled sea ice-ocean mechanism." Climate Dynamics (2011) 37: 2373. https://doi.org/10.1007/s00382-010-0967-z.

428 G.H. Miller et al., 2012, "Abrupt onset of the Little Ice Age triggered by volcanism and sustained by sea-ice/ocean feedbacks." Geophysical Research Letters, 39, L02708, doi:10.1029/2011GL050168.

429 C. Newhall et al., 2018, Anticipating future Volcanic Explosivity Index (VEI) 7 eruptions and their chilling impacts: Geosphere, v. 14, no. 2, p. 1–32, doi:10.1130/GES01513.1.

430 J. Slawinska and A. Robock, 2018, "Impact of Volcanic Eruptions on Decadal to Centennial Fluctuations of Arctic Sea Ice Extent during the Last Millennium and on Initiation of the Little Ice Age." J. Climate, 31, 2145–2167, https://doi.org/10.1175/JCLI-D-16-0498.1.

431 G. H. Miller et al., 2012, "Abrupt onset of the Little Ice Age triggered by volcanism and sustained by sea-ice/ocean feedbacks." Geophysical Research Letters, 39, L02708, doi:10.1029/2011GL050168.

432 F. Lehner et al., 2013, "Amplified inception of European Little Ice Age by sea ice–ocean–atmosphere feedbacks." J. Climate, 26, 7586–7602. https://doi.org/10.1175/JCLI-D-12-00690.1.

433 C. Gao et al., 2008, "Volcanic forcing of climate over the past 1500 years: An improved ice core-based index for climate models." Journal of Geophysical Research, 113, D23111, doi: 10.1029/2008JD010239. [See Figure 2, page 5].

434 Y. Zhong et al., "Centennial-scale climate change from decadally-paced explosive volcanism: a coupled sea ice-ocean mechanism." Climate Dynamics (2011) 37: 2373. https://doi.org/10.1007/s00382-010-0967-z.

435 J. Slawinska and A. Robock, 2018, "Impact of Volcanic Eruptions on Decadal to Centennial Fluctuations of Arctic Sea Ice Extent during the Last Millennium and on Initiation of the Little Ice Age." J. Climate, 31, 2145–2167, https://doi.org/10.1175/JCLI-D-16-0498.1.

436 V. Bucha, "Geomagnetic activity and the North Atlantic Oscillation." Studia Geophysica et Geodaetica. July 2014, Volume 58, Issue 3, 461–472. https://doi.org/10.1007/s11200-014-0508-z.

437 J. G. Pinto and C. C. Raible, 2012, "Past and recent changes in the North Atlantic oscillation." WIREs Climate Change, 3: 79-90. doi:10.1002/wcc.150.

438 Jesper Olsen et al., "Variability of the North Atlantic Oscillation over the past 5,200 years." Nature Geoscience Volume 5, 808–812 (2012). DOI: 10.1038/NGEO1589.

439 P. Thejll et al., "On correlations between the North Atlantic Oscillation, geopotential heights, and geomagnetic activity." Geophysical Research Letters, 30 (6), 1347, 2003. doi:10.1029/2002GL016598.

440 J.W. Hurrell et al., 2013, "An Overview of the North Atlantic Oscillation." In The North Atlantic Oscillation: Climatic Significance and Environmental Impact (eds J. W. Hurrell, Y. Kushnir, G. Ottersen and M. Visbeck). doi:10.1029/134GM01.

441 Jesper Olsen et al., "Variability of the North Atlantic Oscillation over the past 5,200 years." Nature Geoscience Volume 5, 808–812 (2012). DOI: 10.1038/NGEO1589.

442 A. Mazzarella and N. Scafetta, 2012, "Evidences for a quasi 60-year North Atlantic Oscillation since 1700 and its meaning for global climate change." Theoretical Applied Climatology 107, 599-609. DOI: 10.1007/s00704-011-0499-4.

443 T. Bradwell et al., 2006, "The Little Ice Age glacier maximum in Iceland and the North Atlantic Oscillation: evidence from Lambatungnajökull, southeast Iceland." Boreas, 35: 61-80. doi:10.1111/j.1502-3885.2006.tb01113.x.

444 Jesper Olsen et al., "Variability of the North Atlantic Oscillation over the past 5,200 years." Nature Geoscience Volume 5, 808–812 (2012). DOI: 10.1038/NGEO1589.

445 M.H. Ambaum and B.J. Hoskins, 2002, "The NAO Troposphere–Stratosphere Connection." J. Climate, 15, 1969–1978, https://doi.org/10.1175/1520-0442(2002)015<1969:TNTSC>2.0.CO;2.

446 V. Bucha, "Geomagnetic activity and the North Atlantic Oscillation." Studia Geophysica et Geodaetica. July 2014, Volume 58, Issue 3, 461–472. https://doi.org/10.1007/s11200-014-0508-z.

447 P. B. Thejll et al., "On correlations between the North Atlantic Oscillation, geopotential heights, and geomagnetic activity." Geophysical Research Letters, 30 (6), 1347, 2003. doi:10.1029/2002GL016598.

448 H. Lu et al., 2008, "Possible solar wind effect on the northern annular mode and northern hemispheric circulation during winter and spring." Journal of Geophysical Research, 113, D23104, doi: 10.1029/2008JD010848.

449 V. Bucha, "Geomagnetic activity and the North Atlantic Oscillation." Studia Geophysica et Geodaetica. July 2014, Volume 58, Issue 3, 461–472. https://doi.org/10.1007/s11200-014-0508-z.

450 J.W. Hurrell et al., 2013, "An Overview of the North Atlantic Oscillation." In The North Atlantic Oscillation: Climatic Significance and Environmental Impact (eds J. W. Hurrell, Y. Kushnir, G. Ottersen and M. Visbeck). doi:10.1029/134GM01.

451 J.W. Hurrell et al., 2013, "An Overview of the North Atlantic Oscillation." In The North Atlantic Oscillation: Climatic Significance and Environmental Impact (eds J. W. Hurrell, Y. Kushnir, G. Ottersen and M. Visbeck). doi:10.1029/134GM01. [Citing Robock and Mao, 1992; Kodera, 1994; Graf et al., 1994; Kelley et al., 1996.].

452 J. Slawinska and A. Robock, 2018, "Impact of Volcanic Eruptions on Decadal to Centennial Fluctuations of Arctic Sea Ice Extent during the Last Millennium and on Initiation of the Little Ice Age." J. Climate, 31, 2145–2167, https://doi.org/10.1175/JCLI-D-16-0498.1.

453 F. Lehner et al., 2013, "Amplified inception of European Little Ice Age by sea ice–ocean–atmosphere feedbacks." J. Climate, 26, 7586–7602. https://doi.org/10.1175/JCLI-D-12-00690.1.

454 C. Newhall et al., 2018, "Anticipating future Volcanic Explosivity Index (VEI) 7 eruptions and their chilling impacts." Geosphere, v. 14, no. 2, p. 1–32, doi:10.1130/GES01513.1.

455 Odd Helge Otterå et al., "External forcing as a metronome for Atlantic multidecadal variability." Nature Geoscience Volume 3, 688–694 (2010).

456 Y. Zhong et al., "Centennial-scale climate change from decadally-paced explosive volcanism: a coupled sea ice-ocean mechanism." Climate Dynamics (2011) 37: 2373. https://doi.org/10.1007/s00382-010-0967-z.

457 Takuro Kobashi et al., 2017, "Volcanic influence on centennial to millennial Holocene Greenland temperature change." Scientific Reports, 7, 1441. doi: 10.1038/s41598-017-01451-7. Data provided by the National Centers for Environmental Information, NESDIS, NOAA, U.S. Department of Commerce. https://www.ncdc.noaa.gov/paleo-search/study/22057. Data accessed 21/08/2018.

458 Data: (1) Helen Sian Crosweller et al., "Global database on large magnitude explosive volcanic eruptions (LaMEVE)." Journal of Applied Volcanology Society and Volcanoes 20121:4. https://doi.org/10.1186/2191-5040-1-4. Volcano Global Risk Identification and Analysis Project database (VOGRIPA), British Geological Survey. Data Access: http://www.bgs.ac.uk/vogripa/. Data downloaded 07/05/2018. (2) S.K. Solanki et al., 2004, "An unusually active Sun during recent decades compared to the previous 11,000 years." Nature, Volume 431, No. 7012, 1084-1087, 28 October 2004. Data: S.K. Solanki et al., 2005, "11,000 Year Sunspot Number Reconstruction." IGBP PAGES/World Data Center for Paleoclimatology. Data Contribution Series #2005-015. NOAA/NGDC Paleoclimatology Program, Boulder CO, USA. https://www.ncdc.noaa.gov/paleo-search/study/5780. Downloaded 05/06/2018. Personal Research: A total of 67 VEI 6 and 7 eruptions were extracted from the LaMEVE database. These were plotted alongside the above-cited Solanki et al. sunspot numbers. The number of 10-year periods was counted from each eruption to the previous or next sunspot number peak or trough. The data is tabulated above, at the start of the endnotes and referencing this endnote. Results: 82 percent of VEI 6-7 eruptions occurred at or within one decade of a sunspot number peak or trough. This peak and trough occurrence coincides with either a grand solar maximum or minimum, or a smaller peak or sub-trough of sunspot numbers.

459 Data: (1) Takuro Kobashi et al., 2017, "Volcanic influence on centennial to millennial Holocene Greenland temperature change." Scientific Reports, 7, 1441. doi: 10.1038/s41598-017-01451-7. Data provided by the National Centers for Environmental Information, NESDIS, NOAA, U.S. Department of Commerce. https://www.ncdc.noaa.gov/paleo-search/study/22057. Data accessed 21/08/2018. (2) Solanki, S.K., et al. 2004. "An unusually active Sun during recent decades compared to the previous 11,000 years." Nature, Volume 431, No. 7012, 1084-1087, 28 October 2004. Data: Solanki, S.K., et al. 2005. "11,000 Year Sunspot Number Reconstruction." IGBP PAGES/World Data Center for Paleoclimatology. Data Contribution Series #2005-015. NOAA/NGDC Paleoclimatology Program, Boulder CO, USA. https://www.ncdc.noaa.gov/paleo-search/study/5780. Downloaded 05/06/2018. Personal Research: Figure 5.1.A: Using the above-cited climate-forcing volcanic eruption data a quantitative filter was utilized to identify the largest climate forcing eruptions, and to group all eruption events into climate-forcing categories. Each volcanic eruption started with the first data point in a group series, and this group series magnitude was represented by the maximum volcanic forcing magnitude data point (i.e., the most negative Watts/meter-squared value) for that group series (i.e., a 1-year value from within a range of 1-10 years). This was completed for the entire time series (11,054 years). In this manner 403 volcanic events were identified over 11,054 years. The eruption events were preliminarily assigned to groups based on their maximum solar forcing impact,

as follows: Group-1, $\leq$-10 W/m2 (N=23). Group-2, -5 to <-9.99 W/m2 (N=50). Group-3, -2 to <-4.99 W/m2 (N=89). Group-4, 0 to <-1.99 W/m2 (N=241). Volcanic events were then grouped and compiled into 500, 400, and 300 year bin totals spanning the last 5,000, 8,000, and 11,000 years. The average sunspot numbers were calculated for each bin period. A goodness of fit and outlier tests were conducted for all groupings. Pearson and Spearman rank correlations and their significance levels were calculated for each 5,000, 8,000, and 11,000 year periods to help understand if significant relationships existed or not. Results: The 500-year bin totals generated the highest and most significant correlations, and the 8,000 and 5,000 year periods maximized the correlation coefficients. The correlation values were reduced for 11,000-year period versus the 8,000-year period, and were marginally smaller for 400-year bins, and much smaller for 300-year bins (Data not shown) compared with the 500-year bins. On this basis, the 8,000-year duration and 500-year bin totals represented the optimum grouping which maximized the duration of the relationship i.e., since the Holocene Climate Optimum. The 8,000-year data summary is tabulated above (at the start of the endnotes, referencing this endnote). The outcome of this analysis was to compile Groups 1 and 2 into a single group and set the climate forcing eruption threshold at $\leq$ -5.0 Watts/meter-squared i.e., large volcanic eruptions. All 73 large climate-forcing volcanic eruptions were plotted against the above-cited Solanki et al. sunspot numbers to produce Figure 5.1.A's graphic. Figure 5.1.B: The 73 climate-forcing eruptions selected above were tabulated alongside the above-cited Solanki et al. sunspot numbers in the year of the eruption's occurrence. The number of periods (at a 10-year resolution) was counted to the previous or next big (grand solar) and small (sub-) peak or trough for all eruption events. An eruption was then assigned to a big or little peak or trough based on its closest proximity to one of those events. Each eruption was only counted once. The data was used to derive Figure 5.1.B, and is tabulated above at the start of the endnotes (referencing this endnote).

460 R. B. Alley et al., "Holocene climatic instability: A prominent, widespread event 8200 year ago." Geology ; 25 (6): 483–486. doi: https://doi.org/10.1130/0091-7613(1997)025<0483:HCIAP W>2.3.CO;2.
461 Kaarina Sarmaja-Korjonen and H. Seppa, 2007, "Abrupt and consistent responses of aquatic and terrestrial ecosystems to the 8200 cal. year cold event: a lacustrine record from Lake Arapisto, Finland". The Holocene 17 (4): 457–467. doi:10.1177/0959683607077020.
462 D.C. Barber et al., 1999, "Forcing of the cold event of 8,200 years ago by catastrophic drainage of Laurentide lakes." Nature Volume 400, 344–348 (22 July 1999). doi:10.1038/22504.
463 Christopher R W Ellison et al., 2006, "Surface and Deep Ocean Interactions During the Cold Climate Event 8200 Years Ago." Science. 2006 Jun 30;312(5782):1929-32. DOI10.1126/science.1127213.
464 Data: (1) Helen Sian Crosweller et al., "Global database on large magnitude explosive volcanic eruptions (LaMEVE)." Journal of Applied Volcanology Society and Volcanoes 20121:4. https://doi.org/10.1186/2191-5040-1-4. Volcano Global Risk Identification and Analysis Project database (VOGRIPA), British Geological Survey. Data Access: http://www.bgs.ac.uk/vogripa/. Data downloaded 07/05/2018. (2) S.K. Solanki et al., 2004, "An unusually active Sun during recent decades compared to the previous 11,000 years." Nature, Volume 431, No. 7012, 1084-1087, 28 October 2004. Data: Solanki, S.K., et al. 2005. 11,000 Year Sunspot Number Reconstruction. IGBP PAGES/World Data Center for Paleoclimatology. Data Contribution Series #2005-015. NOAA/NGDC Paleoclimatology Program, Boulder CO, USA. https://www.ncdc.noaa.gov/paleo-search/study/5780. Downloaded 05/06/2018. (3) Takuro Kobashi et al., 2017, "Volcanic influence on centennial to millennial Holocene Greenland temperature change." Scientific Reports, 7, 1441. doi: 10.1038/s41598-017-01451-7. Data provided by the National Centers for Environmental Information, NESDIS, NOAA, U.S. Department of Commerce. https://www.ncdc.noaa.gov/paleo-search/study/22057. Data accessed 21/08/2018. Personal Research: (1) Figure 5.2.A: The 11 total VEI 6 and 7 eruptions between 1235 and 1885 were extracted from the LaMEVE database and graphically plotted as discrete events on the above-cited Solanki et al. sunspot number data within this same period. In this manner, the occurrence of VEI 6-7 eruptions can be viewed relative to the grand solar maximum or minimum, or a smaller sub-peak or sub-trough of sunspot numbers going into or coming out of a grand solar minimum trough. (2) Figure 5.2.B: Two periods running from grand solar

maxima-to-minima-to-maxima were extracted from the above-cited Solanki et al. sunspot number data. The corresponding climate forcing volcanic eruptions from the Takuro Kobashi, et al. volcanic eruption data (the same as utilized for Figure 5.1.A) were plotted in the periods that they occurred. This highlights the association of large climate-forcing volcanic eruptions with either a grand solar maximum or minimum, or a smaller sub-peak or sub-trough of sunspot numbers going into or coming out of a grand solar minimum.

465 I.G.M. Usoskin et al., "Solar activity, cosmic rays, and Earth's temperature: A millennium-scale comparison." Journal of Geophysical Research, 110, A10102, doi:10.1029/2004JA010946. [Exposé: See page 1. This tells us cosmogenic isotopes (Beryllium-10, Carbon-14) are used as proxies for solar activity, and that their production is caused by galactic cosmic ray flux, which is influenced by the solar system's (heliospheric) magnetic field and is modulated by solar activity. Comment: Magnetized solar wind modulates the solar system's magnetic shield (i.e., the heliosphere) and the earth's magnetic shield (i.e. the magnetosphere), thereby regulating cosmic ray entry into the solar system and the earth system respectively. Cosmic ray entry into the upper atmosphere from space is modulated by solar activity and geomagnetism. Lower solar activity and lower geomagnetism permit more cosmic ray entry into the atmosphere, and conversely. Increased cosmic ray levels are associated with increased low-cloud formation, which is associated with planetary cooling, and conversely. The cosmic ray and low-cloud cooling effect are concentrated into the polar regions. Cosmogenic isotopes (Carbon-14, Beryllium-10) are generated by cosmic rays in the atmosphere, with more cosmic rays generating more cosmogenic isotopes, and conversely. Cosmogenic isotopes are then embedded in earth repositories (i.e., tree rings, ice cores) and therefore indirectly tell us about solar activity and the resulting magnetized solar wind that contacts the earth's magnetosphere. By utilizing cosmogenic isotopes to assess relationships between the sun and earth systems (i.e., climate, volcanism) we know that the solar activity that is being assessed is magnetism based, and not electromagnetism (i.e. not solar irradiance).].

466 Data: (1) Helen Sian Crosweller et al., "Global database on large magnitude explosive volcanic eruptions (LaMEVE)." Journal of Applied Volcanology Society and Volcanoes 20121:4. https://doi.org/10.1186/2191-5040-1-4. Volcano Global Risk Identification and Analysis Project database (VOGRIPA), British Geological Survey. Data Access: http://www.bgs.ac.uk/vogripa/. Data downloaded 07/05/2018. (2) S.K. Solanki et al., 2004, "An unusually active Sun during recent decades compared to the previous 11,000 years." Nature, Volume 431, No. 7012, 1084-1087, 28 October 2004. Data: S.K. Solanki et al., 2005, "11,000 Year Sunspot Number Reconstruction." IGBP PAGES/World Data Center for Paleoclimatology. Data Contribution Series #2005-015. NOAA/NGDC Paleoclimatology Program, Boulder CO, USA. https://www.ncdc.noaa.gov/paleo-search/study/5780. Downloaded 05/06/2018. Personal Research: Utilizing VOGRIPA's LaMEVE VEI 4-7 eruption events, these were grouped into 500-year bins from 1899 and back over the prior 5,000 years. The above cited Solanki, S.K., et al. was used to calculate 500-year average sunspot numbers. Using Microsoft Excel scatter plots were created, and various trend lines were fitted to the data. The power trend best optimized the R-squared value; (1) Power 0.803 versus (2) Exponential 0.748, (3) Logarithmic 0.713, and (4) Linear 0.639. The significant non-linear expansion in the number of large magnitude volcanic eruptions observed during the period 1400 to 1899 CE (i.e., the Little Ice Age) corresponded with the lowest 500-year average sunspot number in 7,000 years (mean of 15 sunspots). Cautionary Note: See the following citation (S.K. Brown et al., 2014) for an analysis-critique of the VOGRIPA database's recognized underreporting bias. This inadvertent underreporting of eruptions theoretically skews the data, so a higher incidence of volcanic eruptions or a growing frequency is more "apparent" over the last millennium. The VOGRIPA database represents the best of its kind and compiles numerous other databases. This LaMEVE database skewing gives the impression of an acceleration effect in the frequency of VEI 4-7 eruptions over the last 1,000 years compared with the prior 10,000 years and 2.6 million year period. This theoretically confounds the interpretation of the result, meriting caution with its interpretation. However, given that the VOGRIPA data is paired with a declining long-term trend in solar activity to derive this high R-squared non-linear correlation, the VOGRIPA data derived result should not be fully dismissed.].

467   S.K. Brown et al., "Characterization of the Quaternary eruption record: analysis of the Large Magnitude Explosive Volcanic Eruptions (LaMEVE) database." J Appl. Volcanology. (2014) 3: 5. https://doi.org/10.1186/2191-5040-3-5.

468   <u>Data</u>: (1) Takuro Kobashi et al., 2017, "Volcanic influence on centennial to millennial Holocene Greenland temperature change." Scientific Reports, 7, 1441. doi: 10.1038/s41598-017-01451-7. Data provided by the National Centers for Environmental Information, NESDIS, NOAA, U.S. Department of Commerce. https://www.ncdc.noaa.gov/paleo-search/study/22057. Data accessed 21/08/2018. (2) S.K. Solanki et al., 2004, "An unusually active Sun during recent decades compared to the previous 11,000 years." Nature, Volume 431, No. 7012, 1084-1087, 28 October 2004. Data: S.K. Solanki et al., 2005, "11,000 Year Sunspot Number Reconstruction." IGBP PAGES/World Data Center for Paleoclimatology. Data Contribution Series #2005-015. NOAA/NGDC Paleoclimatology Program, Boulder CO, USA. https://www.ncdc.noaa.gov/paleo-search/study/5780. Downloaded 05/06/2018. <u>Personal Research</u>: Using the above-cited data and the methodology cited in Figure 5.1.A, the largest climate-forcing volcanic eruptions ($\leq$5 Watts/meter-squared) were grouped into 500-year bin totals starting in 1895 and extending back 8,000 years. Five hundred-year average sunspot numbers were generated for the corresponding periods. Figure 5.3.A: Both previously derived parameters were plotted against one another and a two-period moving average created to highlight the inverse relationship. Figure 5.3.B: Both previously derived parameters were plotted using a scatter plot (Microsoft Excel) and a linear trend line fitted. Pearson and Spearman rank correlations were calculated, with both yielding a correlation coefficient $r = -0.72$, two-tailed P-value 0.002 (N=43 eruptions organized into 16 groups). There were no outliers for either parameter. A goodness of fit using the Shapiro-Wilks test indicated the 500-year sunspot number averages were normally distributed. The 500-year bin totals of volcanic eruptions yielded a $P = 0.031$ indicating a non-normal distribution, hence the Spearman rank correlation inclusion.

469   J. Slawinska and A. Robock, 2018, "Impact of Volcanic Eruptions on Decadal to Centennial Fluctuations of Arctic Sea Ice Extent during the Last Millennium and on Initiation of the Little Ice Age." J. Climate, 31, 2145–2167, https://doi.org/10.1175/JCLI-D-16-0498.1.

470   See endnote 461.

471   S.K. Solanki et al., 2004, "An unusually active Sun during recent decades compared to the previous 11,000 years." Nature, Volume 431, No. 7012, 1084-1087, 28 October 2004. Data: S.K. Solanki et al., 2005, 11,000 Year Sunspot Number Reconstruction. IGBP PAGES/World Data Center for Paleoclimatology. Data Contribution Series #2005-015. NOAA/NGDC Paleoclimatology Program, Boulder CO, USA. https://www.ncdc.noaa.gov/paleo-search/study/5780. Downloaded 05/06/2018.

472   J. Slawinska and A. Robock, 2018, "Impact of Volcanic Eruptions on Decadal to Centennial Fluctuations of Arctic Sea Ice Extent during the Last Millennium and on Initiation of the Little Ice Age." J. Climate, 31, 2145–2167, https://doi.org/10.1175/JCLI-D-16-0498.1.

473   F. Lehner et al., 2013, "Amplified inception of European Little Ice Age by sea ice–ocean–atmosphere feedbacks." J. Climate, 26, 7586–7602. https://doi.org/10.1175/JCLI-D-12-00690.1.

474   C. Newhall et al., 2018, "Anticipating future Volcanic Explosivity Index (VEI) 7 eruptions and their chilling impacts." Geosphere, v. 14, no. 2, p. 1–32, doi:10.1130/GES01513.1.

475   Odd Helge Otterå et al., "External forcing as a metronome for Atlantic multidecadal variability." Nature Geoscience Volume 3, 688–694 (2010).

476   Y. Zhong et al, "Centennial-scale climate change from decadally-paced explosive volcanism: a coupled sea ice-ocean mechanism." Climate Dynamics (2011) 37: 2373. https://doi.org/10.1007/s00382-010-0967-z.

477   Nils-Axel Mörner, "Solar Wind, Earth's Rotation and Changes in Terrestrial Climate." Physical Review & Research International 3(2): 117-136, 2013. [Citing multiple sources].

478   Adriano Mazzarella, "Solar Forcing of Changes in Atmospheric Circulation, Earth's Rotation and Climate." The Open Atmospheric Science Journal, 2008, 2, 181-184.

479   George C. Reid, "Solar Variability and the Earth's Climate: Introduction and Overview." Space Science Reviews (2000) 94: 1. https://doi.org/10.1023/A:1026797127105. [See pages 2-4, Figure 1].

480   E. Friis-Christensen, K. Lassen, "Length of the Solar Cycle: An Indicator of Solar Activity Closely Associated with Climate." Science, New Series, Volume 254, No. 5032. (Nov. 1, 1991), 698-700. [See Figures 1-3, 699-700].

481 V. Bucha, "Geomagnetic activity and the North Atlantic Oscillation." Studia Geophysica et Geodaetica. July 2014, Volume 58, Issue 3, 461–472. https://doi.org/10.1007/s11200-014-0508-z.

482 J. G. Pinto and C.C. Raible, 2012, "Past and recent changes in the North Atlantic oscillation." WIREs Climate Change, 3: 79-90. doi:10.1002/wcc.150.

483 J.W. Hurrell et al., 2013, "An Overview of the North Atlantic Oscillation." In The North Atlantic Oscillation: Climatic Significance and Environmental Impact (eds J. W. Hurrell, Y. Kushnir, G. Ottersen and M. Visbeck). doi:10.1029/134GM01.

484 A. Mazzarella and N. Scafetta, 2012, "Evidences for a quasi 60-year North Atlantic Oscillation since 1700 and its meaning for global climate change." Theoretical Applied Climatology 107, 599-609. DOI: 10.1007/s00704-011-0499-4.

485 M.H. Ambaum and B.J. Hoskins, 2002, "The NAO Troposphere–Stratosphere Connection." J. Climate, 15, 1969–1978, https://doi.org/10.1175/1520-0442(2002)015<1969:TNTSC>2.0.CO;2.

486 P. Thejll et al., On correlations between the North Atlantic Oscillation, geopotential heights, and geomagnetic activity. Geophysical Research Letters, 30 (6), 1347, 2003. doi:10.1029/2002GL016598.

487 H. Lu et al., 2008, "Possible solar wind effect on the northern annular mode and northern hemispheric circulation during winter and spring." Journal of Geophysical Research, 113, D23104, doi: 10.1029/2008JD010848.

488 V. Bucha, "Geomagnetic activity and the North Atlantic Oscillation." Studia Geophysica et Geodaetica. July 2014, Volume 58, Issue 3, 461–472. https://doi.org/10.1007/s11200-014-0508-z.

489 Boian Kirov and Katya Georgieva, "Long-term variations and interrelations of ENSO, NAO and solar activity." Physics and Chemistry of the Earth Parts. 2002. A/B/C 27(6-8):441-448. DOI10.1016/S1474-7065(02)00024-4.

490 S. V. Belova et al., "On the Relation between Endogenic Activity of the Earth and Solar and Geomagnetic Activity." Geophysics. DOI: 10.1134/S1028334X0907023X. Original Russian Text © S.V. Belov et al., 2009, published in Doklady Akademii Nauk, 2009, Volume 428, No. 1, 104–108.

491 V.E. Khain and E.N. Khalilov, "About possible influence of solar activity upon seismic and volcanic activities: Long-term Forecast." Science Without Borders. Transactions of the International Academy of Science H & E. Volume .3. 2007/2008, SWB, Innsbruck, 2008.

492 M. Tavares and A. Azevedo, 2011, "Influences of Solar Cycles on Earthquakes." Natural Science, 3, 436-443. Scientific Research. doi: 10.4236/ns.2011.36060. The data provided within this cited publication was extracted and presented in this graphical format without modification to the data. https://creativecommons.org/licenses/by/4.0/.

493 S. Odintsov et al., "Long-period trends in global seismic and geomagnetic activity and their relation to solar activity." 2006 Physics and Chemistry of the Earth Parts A/B/C 31(1):88-93. DOI10.1016/j.pce.2005.03.004.

494 I. P. Shestopalov and E. P. Kharin, "Relationship between solar activity and global seismicity and neutrons of terrestrial origin." 2014 Russian Journal of Earth Sciences 14(1):1-10. DOI10.2205/2014ES000536.

495 Bijan Nikouravan, 2012, "Do Solar Activities Cause Local Earthquakes?" International Journal of Fundamental Physical Sciences. http://fundamentaljournals.org/ijfps/downloads/29-Bijan%20150612.pdf, 2 (2). 20-23.

496 M. N. Gousheval et al., "On the relation between solar activity and seismicity." 0-7803-8 142-4/03/$17.0002003 IEEE 236. Recent Advances in Space Technologies, 2003. RAST '03. International Conference on. Proceedings of. DOI: 10.1109/RAST.2003.1303913.

497 G. Anagnostopoulos et al., "Solar wind triggering of geomagnetic disturbances and strong (M>6.8) earthquakes during the November – December 2004 period." Geophysics. arXiv:1012.3585. https://arxiv.org/abs/1012.3585.

498 E. Yu Aleshkina, "On correlation between variations in Earth rotation and frequency of earthquakes." Proceedings of the Journées 2010 "Systèmes de référence spatio-temporels" (JSR2010). 2011jsrs.conf..192A.

499 L. Ostřihanský, "Earth's rotation variations and earthquakes 2010–2011." Solid Earth Discuss., 4, 33–130, 2012. https://www.solid-earth-discuss.net/se-2011-36/. doi:10.5194/sed-4-33-2012.

500 P. Varga et al. "The relationship between the global seismicity and the rotation of the Earth." Journées 2004 - systèmes de référence spatio-temporels. Fundamental astronomy: new

concepts and models for high accuracy observations, Paris, 20-22 September 2004, edited by N. Capitaine, Paris: Observatoire de Paris, ISBN 2-901057-51-9, 2005, p. 115 – 120.

501   Elena Sasoroval and Boris Levin, "Relationship between Seismic Activity and Variations in the Earth's Rotation Angular Velocity." Journal of Geography and Geology; Volume 10, No. 2; 2018. doi:10.5539/jgg.v10n2p43.

502   P. Varga et al., "Global pattern of earthquakes and seismic energy distributions: Insights for the mechanisms of plate tectonics." Tectonophysics (2011), doi: 10.1016/j.tecto.2011.10.014.

503   N.A. Mörner et al., "General conclusions regarding the planetary–solar–terrestrial interaction." Pattern Recognition Physics, 1, 205–206, 2013. www.pattern-recogn-phys.net/1/205/2013/. doi:10.5194/prp-1-205-2013.

504   J.E. Solheim, "The sunspot cycle length – modulated by planets?" Pattern Recognition Physics, 1, 159–164, 2013. www.pattern-recogn-phys.net/1/159/2013/. doi:10.5194/prp-1-159-2013.

505   I.R.G. Wilson et al., "Does a Spin-Orbit Coupling Between the Sun and the Jovian Planets Govern the Solar Cycle?" Astronomical Society of Australia, Volume 25, Issue 2, 85-93. DOI:10.1071/AS06018.

506   R. Tattersall, 2013, "Apparent relations between planetary spin, orbit, and solar differential rotation." Pattern Recognition in Physics, 1 (1). 199 - 202. https://doi.org/10.5194/prp-1-199-2013.

507   Katya Georgieva, "Effects of interplanetary disturbances on the Earth's atmosphere and climate." http://www.issibern.ch/teams/interplanetarydisturb/wp-content/uploads/2014/01/proposal.pdf.

508   European Space Agency, "SOHO, New Solar Cycle Starts with a Bang." http://www.esa.int/Our_Activities/Space_Science/SOHO_the_new_solar_cycle_starts_with_a_bang.

509   Sunspot data from the World Data Center SILSO, Royal Observatory of Belgium, Brussels. http://sidc.be/silso/datafiles#total. [Data: Based on the Yearly mean sunspot number. The 1980 (1979.5) peak sunspot number was 220, 1990 (1989.5) peak 211, 2001 (2000.5) peak 174, 2015 (2014.5) peak 113. Downloaded 05/05/2018.].

510   M.G. Ogurtsov et al., Long-Period Cycles of the Sun's Activity Recorded in Direct Solar Data and Proxies. Solar Physics (2002) 211: 371. https://doi.org/10.1023/A:1022411209257.

511   I.G. Usoskin et al., "Grand minima and maxima of solar activity: new observational constraints." Astron.Astrophys.471:301-309,2007. DOI:10.1051/0004-6361:20077704.

512   Sunspot data from the World Data Center SILSO, Royal Observatory of Belgium, Brussels. http://sidc.be/silso/datafiles#total. [Data: Yearly mean sunspot numbers from 1700 to the present. Downloaded 05/05/2018.].

513   Solar Wind. https://www.jpl.nasa.gov/nmp/st5/SCIENCE/solarwind.html.

514   Theodor Landscheidt, "New Little Ice Age, Instead of Global Warming?" Schroeter Institute for Research in Cycles of Solar Activity, Klammerfelsweg 5, 93449 Waldmunchen, Germany. http://www.schulphysik.de/klima/landscheidt/iceage.htm.

515   Sami K Solanki, "Solar variability and climate change: is there a link?" Sami K Solanki presents the Harold Jeffrey's Lecture on the links between our climate and the behavior of the Sun, from the perspective of a solar physicist. https://bit.ly/2pGR2LH.

516   Sun-Earth Connections. https://www.nasa.gov/mission_pages/themis/auroras/sun_earth_connect.html

517   Earth's ionospheric and atmospheric layers in visual display. https://www.nasa.gov/mission_pages/sunearth/science/atmosphere-layers2.html.

518   Katya Georgieva, "Effects of interplanetary disturbances on the Earth's atmosphere and climate." http://www.issibern.ch/teams/interplanetarydisturb/wp-content/uploads/2014/01/proposal.pdf. [Citing: (1) C.E. Randall et al., Enhanced NOx in 2006 linked to strong upper stratospheric Arctic vortex, Geophysical Research Letters 33 (18), CiteID L18811, 2006. (2) S.V. Veretenenko and M.G. Ogurtsov, Stratospheric circumpolar vortex as a link between solar activity and circulation of the lower atmosphere. Geomagnetism and Aeronomy 52 (7) 937-943, 2012. (3) E.W. Kolstad et al., The association between stratospheric weak polar vortex events and cold air outbreaks in the Northern hemisphere. Quarterly Journal of the Royal Meteorological Society, 136, 886-893, 2010. (4) V. Bucha and V. Bucha Junior, Geomagnetic forcing of changes in climate and in the atmospheric circulation. Journal of Atmospheric and Solar-Terrestrial Physics 60, (2), 145-169, 1998.].

519 Le Mouël et al., 2010, "Solar forcing of the semi-annual variation of length-of-day." Geophysical Research Letters, 37, L15307, doi:10.1029/2010GL043185.

520 Nils-Axel Mörner, "Solar Wind, Earth's Rotation and Changes in Terrestrial Climate." Physical Review & Research International 3(2): 117-136, 2013.

521 S. Odintsov et al., "Long-period trends in global seismic and geomagnetic activity and their relation to solar activity." 2006 Physics and Chemistry of the Earth Parts A/B/C 31(1):88-93. DOI10.1016/j.pce.2005.03.004. [Citing: V. Bucha and V. Bucha Junior, Geomagnetic forcing of changes in climate and in the atmospheric circulation. Journal of Atmospheric and Solar-Terrestrial Physics 60, (2), 145-169, 1998].

522 See endnote 514.

523 Nils-Axel Mörner, "Solar Wind, Earth's Rotation and Changes in Terrestrial Climate." Physical Review & Research International 3(2): 117-136, 2013.

524 M. N. Gousheval et al., "On the relation between solar activity and seismicity." 0-7803-8 142-4/03/$17.0002003 IEEE 236. Recent Advances in Space Technologies, 2003. RAST '03. International Conference on. Proceedings of. DOI: 10.1109/RAST.2003.1303913.

525 Darryn W. Waugh and Lorenzo M. Polvani, "Stratospheric polar vortices." The Stratosphere: Dynamics, Transport, and Chemistry, Geophysical Monograph Series 190 (2010): 43-57.

526 V. Bucha, "Geomagnetic activity and the North Atlantic Oscillation." Studia Geophysica et Geodaetica. July 2014, Volume 58, Issue 3, 461–472. https://doi.org/10.1007/s11200-014-0508-z.

527 R.X. Black, 2002, "Stratospheric Forcing of Surface Climate in the Arctic Oscillation." J. Climate, 15, 268–277, https://doi.org/10.1175/1520-0442(2002)015<0268:SFOSCI>2.0.CO;2.

528 J. Perlwitz and N. Harnik, 2004, "Downward Coupling between the Stratosphere and Troposphere: The Relative Roles of Wave and Zonal Mean Processes." J. Climate, 17, 4902–4909, https://doi.org/10.1175/JCLI-3247.1.

529 Nils-Axel Mörner, "Solar Wind, Earth's Rotation and Changes in Terrestrial Climate." Physical Review & Research International 3(2): 117-136, 2013.

530 Adriano Mazzarella, "Solar Forcing of Changes in Atmospheric Circulation, Earth's Rotation and Climate." The Open Atmospheric Science Journal, 2008, 2, 181-184.

531 Nils-Axel Mörner, "Solar Wind, Earth's Rotation and Changes in Terrestrial Climate." Physical Review & Research International 3(2): 117-136, 2013. [Multiple citations].

532 Adriano Mazzarella, "Solar Forcing of Changes in Atmospheric Circulation, Earth's Rotation and Climate." The Open Atmospheric Science Journal, 2008, 2, 181-184.

533 Nils-Axel Mörner, "Solar Minima, Earth's rotation and Little Ice Ages in the past and in the future." The North Atlantic–European case. Global and Planetary Change 72 (2010) 282–293. doi:10.1016/j.gloplacha.2010.01.004. [Multiple citations].

534 M.A. Vukcevic, "Evidence of Length of Day (LOD) Bidecadal Variability Concurrent with the Solar Magnetic cycles." [Research Report] STAR. 2014.<hal-01071375v2>.

535 R. G. Currie, 1980, "Detection of the 11 year sunspot cycle signal in Earth rotation." Geophysical Journal of the Royal Astronomical Society, 61: 131-140. doi:10.1111/j.1365-246X.1980.tb04309.x.

536 T. Barlyaeval et al., "Rotation of the Earth, solar activity and cosmic ray intensity." July 2014 Annales Geophysicae 32(7):761-771. DOI: 10.5194/angeo-32-761-2014.

537 J.L. Le Mouël et al., 2010, "Solar forcing of the semi-annual variation of length-of-day." Geophysical Research Letters, 37, L15307, doi:10.1029/2010GL043185.

538 H. Wanner et al., "Structure and origin of Holocene cold events." Quaternary Science Reviews (2011), doi:10.1016/j.quascirev.2011.07.010.

539 D.S. Kaufman et al., "Holocene thermal maximum in the western Arctic (0–180°W)." Quaternary Science Reviews, Volume 23, Issues 5–6, 2004, 529-560. https://doi.org/10.1016/j.quascirev.2003.09.007.

540 Miller, G. H et al., 2012, "Abrupt onset of the Little Ice Age triggered by volcanism and sustained by sea-ice/ocean feedbacks." Geophysical Research Letters, 39, L02708, doi:10.1029/2011GL050168.

541 Y. Zhong et al., "Centennial-scale climate change from decadally-paced explosive volcanism: a coupled sea ice-ocean mechanism." Climate Dynamics (2011) 37: 2373. https://doi.org/10.1007/s00382-010-0967-z.

542   Darrell Kaufman et al., "Recent Warming Reverses Long-Term Arctic Cooling." September 2009. Science 325(5945):1236-1239. DOI: 10.1126/science.1173983.
543   Sami K. Solanki, "Solar variability and climate change: is there a link?" Astronomy & Geophysics, Volume 43, Issue 5, 1 October 2002, 5.9–5.13, https://doi.org/10.1046/j.1468-4004.2002.43509.x.
544   Sami K. Solanki, "Solar variability and climate change: is there a link?" Astronomy & Geophysics, Volume 43, Issue 5, 1 October 2002, 5.9–5.13, https://doi.org/10.1046/j.1468-4004.2002.43509.x.
545   "Global Radiative Forcing, CO2-equivalent mixing ratio" the AGGI 1979-2016. www.esrl.noaa.gov/gmd/aggi/aggi.html. [See table 2 for 2016's global radiative forcing of 3.027 W/m2].
546   George C. Reid, "Solar Variability and the Earth's Climate: Introduction and Overview." Space Science Reviews (2000) 94: 1. https://doi.org/10.1023/A:1026797127105. [See pages 2-4, Figure 1].
547   E. Friis-Christensen and K. Lassen, "Length of the Solar Cycle: An Indicator of Solar Activity Closely Associated with Climate." Science, New Series, Volume 254, No. 5032. (Nov. 1, 1991), 698-700. [See Figures 1-3, 699-700].
548   N.A. Mörner et al., "General conclusions regarding the planetary-solar-terrestrial interaction." Pattern Recognition Physics, 1, 205–206, 2013. www.pattern-recogn-phys.net/1/205/2013/. doi:10.5194/prp-1-205-2013.
549   J.E. Solheim, "The sunspot cycle length – modulated by planets?" Pattern Recognition Physics, 1, 159–164, 2013. www.pattern-recogn-phys.net/1/159/2013/. doi:10.5194/prp-1-159-2013.
550   I.R.G. Wilson et al., "Does a Spin-Orbit Coupling Between the Sun and the Jovian Planets Govern the Solar Cycle?" Astronomical Society of Australia, Volume 25, Issue 2, 85-93. DOI:10.1071/AS06018.
551   See endnote 461.
552   See endnote 461.
553   S.K. Solanki et al., 2004, "An unusually active Sun during recent decades compared to the previous 11,000 years." Nature, Volume 431, No. 7012, 1084-1087, 28 October 2004. Data: Solanki, S.K., et al. 2005. 11,000 Year Sunspot Number Reconstruction. IGBP PAGES/World Data Center for Paleoclimatology. Data Contribution Series #2005-015. NOAA/NGDC Paleoclimatology Program, Boulder CO, USA. https://www.ncdc.noaa.gov/paleo-search/study/5780. Downloaded 05/06/2018.
554   J. Slawinska and A. Robock, 2018, "Impact of Volcanic Eruptions on Decadal to Centennial Fluctuations of Arctic Sea Ice Extent during the Last Millennium and on Initiation of the Little Ice Age." J. Climate, 31, 2145–2167, https://doi.org/10.1175/JCLI-D-16-0498.1.
555   F. Lehner et al., 2013, "Amplified inception of European Little Ice Age by sea ice–ocean–atmosphere feedbacks." J. Climate, 26, 7586–7602. https://doi.org/10.1175/JCLI-D-12-00690.1.
556   C. Newhall et al., 2018, "Anticipating future Volcanic Explosivity Index (VEI) 7 eruptions and their chilling impacts." Geosphere, v. 14, no. 2, p. 1–32, doi:10.1130/GES01513.1.
557   Odd Helge Otterå et al., "External forcing as a metronome for Atlantic multidecadal variability." Nature Geoscience Volume 3, 688–694 (2010).
558   Y. Zhong et al., "Centennial-scale climate change from decadally-paced explosive volcanism: a coupled sea ice-ocean mechanism." Climate Dynamics (2011) 37: 2373. https://doi.org/10.1007/s00382-010-0967-z.
559   Data: (1) A.M Berggren et al., 2009, "A 600-year annual 10Be record from the NGRIP ice core, Greenland." Geophysical Research Letters, 36, L11801, doi:10.1029/2009GL038004. National Centers for Environmental Information, NESDIS, NOAA, U.S. Department of Commerce. North GRIP - 600 Year Annual 10Be Data. https://www.ncdc.noaa.gov/paleo-search/study/8618. Downloaded 05/05/2018. (2) T. Kobashi et al., 2013, "Causes of Greenland temperature variability over the past 4000 year: implications for northern hemispheric temperature changes." Climate of the Past, 9(5), 2299-2317. doi: 10.5194/cp-9-2299-2013. National Centers for Environmental Information, NESDIS, NOAA, U.S. Department of Commerce. Northern Hemisphere 4000 Year Temperature Reconstructions. https://www.ncdc.noaa.gov/paleo/study/15535. Downloaded 05/05/2018. Statistics Software Utilized: The Spearman rank calculator that was utilized: Wessa P., (2017), Spearman Rank Correlation (v1.0.3) in Free Statistics Software (v1.2.1), Office for Research Development and Education, URL https://www.wessa.net/rwasp_spearman.wasp/. Personal Research: Figure 5.4.A: The

above-cited Northern Hemisphere temperature anomaly and a 18-year trailing average Beryllium-10 concentration anomaly data were plotted against one another (1889-1985). An 18-year moving average of the Beryllium-10 concentration anomaly was selected using a Microsoft Excel scatterplot of this data, which utilized a linear trend line to maximize the R-squared (compared 18, 11, and 5-year moving averages, and the anomaly data; data not shown). Figure 5.4.B: A scatter plot of Figure 5.4.A's data was created to reveal the relationship. The 18-year moving average data for Beryllium-10 concentration anomaly was not normally distributed and contained outliers (Iglewicz and Hoaglin's robust test, two-sided test, using a Modified Z score ≥ 3.5). As such the Spearman rank correlation calculation was utilized. Results: Spearman rank correlation -0.895, two-tailed P-value = <0.00001, N=97 annual data pairings. Furthermore, the scatter plot graphic best fits a non-linear trend line (R-squared 0.81) versus a linear trend (R-squared 0.75). The data used in this analysis and graphic continues on from that used to derive Figure 4.4.A (Spearman rank correlation r=-0.76, two-tailed P-value <0.00001, N=484 years). The correlation value in this figure 5.4.B was similar to the correlation value of the preceding four grand solar minima. This would indicate, in both cases, that the correlation appears to strength during grand solar minimum and maximum phases.

560 Nils-Axel Mörner, "Solar Wind, Earth's Rotation and Changes in Terrestrial Climate." Physical Review & Research International 3(2): 117-136, 2013. [Multiple citations].

561 Adriano Mazzarella, "Solar Forcing of Changes in Atmospheric Circulation, Earth's Rotation and Climate." The Open Atmospheric Science Journal, 2008, 2, 181-184.

562 Nils-Axel Mörner, "Solar Minima, Earth's rotation and Little Ice Ages in the past and in the future. The North Atlantic–European case." Global and Planetary Change 72 (2010) 282–293. doi:10.1016/j.gloplacha.2010.01.004. [Multiple citations].

563 M.A. Vukcevic, "Evidence of Length of Day (LOD) Bidecadal Variability Concurrent with the Solar Magnetic cycles." [Research Report] STAR. 2014.<hal-01071375v2>.

564 R.G. Currie, 1980, "Detection of the 11 year sunspot cycle signal in Earth rotation." Geophysical Journal of the Royal Astronomical Society, 61: 131-140. doi:10.1111/j.1365-246X.1980.tb04309.x.

565 T. Barlyaeval et al., "Rotation of the Earth, solar activity and cosmic ray intensity." July 2014 Annales Geophysicae 32(7):761-771. DOI: 10.5194/angeo-32-761-2014.

566 J.L. Le Mouël et al., 2010, "Solar forcing of the semi-annual variation of length-of-day." Geophysical Research Letters, 37, L15307, doi:10.1029/2010GL043185.

567 T. Barlyaeval et al., "Rotation of the Earth, solar activity and cosmic ray intensity." July 2014 Annales Geophysicae 32(7):761-771. DOI: 10.5194/angeo-32-761-2014. [Citing multiple references].

568 M.A. Vukcevic., "Evidence of Length of Day (LOD) Bidecadal Variability Concurrent with the Solar Magnetic cycles." [Research Report] STAR. 2014.<hal-01071375v2>.

569 K. Scherer et al., "Interstellar-Terrestrial Relations: Variable Cosmic Environments, The Dynamic Heliosphere, and Their Imprints on Terrestrial Archives and Climate." Space Science Reviews (2006) 127: 327. https://doi.org/10.1007/s11214-006-9126-6.

570 Boian Kirov and Katya Georgieva, "Long-term variations and interrelations of ENSO, NAO and solar activity." Physics and Chemistry of the Earth Parts. 2002. A/B/C 27(6-8):441-448. DOI10.1016/S1474-7065(02)00024-4.

571 Nils-Axel Mörner, "Solar Minima, Earth's rotation and Little Ice Ages in the past and in the future. The North Atlantic–European case." Global and Planetary Change 72 (2010) 282–293. doi:10.1016/j.gloplacha.2010.01.004.

572 V. Bucha, "Geomagnetic activity and the North Atlantic Oscillation." Studia Geophysica et Geodaetica. July 2014, Volume 58, Issue 3, 461–472. https://doi.org/10.1007/s11200-014-0508-z.

573 J. G. Pinto and C.C. Raible, 2012, "Past and recent changes in the North Atlantic oscillation." WIREs Climate Change, 3: 79-90. doi:10.1002/wcc.150.

574 Jesper Olsen et al., "Variability of the North Atlantic Oscillation over the past 5,200 years." Nature Geoscience volume 5, 808–812 (2012). DOI: 10.1038/NGEO1589.

575 P. Thejll et al., "On correlations between the North Atlantic Oscillation, geopotential heights, and geomagnetic activity." Geophysical Research Letters, 30 (6), 1347, 2003. doi:10.1029/2002GL016598.

576 J.W. Hurrell et al., 2013, "An Overview of the North Atlantic Oscillation. In The North Atlantic Oscillation: Climatic Significance and Environmental Impact" (eds J. W. Hurrell, Y. Kushnir, G. Ottersen and M. Visbeck). doi:10.1029/134GM01.

577 A. Mazzarella and N. Scafetta, 2012, "Evidences for a quasi 60-year North Atlantic Oscillation since 1700 and its meaning for global climate change." Theoretical Applied Climatology 107, 599-609. DOI: 10.1007/s00704-011-0499-4.

578 M.H. Ambaum and B.J. Hoskins, 2002, "The NAO Troposphere–Stratosphere Connection." J. Climate, 15, 1969–1978, https://doi.org/10.1175/1520-0442(2002)015<1969:TNTSC>2.0.CO;2.

579 V. Bucha, "Geomagnetic activity and the North Atlantic Oscillation." Studia Geophysica et Geodaetica. July 2014, Volume 58, Issue 3, 461–472. https://doi.org/10.1007/s11200-014-0508-z.

580 P. Thejll et al., "On correlations between the North Atlantic Oscillation, geopotential heights, and geomagnetic activity." Geophysical Research Letters, 30 (6), 1347, 2003. doi:10.1029/2002GL016598.

581 H. Lu et al., 2008, "Possible solar wind effect on the northern annular mode and northern hemispheric circulation during winter and spring." Journal of Geophysical Research, 113, D23104, doi: 10.1029/2008JD010848.

582 V. Bucha, "Geomagnetic activity and the North Atlantic Oscillation." Studia Geophysica et Geodaetica. July 2014, Volume 58, Issue 3, 461–472. https://doi.org/10.1007/s11200-014-0508-z.

583 A. Mazzarella and N. Scafetta, 2012, "Evidences for a quasi 60-year North Atlantic Oscillation since 1700 and its meaning for global climate change." Theoretical Applied Climatology 107, 599-609. DOI: 10.1007/s00704-011-0499-4.

584 Boian Kirov and Katya Georgieva, "Long-term variations and interrelations of ENSO, NAO and solar activity." Physics and Chemistry of the Earth Parts. 2002. A/B/C 27(6-8):441-448. DOI10.1016/S1474-7065(02)00024-4.

585 "Frequently asked questions about El Niño and La Niña." National Weather Service. Climate Prediction Center. http://www.cpc.ncep.noaa.gov/products/analysis_monitoring/ensostuff/ensofaq.shtml#NINO.

586 Boian Kirov and Katya Georgieva, "Long-term variations and interrelations of ENSO, NAO and solar activity." Physics and Chemistry of the Earth Parts. 2002. A/B/C 27(6-8):441-448. DOI10.1016/S1474-7065(02)00024-4.

587 Vincent Courtillot et al., "Are there connections between the Earth's magnetic field and climate?" Earth and Planetary Science Letters 253 (2007) 328–339.

588 K. Scherer et al., "Interstellar-Terrestrial Relations: Variable Cosmic Environments, The Dynamic Heliosphere, and Their Imprints on Terrestrial Archives and Climate." 2006. Space Science Reviews. 127(1): 327-465. DOI10.1007/s11214-006-9126-6.

589 Dimitar Todorov Valev, "Statistical relationships between the surface air temperature anomalies and the solar and geomagnetic activity indices." 2006. Physics and Chemistry of the Earth Parts A/B/C 31(1):109-112. DOI10.1016/j.pce.2005.03.005.

590 Theodor Landscheidt, "New Little Ice Age, Instead of Global Warming?" Schroeter Institute for Research in Cycles of Solar Activity, Klammerfelsweg 5, 93449 Waldmünchen, Germany. http://www.schulphysik.de/klima/landscheidt/iceage.htm.

591 V. Bucha and V. Bucha Junior, "Geomagnetic forcing of changes in climate and in the atmospheric circulation." Journal of Atmospheric and Solar-Terrestrial Physics, Volume 60, Issue 2, p. 145-169. DOI:10.1016/S1364-6826(97)00119-3. http://adsabs.harvard.edu/abs/1998JASTP..60..145B.

592 Yves Gallet et al., "Possible impact of the Earth's magnetic field on the history of ancient civilizations." Earth and Planetary Science Letters. Volume 246, Issues 1–2, 15 June 2006, 17-26. https://doi.org/10.1016/j.epsl.2006.04.001.

593 K. S. Carslaw et al., "Cosmic Rays, Clouds, and Climate." Science 29 Nov 2002: Volume 298, Issue 5599, 1732-1737. DOI: 10.1126/science.1076964. http://science.sciencemag.org/content/298/5599/1732.full.

594 F. Yu, "Altitude variations of cosmic ray induced production of aerosols: Implications for global cloudiness and climate." Journal of Geophysical Research, 107(A7), doi:10.1029/2001JA000248, 2002.

595 J.E. Kristjánsson and J. Kristiansen, 2000, "Is there a cosmic ray signal in recent variations in global cloudiness and cloud radiative forcing?" Journal of Geophysical Research, 105(D9), 11851–11863, doi:10.1029/2000JD900029.

596 L.E.A. Vieira, and L.A. da Silva, 2006, "Geomagnetic modulation of clouds effects in the Southern Hemisphere Magnetic Anomaly through lower atmosphere cosmic ray effects." Geophysical Research Letters, 33, L14802, doi:10.1029/2006GL026389.

597 Le Mouël et al., 2010, "Solar forcing of the semi-annual variation of length-of-day." Geophysical Research Letters, 37, L15307, doi:10.1029/2010GL043185. [See Figure 1, page 2].

598 G.A. Kovaltsov and I.G. Usoskin, "Regional cosmic ray induced ionization and geomagnetic field changes." Advances in Geosciences, European Geosciences Union, 2007, 13, 31-35.

599 K. S. Carslaw et al., "Cosmic Rays, Clouds, and Climate." Science 29 Nov 2002: Volume 298, Issue 5599, 1732-1737. DOI: 10.1126/science.1076964. http://science.sciencemag.org/content/298/5599/1732.full.

600 S. V. Belova et al., "On the Relation between Endogenic Activity of the Earth and Solar and Geomagnetic Activity." GEOPHYSICS. DOI: 10.1134/S1028334X0907023X. Original Russian Text © S.V. Belov et al., 2009, published in Doklady Akademii Nauk, 2009, Volume 428, No. 1, 104–108. [See page 209, Figures 1-2].

601 R.B. Stothers, 1989, "Volcanic eruptions and solar activity." Journal of Geophysical Research, 94(B12), 17371–17381, doi:10.1029/JB094iB12p17371.

602 V.E. Khain and E.N. Khalilov, "About possible influence of solar activity upon seismic and volcanic activities: Long-term Forecast." Science Without Borders. Transactions of the International Academy of Science H & E. Volume 3. 2007/2008, SWB, Innsbruck, 2008 ISBN 978-9952-451-01-6 ISSN 2070-0334.

603 M. Tavares and A. Azevedo, 2011, "Influences of Solar Cycles on Earthquakes." Natural Science, 3, 436-443. Scientific Research. doi: 10.4236/ns.2011.36060.

604 M. Tavares and A. Azevedo, 2011, "Influences of Solar Cycles on Earthquakes." Natural Science, 3, 436-443. Scientific Research. doi: 10.4236/ns.2011.36060. [See Tables 1-2 for the Maunder and Dalton minimum].

605 S.V. Belova, et al., "On the Relation between Endogenic Activity of the Earth and Solar and Geomagnetic Activity." Geophysics. DOI: 10.1134/S1028334X0907023X. Original Russian Text © S.V. Belov, I.P. Shestopalov, E.P. Kharin, 2009, published in Doklady Akademii Nauk, 2009, Volume 428, No. 1, 104–108.

606 V.E. Khain and E.N. Khalilov, "About possible influence of solar activity upon seismic and volcanic activities: Long-term Forecast." Science Without Borders. Transactions of the International Academy of Science H & E. Volume 3. 2007/2008, SWB, Innsbruck, 2008.

607 M. Tavares and A. Azevedo, 2011, "Influences of Solar Cycles on Earthquakes." Natural Science, 3, 436-443. Scientific Research. doi: 10.4236/ns.2011.36060. The data provided within this cited publication was extracted and presented in this graphical format without modification to the data. https://creativecommons.org/licenses/by/4.0/.

608 V.E. Khain and E.N. Khalilov, "About possible influence of solar activity upon seismic and volcanic activities: Long-term Forecast." Science Without Borders. Transactions of the International Academy of Science H & E. Volume 3. 2007/2008, SWB, Innsbruck.

609 R. B. Stothers, 1989, "Volcanic eruptions and solar activity." Journal of Geophysical Research, 94(B12), 17371–17381, doi:10.1029/JB094iB12p17371.

610 M.G. Ogurtsov et al., "Long-Period Cycles of the Sun's Activity Recorded in Direct Solar Data and Proxies." Solar Physics (2002) 211: 371. https://doi.org/10.1023/A:1022411209257.

611 Toshikazu Ebisuzaki et al., "Explosive volcanic eruptions triggered by cosmic rays: Volcano as a bubble chamber." Gondwana Research 19 (2011) 1054–1061. https://doi.org/10.1016/j.gr.2010.11.004.

612 S. V. Belova et al., "On the Relation between Endogenic Activity of the Earth and Solar and Geomagnetic Activity." Geophysics. DOI: 10.1134/S1028334X0907023X. Original Russian Text © S.V. Belov, I.P. Shestopalov, E.P. Kharin, 2009, published in Doklady Akademii Nauk, 2009, Volume 428, No. 1, 104–108.

613 Gerald Duma, "Earthquake activity controlled by the regular induced telluric currents." Central Institute for Meteorology and Geodynamics, Department of Geophysics. http://www.isfep.com/website%2014_013-G.Duma.pdf.

614 S. Odintsov et al., "Long-period trends in global seismic and geomagnetic activity and their relation to solar activity." 2006 Physics and Chemistry of the Earth Parts A/B/C 31(1):88-93. DOI10.1016/j.pce.2005.03.004.

615 I.P. Shestopalov and E.P. Kharin, "Relationship between solar activity and global seismicity and neutrons of terrestrial origin." 2014 Russian Journal of Earth Sciences 14(1):1-10. DOI10.2205/2014ES000536.

616 Bijan Nikouravan, 2012, "Do Solar Activities Cause Local Earthquakes?" International Journal of Fundamental Physical Sciences. http://fundamentaljournals.org/ijfps/downloads/29-Bijan%20150612.pdf, 2 (2). 20-23.

617 M.N. Gousheval et al., "On the relation between solar activity and seismicity." 0-7803-8 142-4/03/$17.0002003 IEEE 236. Recent Advances in Space Technologies, 2003. RAST '03. International Conference on. Proceedings of. DOI: 10.1109/RAST.2003.1303913.

618 G. Anagnostopoulos et al., "Solar wind triggering of geomagnetic disturbances and strong (M>6.8) earthquakes during the November – December 2004 period." Geophysics. arXiv:1012.3585. https://arxiv.org/abs/1012.3585.

619 S. Odintsov et al., "Long-period trends in global seismic and geomagnetic activity and their relation to solar activity." 2006 Physics and Chemistry of the Earth Parts A/B/C 31(1):88-93. DOI10.1016/j.pce.2005.03.004.

620 M.N. Gousheval et al., "On the relation between solar activity and seismicity." 0-7803-8 142-4/03/$17.0002003 IEEE 236. Recent Advances in Space Technologies, 2003. RAST '03. International Conference on. Proceedings of. DOI:10.1109/RAST.2003.1303913.

621 E. Yu. Aleshkina, "On correlation between variations in Earth rotation and frequency of earthquakes." Proceedings of the Journées 2010 "Systèmes de référence spatio-temporels" (JSR2010). 2011jsrs.conf..192A.

622 L. Ostřihanský, "Earth's rotation variations and earthquakes 2010–2011." Solid Earth Discuss., 4, 33–130, 2012. www.solid-earth-discuss.net/4/33/2012/. doi:10.5194/sed-4-33-2012.

623 P. Varga et al., "The relationship between the global seismicity and the rotation of the Earth." Journées 2004 - systèmes de référence spatio-temporels. Fundamental astronomy: new concepts and models for high accuracy observations, Paris, 20-22 September 2004, edited by N. Capitaine, Paris: Observatoire de Paris, ISBN 2-901057-51-9, 2005, p. 115 – 120.

624 Elena Sasoroval and Boris Levin, "Relationship between Seismic Activity and Variations in the Earth's Rotation Angular Velocity." Journal of Geography and Geology; Volume 10, No. 2; 2018. doi:10.5539/jgg.v10n2p43.

625 Elena Sasoroval and Boris Levin, "Relationship between Seismic Activity and Variations in the Earth's Rotation Angular Velocity." Journal of Geography and Geology; Volume 10, No. 2; 2018. doi:10.5539/jgg.v10n2p43.

626 L. Ostřihanský, "Earth's rotation variations and earthquakes 2010–2011." Solid Earth Discuss., 4, 33–130, 2012. www.solid-earth-discuss.net/4/33/2012/. doi:10.5194/sed-4-33-2012.

627 P. Varga et al., "Global pattern of earthquakes and seismic energy distributions: Insights for the mechanisms of plate tectonics." Tectonophysics (2011), doi: 10.1016/j.tecto.2011.10.014.

628 L. Ostřihanský, "Earth's rotation variations and earthquakes 2010–2011." Solid Earth Discuss., 4, 33–130, 2012. www.solid-earth-discuss.net/4/33/2012/. doi:10.5194/sed-4-33-2012.

629 P. Varga et al., "The relationship between the global seismicity and the rotation of the Earth." Journées 2004 - systèmes de référence spatio-temporels. Fundamental astronomy: new concepts and models for high accuracy observations, Paris, 20-22 September 2004, edited by N. Capitaine, Paris: Observatoire de Paris, ISBN 2-901057-51-9, 2005, p. 115 – 120. [Exposé: "Annual energy budgets: Rotational energy ≈1.6 × 1019 Joules/annum, Energy of earthquakes ≈ 9.5 × 1018 Joules/annum, Volcanic energy ≈ 2.0 × 1018 Joules/annum, and Atmospheric circulations ≈6.3 × 1022 Joules/annum."].

630 L. Ostřihanský, "Earth's rotation variations and earthquakes 2010–2011." Solid Earth Discuss., 4, 33–130, 2012. https://www.solid-earth-discuss.net/se-2011-36/. doi:10.5194/sed-4-33-2012.

631 P. Varga et al., "The relationship between the global seismicity and the rotation of the Earth." Journées 2004 - systèmes de référence spatio-temporels. Fundamental astronomy: new concepts and models for high accuracy observations, Paris, 20-22 September 2004, edited by N. Capitaine, Paris: Observatoire de Paris, ISBN 2-901057-51-9, 2005, p. 115 – 120. [Exposé: "Annual energy budgets: Rotational energy ≈1.6 × 1019 Joules/annum, Energy of earthquakes

≈ 9.5 × 1018 Joules/annum, Volcanic energy ≈ 2.0 × 1018 Joules/annum, and Atmospheric circulations ≈6.3 × 1022 Joules/annum."].

632   S. Odintsov et al., "Long-period trends in global seismic and geomagnetic activity and their relation to solar activity." 2006 Physics and Chemistry of the Earth Parts A/B/C 31(1):88-93. DOI10.1016/j.pce.2005.03.004. [Citing: Sytinskii, A.D., 1997. Influence of interplanetary disturbances on the seismicity and atmosphere of the Earth. Geomagnetism and Aeronomy 37 (2), 138–141.]

633   M. N. Gousheval et al., "On the relation between solar activity and seismicity." 0-7803-8 142-4/03/$17.0002003 IEEE 236. Recent Advances in Space Technologies, 2003. RAST '03. International Conference on. Proceedings of. DOI: 10.1109/RAST.2003.1303913.

634   See endnote 114.

635   IPCC Risk mitigation. IPCC, 2014: Climate Change 2014: Impacts, Adaptation, and Vulnerability. Part A: Global and Sectoral Aspects. Contribution of Working Group II to the Fifth Assessment Report of the Intergovernmental Panel on Climate Change [Field, C.B., V.R. Barros, D.J. Dokken, K.J. Mach, M.D. Mastrandrea, T.E. Bilir, M. Chatterjee, K.L. Ebi, Y.O. Estrada, R.C. Genova, B. Girma, E.S. Kissel, A.N. Levy, S. MacCracken, P.R. Mastrandrea, and L.L. White (eds.)]. Cambridge University Press, Cambridge, United Kingdom and New York, NY, USA, 1132 pages [Exposé: See page 14, second paragraph. This paragraph tells us that climate change risks can be reduced by reducing greenhouse gas emissions.].

636   N. Scafetta, "Multi-scale harmonic model for solar and climate cyclical variation throughout the Holocene based on Jupiter-Saturn tidal frequencies plus the 11-year solar dynamo cycle." Journal of Atmospheric and Solar-Terrestrial Physics (2012). doi:10.1016/j.jastp.2012.02.016.

637   Theodor Landscheidt, "New Little Ice Age Instead of Global Warming?" Energy & Environment. 2003. Volume 14, Issue 2, 327 – 350. https://doi.org/10.1260/095830503765184646.

638   R.J. Salvador, "A mathematical model of the sunspot cycle for the past 1000 years." Pattern Recognition Physics, 1, 117-122, doi:10.5194/prp-1-117-2013, 2013.

639   Habibullo Abdussamatov, "Current Long-Term Negative Average Annual Energy Balance of the Earth Leads to the New Little Ice age." Thermal Science. 2015 Supplement, Volume 19, S279-S288.

640   Jan-Erik Solheim, https://www.mwenb.nl/wp-content/uploads/2014/10/Blog-Jan-Erik-Solheim-def.pdf. Referred from http://www.climatedialogue.org/what-will-happen-during-a-new-maunder-minimum/. Citing blog for 4-5 solar-climate experts.

641   Nils-Axel Mörner, "Solar Wind, Earth's Rotation and Changes in Terrestrial Climate." Physical Review & Research International 3 (2): 117-136, 2013.

642   Boncho P. Bonev et al., "Long-Term Solar Variability and the Solar Cycle in the 21st Century." The Astrophysical Journal, 605:L81–L84, 2004 April 10.

643   See endnote 181.

644   See endnote 8.

645   See endnote 9.

646   David D. Zhang et al., "Global climate change, war, and population decline in recent human history." Proceedings of the National Academy of Sciences Dec 2007, 104 (49) 19214-19219; DOI: 10.1073/pnas.0703073104.

647   Dian Zhang et al., "Climate change, social unrest and dynastic transition in ancient China." China Science Bulletin January 2005, Volume 50, Issue 2, 137–144. https://doi.org/10.1007/BF02897517.

648   Anthony J. McMichael, "Insights from past millennia into climatic impacts on human health and survival." Proceedings of the National Academy of Sciences Mar 2012, 109 (13) 4730-4737; DOI: 10.1073/pnas.1120177109. [See page 4734, column 2, second paragraph].

649   Geoffrey Parker, "Crisis and Catastrophe: The Global Crisis of the Seventeenth Century Reconsidered." The American Historical Review, Volume 113, No. 4 (Oct., 2008), 1053-1079. http://www.jstor.org/stable/30223245.

650   D. Collet and M. Schuh (eds.), "Famines During the 'Little Ice Age' (1300–1800) ." DOI 10.1007/978-3-319-54337-6_2. [See page 21].

651 Michael J. Puma et al., "Exploring the potential impacts of historic volcanic eruptions on the contemporary global food system." PAGES Magazine. Science Highlights. Volcanoes and Climate. Volume 23, No 2, December 2015.

652 Clive Oppenheimer, "Climatic, environmental and human consequences of the largest known historic eruption: Tambora volcano (Indonesia) 1815." Progress in Physical Geography: Earth and Environment (2003). Volume 27, Issue 2, 230 – 259. https://doi.org/10.1191/0309133303pp379ra.

653 Anthony J. McMichael, "Insights from past millennia into climatic impacts on human health and survival." Proceedings of the National Academy of Sciences Mar 2012, 109 (13) 4730-4737; DOI: 10.1073/pnas.1120177109. [See page 4735, column 2, paragraph 2].

654 R.B. Stothers, "Climatic and Demographic Consequences of the Massive Volcanic Eruption of 1258." Climatic Change (2000) 45: 361. https://doi.org/10.1023/A:1005523330643.

655 C. Oppenheimer, 2003, "Ice core and paleoclimate evidence for the timing and nature of the great mid-13th century volcanic eruption." International Journal of Climatology, 23: 417-426. doi:10.1002/joc.891.

656 J. Slawinska and A. Robock, 2018, "Impact of Volcanic Eruptions on Decadal to Centennial Fluctuations of Arctic Sea Ice Extent during the Last Millennium and on Initiation of the Little Ice Age." J. Climate, 31, 2145–2167, https://doi.org/10.1175/JCLI-D-16-0498.1.

657 F. Lehner et al., 2013, "Amplified inception of European Little Ice Age by sea ice–ocean–atmosphere feedbacks." J. Climate, 26, 7586–7602. https://doi.org/10.1175/JCLI-D-12-00690.1.

658 C. Newhall et al., 2018, "Anticipating future Volcanic Explosivity Index (VEI) 7 eruptions and their chilling impacts." Geosphere, v. 14, no. 2, p. 1–32, doi:10.1130/GES01513.1.

659 Odd Helge Otterå et al., "External forcing as a metronome for Atlantic multidecadal variability." Nature Geoscience Volume 3, 688–694 (2010).

660 Y. Zhong et al., "Centennial-scale climate change from decadally-paced explosive volcanism: a coupled sea ice-ocean mechanism." Climate Dynamics (2011) 37: 2373. https://doi.org/10.1007/s00382-010-0967-z.

661 See endnote 114.

662 United Nations Framework Convention on Climate Change. United Nations 1992. FCCC/INFORMAL/84, GE.05-62220 (E), 200705. [See page 4]. Accessed 14/05/2018.

663 IPCC, 2014: Climate Change 2014: Impacts, Adaptation, and Vulnerability. Part A: Global and Sectoral Aspects. Contribution of Working Group II to the Fifth Assessment Report of the Intergovernmental Panel on Climate Change [Field, C.B., V.R. Barros, D.J. Dokken, K.J. Mach, M.D. Mastrandrea, T.E. Bilir, M. Chatterjee, K.L. Ebi, Y.O. Estrada, R.C. Genova, B. Girma, E.S. Kissel, A.N. Levy, S. MacCracken, P.R. Mastrandrea, and L.L. White (eds.)]. Cambridge University Press, Cambridge, United Kingdom and New York, NY, USA, 1132 pages [See page 13].

664 IPCC, 2014: Climate Change 2014: Synthesis Report. Contribution of Working Groups I, II and III to the Fifth Assessment Report of the Intergovernmental Panel on Climate Change [Core Writing Team, R.K. Pachauri and L.A. Meyer (eds.)]. IPCC, Geneva, Switzerland, 151 pages [See page 16].

665 IPCC, 2014: Climate Change 2014: Synthesis Report. Contribution of Working Groups I, II and III to the Fifth Assessment Report of the Intergovernmental Panel on Climate Change [Core Writing Team, R.K. Pachauri and L.A. Meyer (eds.)]. IPCC, Geneva, Switzerland, 151 pages [See page 13].

666 IPCC, 2014: Climate Change 2014: Impacts, Adaptation, and Vulnerability. Part A: Global and Sectoral Aspects. Contribution of Working Group II to the Fifth Assessment Report of the Intergovernmental Panel on Climate Change [Field, C.B., V.R. Barros, D.J. Dokken, K.J. Mach, M.D. Mastrandrea, T.E. Bilir, M. Chatterjee, K.L. Ebi, Y.O. Estrada, R.C. Genova, B. Girma, E.S. Kissel, A.N. Levy, S. MacCracken, P.R. Mastrandrea, and L.L. White (eds.)]. Cambridge University Press, Cambridge, United Kingdom and New York, NY, USA, 1132 pages [See page 14].

667 IPCC, 2013: Climate Change 2013: The Physical Science Basis. Contribution of Working Group I to the Fifth Assessment Report of the Intergovernmental Panel on Climate Change [Stocker, T.F., D. Qin, G.-K. Plattner, M. Tignor, S.K. Allen, J. Boschung, A. Nauels, Y. Xia, V. Bex and

P.M. Midgley (eds.)]. Cambridge University Press, Cambridge, United Kingdom and New York, NY, USA, 1535 pages.

668 IPCC, 2014: Climate Change 2014: Impacts, Adaptation, and Vulnerability. Part A: Global and Sectoral Aspects. Contribution of Working Group II to the Fifth Assessment Report of the Intergovernmental Panel on Climate Change [Field, C.B., V.R. Barros, D.J. Dokken, K.J. Mach, M.D. Mastrandrea, T.E. Bilir, M. Chatterjee, K.L. Ebi, Y.O. Estrada, R.C. Genova, B. Girma, E.S. Kissel, A.N. Levy, S. MacCracken, P.R. Mastrandrea, and L.L. White (eds.)]. Cambridge University Press, Cambridge, United Kingdom and New York, NY, USA, 1132 pages.

669 IPCC, 2014: Climate Change 2014: Impacts, Adaptation, and Vulnerability. Part B: Regional Aspects. Contribution of Working Group II to the Fifth Assessment Report of the Intergovernmental Panel on Climate Change [Barros, V.R., C.B. Field, D.J. Dokken, M.D. Mastrandrea, K.J. Mach, T.E. Bilir, M. Chatterjee, K.L. Ebi, Y.O. Estrada, R.C. Genova, B. Girma, E.S. Kissel, A.N. Levy, S. MacCracken, P.R. Mastrandrea, and L.L. White (eds.)]. Cambridge University Press, Cambridge, United Kingdom and New York, NY, USA, pages 688.

670 IPCC, 2014: Climate Change 2014: Mitigation of Climate Change. Contribution of Working Group III to the Fifth Assessment Report of the Intergovernmental Panel on Climate Change [Edenhofer, O., R. Pichs-Madruga, Y. Sokona, E. Farahani, S. Kadner, K. Seyboth, A. Adler, I. Baum, S. Brunner, P. Eickemeier, B. Kriemann, J. Savolainen, S. Schlömer, C. von Stechow, T. Zwickel and J.C. Minx (eds.)]. Cambridge University Press, Cambridge, United Kingdom and New York, NY, USA.

671 IPCC, 2012: Managing the Risks of Extreme Events and Disasters to Advance Climate Change Adaptation. A Special Report of Working Groups I and II of the Intergovernmental Panel on Climate Change [Field, C.B., V. Barros, T.F. Stocker, D. Qin, D.J. Dokken, K.L. Ebi, M.D. Mastrandrea, K.J. Mach, G.-K. Plattner, S.K. Allen, M. Tignor, and P.M. Midgley (eds.)]. Cambridge University Press, Cambridge, UK, and New York, NY, USA, 582 pages.

672 IPCC, 2014: Climate Change 2014: Impacts, Adaptation, and Vulnerability. Part A: Global and Sectoral Aspects. Contribution of Working Group II to the Fifth Assessment Report of the Intergovernmental Panel on Climate Change [Field, C.B., V.R. Barros, D.J. Dokken, K.J. Mach, M.D. Mastrandrea, T.E. Bilir, M. Chatterjee, K.L. Ebi, Y.O. Estrada, R.C. Genova, B. Girma, E.S. Kissel, A.N. Levy, S. MacCracken, P.R. Mastrandrea, and L.L. White (eds.)]. Cambridge University Press, Cambridge, United Kingdom and New York, NY, USA, 1132 pages [See page 421].

673 David D. Zhang et al., "Global climate change, war, and population decline in recent human history." Proceedings of the National Academy of Sciences Dec 2007, 104 (49) 19214-19219; DOI: 10.1073/pnas.0703073104. [See 19216, column 2, China's high mortality and Europe's Black Death].

674 Anthony J. McMichael, "Insights from past millennia into climatic impacts on human health and survival." Proceedings of the National Academy of Sciences Mar 2012, 109 (13) 4730-4737; DOI: 10.1073/pnas.1120177109. [See page 4734, column 2, second paragraph].

675 David D. Zhang et al., "The causality analysis of climate change and large-scale human crisis." Proceedings of the National Academy of Sciences Oct 2011, 108 (42) 17296-17301; DOI: 10.1073/pnas.1104268108.

676 Geoffrey Parker, "Crisis and Catastrophe: The Global Crisis of the Seventeenth Century Reconsidered." The American Historical Review, Volume 113, No. 4 (Oct., 2008), 1053-1079. http://www.jstor.org/stable/30223245.

677 The Little Ice Age's impact on wars. IPCC, 2014: Climate Change 2014: Impacts, Adaptation, and Vulnerability. Part A: Global and Sectoral Aspects. Contribution of Working Group II to the Fifth Assessment Report of the Intergovernmental Panel on Climate Change [Field, C.B., V.R. Barros, D.J. Dokken, K.J. Mach, M.D. Mastrandrea, T.E. Bilir, M. Chatterjee, K.L. Ebi, Y.O. Estrada, R.C. Genova, B. Girma, E.S. Kissel, A.N. Levy, S. MacCracken, P.R. Mastrandrea, and L.L. White (eds.)]. Cambridge University Press, Cambridge, United Kingdom and New York, NY, USA, 1132 pages [Exposé: (1) See page 772. According to the IPCC, because the exact causal pathways linking historical climate change to issues faced by societies of the past, i.e., wars during the Little Ice Age, no inference can be made for climate change on today's global society. (2) See page 1001, section 18.4.5. This section details the IPCC's low

confidence in the historical data regarding the climate change link with wars. Critique: This dismissal of the Little Ice Age climate change risk on wars exemplifies confirmation bias. When the IPCC wishes to disprove something counter or inconvenient to their theory, then there are problems with a lack of confidence in the historical data, or there is a low certainty or confidence in the science or historical precedent. Of course, that same lack of understanding of the control mechanisms for the climate system (while ignoring natural climate change), and the large uncertainties associated with climate forecasting, never stopped the IPCC from inaccurately forecasting the climate with a high confidence for three decades.].

678 See endnote 8.

679 Not detailing abrupt or rapid climate change. IPCC, Climate Change 2013: The Physical Science Basis. Contribution of Working Group I to the Fifth Assessment Report of the Intergovernmental Panel on Climate Change [Stocker, T.F., D. Qin, G.-K. Plattner, M. Tignor, S.K. Allen, J. Boschung, A. Nauels, Y. Xia, V. Bex and P.M. Midgley (eds.)]. Cambridge University Press, Cambridge, United Kingdom and New York, NY, USA, 1535 pages [Exposé and Critique: (1) See pages 70, Section TFE.5. This section is supposed to detail abrupt climate change and so-called climate tipping points. We are told there is some information (when there is much information) on this subject matter but there is little consensus and a low confidence that such an event would take place during the 21st century. (2) See page 115, TS.6.4, and page 1033. These sections reiterate the message contained in section TFE.5 on page 70. Critique: This dismissal of abrupt or rapid climate change is despite the history of catastrophic climate change associated with the Little Ice Age (and glacier expansions), climate-forcing volcanism, and the specific rapid climate events known to have taken place at or since the Holocene Climate Optimum, i.e., the 8.2, 5.9 and 4.2 kiloyear rapid climate change events. Once again this exemplifies the IPCC's confirmation bias. When the IPCC wish to disprove science that would undermine its theory or forecasts, it has a low confidence and certainty in the predictions or the underlying science, or there is little consensus in scientists' predictions.].

680 See endnote 6.

681 See endnote 7.

682 Leszek Starkel, "Extreme rainfalls and river floods in Europe during the last millennium." Geographia Polonica (2001) Volume 74, issue 2, 69-79.

683 Martín-Puertas et al., "Hydrological evidence for a North Atlantic oscillation during the Little Ice Age outside its range observed since 1850." Climate of the Past Discuss., 7, 4149-4171, https://doi.org/10.5194/cpd-7-4149-2011, 2011.

684 B. Stefanie et al., "Holocene flood frequency across the Central Alps - solar forcing and evidence for variations in North Atlantic atmospheric circulation." Quaternary Science Reviews 80 (2013) 112e128.

685 Laurent Fouinat et al., "Paleoflood activity and climate change over the last 2000 years recorded by high altitude alpine lake sediments in Western French Alps." Geophysical Research Abstracts. Volume 17, EGU2015-11555, 2015 EGU General Assembly 2015 © Author(s) 2015. CC Attribution 3.0 License.

686 B. Wilhelm et al., 2012, "1400 years of extreme precipitation patterns over the Mediterranean French Alps and possible forcing mechanisms." Quaternary Research, 78(1), 1-12. doi:10.1016/j.yqres.2012.03.003.

687 B. Stefanie et al., "A 2000 year long seasonal record of floods in the southern European Alps." Geophysical Research Letters, Volume 40, 4025–4029, doi:10.1002/grl.50741, 2013.

688 O.N. Solomina et al., 2016, "Glacier fluctuations during the past 2000 years." Quaternary Science Reviews, 149, 61-90. DOI: 10.1016/j.quascirev.2016.04.008. [See Figure 5, page 276. This figure collates a stacked time series of the number of glacier advances and recessions in each region into a global total.].

689 Zicheng Yu and Emi Ito, "Possible solar forcing of century-scale drought frequency in the northern Great Plains." Geology ; 27 (3): 263–266. doi: https://doi.org/10.1130/0091-7613(1999)027<0263:PSFOCS>2.3.CO;2.

690 J.E. Nichols and Y. Huang, 2012, "Hydroclimate of the northeastern United States is highly sensitive to solar forcing." Geophysical. Research. Letters., Volume 39, L04707, doi:10.1029/2011GL050720, 2012.

691 H. Xu et al., 2015, "Late Holocene Indian Summer Monsoon Variations Recorded at Lake Erhai, Southwestern China." Quaternary Research, 83(2), 307-314. doi:10.1016/j.yqres.2014.12.004.

692 Shangbin Xiao et al., "Coherence between solar activity and the East Asian winter monsoon variability in the past 8000 years from Yangtze River-derived mud in the East China Sea." Palaeogeography, Palaeoclimatology, Palaeoecology 237 (2006) 293–304. doi:10.1016/j. palaeo.2005.12.003.

693 Liangcheng Tan et al., "Precipitation variations of Longxi, northeast margin of Tibetan Plateau since AD 960 and their relationship with solar activity." Climate of the Past, 4, 19–28, 2008, https://doi.org/10.5194/cp-4-19-2008, 2008.

694 Wenfeng Deng et al., "A comparison of the climates of the Medieval Climate Anomaly, Little Ice Age, and Current Warm Period reconstructed using coral records from the northern South China Sea." December 2016. Journal of Geophysical Research: Oceans 122(1). DOI.10.1002/2016JC012458.

695 J.J. Yin et al., "Variation in the Asian monsoon intensity and dry–wet conditions since the Little Ice Age in central China revealed by an aragonite stalagmite." Climate of the Past, 10, 1803-1816, https://doi.org/10.5194/cp-10-1803-2014, 2014.

696 Wang Shaowu et al., "Climate in China During the Little Ice Age." Department of Geophysics, Peking University, Beijing 100871. http://en.cnki.com.cn/Article_en/CJFDTOTAL-DSJJ199801007.htm.

697 C. Uberoi, "Little Ice Age in Mughal India: Solar Minima Linked to Droughts?" Volume 93 Number 44 30 October 2012 EOS, Transactions, American Geophysical Union. 437–452.

698 Rajesh Agnihotri et al., "Evidence for solar forcing on the Indian monsoon during the last millennium." Earth and Planetary Science Letters 198 (2002) 521-527.

699 Vishwas Kale and Victor R. Baker, "An Extraordinary Period of Low-magnitude Floods Coinciding with the Little Ice Age: Palaeoflood Evidence from Central and Western India." Journal of the Geological Society of India 68(3):477-483.

700 Feng Shi et al., "A tree-ring reconstruction of the South Asian summer monsoon index over the past millennium." Scientific Reports Volume 4, Article number: 6739 (2014). DOI: 10.1038/srep06739.

701 J.M. Russell, T.C. Johnson, "Little Ice Age drought in equatorial Africa: Intertropical Convergence Zone migrations and El Niño–Southern Oscillation variability." Geology (2007) 35 (1): 21-24. DOI: https://doi.org/10.1130/G23125A.1.

702 Dirk Verschuren et al., Cumming. "Rainfall and drought in equatorial east Africa during the past 1,100 years." Nature Volume 403, 410–414 (27 January 2000). doi:10.1038/35000179.

703 James M. Russell et al., "Spatial complexity of 'Little Ice Age' climate in East Africa: sedimentary records from two crater lake basins in western Uganda." The Holocene. Volume 17, Issue 2, 183 – 193. 2007. https://doi.org/10.1177/0959683607075832.

704 P D Tyson et al., "The Little Ice Age and medieval warming in South Africa." March 2000South African Journal of Science 96(3):121-126.

705 Justin Reuter et al., "A new perspective on the hydroclimate variability in northern South America during the Little Ice Age." December 2009 Geophysical Research Letters 36(21). DOI10.1029/2009GL041051.

706 Alexandra Haase-Schramm et al., "Sr/Ca ratios and oxygen isotopes from sclerosponges: Temperature history of the Caribbean mixed layer and thermocline during the Little Ice Age." Paleoceanography, 18(3), 1073, doi:10.1029/2002PA000830, 2003.

707 Juan Pablo Milana and Daniela Kröhling, "Climate changes and solar cycles recorded at the Holocene Paraná Delta, and their impact on human population." August 2015Scientific Reports 5(12851):1-8. DOI10.1038/srep12851.

708 Pablo Mauas et al., "Long-term solar activity influences on South American rivers." Journal of Atmospheric and Solar-Terrestrial Physics. arXiv:1003.0414 [astro-ph.SR]. 10.1016/j. jastp.2010.02.019.

709 Michael J Burn et al., "A sediment-based reconstruction of Caribbean effective precipitation during the Little Ice Age from Freshwater Pond, Barbuda." The Holocene. Volume: 26 issue: 8, 1237-1247. https://doi.org/10.1177/0959683616638418.

710   Amos Winter et al., "Caribbean sea surface temperatures: Two-to-three degrees cooler than present during the Little Ice Age." October 2000 Geophysical Research Letters 27(20):3365-3368. DOI10.1029/2000GL011426.

711   C. Lane et al., 2011, "Oxygen isotope evidence of Little Ice Age aridity on the Caribbean slope of the Cordillera Central, Dominican Republic." Quaternary Research, 75(3), 461-470. doi:10.1016/j.yqres.2011.01.002.

712   D.A. Hodell et al., 2005, "Climate change on the Yucatan Peninsula during the little ice age." Quaternary Research, 63 (2). 109-121. ISSN 0033-5894. DOI.10.1016/j.yqres.2004.11.004.

713   David D. Zhang et al., "Global climate change, war, and population decline in recent human history." Proceedings of the National Academy of Sciences Dec 2007, 104 (49) 19214-19219; DOI: 10.1073/pnas.0703073104.

714   Dian Zhang et al., "Climate change, social unrest and dynastic transition in ancient China." China Science Bulletin January 2005, Volume 50, Issue 2, 137–144. https://doi.org/10.1007/BF02897517.

715   Anthony J. McMichael, "Insights from past millennia into climatic impacts on human health and survival." Proceedings of the National Academy of Sciences Mar 2012, 109 (13) 4730-4737; DOI: 10.1073/pnas.1120177109.

716   David D. Zhang et al., "Global climate change, war, and population decline in recent human history." Proceedings of the National Academy of Sciences Dec 2007, 104 (49) 19214-19219; DOI: 10.1073/pnas.0703073104. [Exposé: China's high mortality. See 19216, column 2, first paragraph].

717   D. Collet and M. Schuh (eds.), "Famines During the 'Little Ice Age' (1300–1800) ." DOI 10.1007/978-3-319-54337-6_2. [See page 21].

718   Anthony J. McMichael, "Insights from past millennia into climatic impacts on human health and survival." Proceedings of the National Academy of Sciences Mar 2012, 109 (13) 4730-4737; DOI: 10.1073/pnas.1120177109. [See page 4734, column 2, second paragraph].

719   David D. Zhang et al., "Global climate change, war, and population decline in recent human history." Proceedings of the National Academy of Sciences Dec 2007, 104 (49) 19214-19219; DOI: 10.1073/pnas.0703073104. [Europe's Black Death. See 19216, column 2, first paragraph].

720   David D. Zhang et al., "Global climate change, war, and population decline in recent human history." Proceedings of the National Academy of Sciences Dec 2007, 104 (49) 19214-19219; DOI: 10.1073/pnas.0703073104. [See 19216, column 2, second paragraph].

721   Geoffrey Parker, "Crisis and Catastrophe: The Global Crisis of the Seventeenth Century Reconsidered." The American Historical Review, Volume 113, No. 4 (Oct., 2008), 1053-1079. http://www.jstor.org/stable/30223245.

722   Geoffrey Parker, "Crisis and Catastrophe: The Global Crisis of the Seventeenth Century Reconsidered." The American Historical Review, Volume 113, No. 4 (Oct., 2008), 1053-1079. http://www.jstor.org/stable/30223245.

723   David D. Zhang et al., "The causality analysis of climate change and large-scale human crisis." Proceedings of the National Academy of Sciences Oct 2011, 108 (42) 17296-17301; DOI: 10.1073/pnas.1104268108.

724   Anthony J. McMichael, "Insights from past millennia into climatic impacts on human health and survival." Proceedings of the National Academy of Sciences Mar 2012, 109 (13) 4730-4737; DOI: 10.1073/pnas.1120177109. [Page 4734, column 1, paragraph 3].

725   M.E. Brown et al., 2015, "Climate Change, Global Food Security, and the U.S. Food System." http://www.usda.gov/oce/climate_change/FoodSecurity2015Assessment/FullAssessment.pdf.

726   Tony Fischer et al., "Crop yields and global food security. Will yield increase continue to feed the world?" Australia Centre for International Agricultural Research. Grains Research and Development Corporation. https://www.aciar.gov.au/node/12101.

727   Michael J Puma et al., "Assessing the evolving fragility of the global food system." Environment Research Letter 10 (2015) 024007 doi:10.1088/1748-9326/10/2/024007.

728   Philippe Marchand et al., 2016, "Reserves and trade jointly determine exposure to food supply shocks." Environment Research Letter 11 095009. Environment Research Letter 11 (2016) 095009 doi:10.1088/1748-9326/11/9/095009.

729   Christopher Bren d'Amour et al., "Teleconnected food supply shocks." Environment Research Letter 11 (2016) 035007 doi:10.1088/1748-9326/11/3/035007.

730   Michael J. Puma et al., "Exploring the potential impacts of historic volcanic eruptions on the contemporary global food system." PAGES Magazine. Science Highlights. Volcanoes and Climate. Volume 23, No 2, December 2015.

731   R. B. Alley et al., "Holocene climatic instability: A prominent, widespread event 8200 year ago." Geology ; 25 (6): 483–486. doi: https://doi.org/10.1130/0091-7613(1997)025<0483:HCIAP W>2.3.CO;2.

732   Kaarina Sarmaja-Korjonen and H. Seppa, 2007, "Abrupt and consistent responses of aquatic and terrestrial ecosystems to the 8200 cal. year cold event: a lacustrine record from Lake Arapisto, Finland". The Holocene 17 (4): 457–467. doi:10.1177/0959683607077020.

733   D.C. Barber et al., 1999, "Forcing of the cold event of 8,200 years ago by catastrophic drainage of Laurentide lakes." Nature Volume 400, 344–348 (22 July 1999). doi:10.1038/22504.

734   Christopher R W Ellison et al., 2006, "Surface and Deep Ocean Interactions During the Cold Climate Event 8200 Years Ago." Science. 2006 Jun 30;312(5782):1929-32. DOI10.1126/science.1127213.

735   A. Parker et al., 2006, "A Record of Holocene Climate Change from Lake Geochemical Analyses in Southeastern Arabia." Quaternary Research, 66(3), 465-476. doi:10.1016/j.yqres.2006.07.001.

736   Peter B. deMenocal, "Cultural Responses to Climate Change During the Late Holocene." Science. 2001: Volume 292, Issue 5517, 667-673. DOI: 10.1126/science.1059287.

737   Robert K. Booth et al., "A severe centennial-scale drought in midcontinental North America 4200 years ago and apparent global linkages." The Holocene. Volume15, Issue 3, 321 – 328. 2005. https://doi.org/10.1191/0959683605hl825ft.

738   J. Wang et al., "The abrupt climate change near 4,400 year BP on the cultural transition in Yuchisi, China and its global linkage." Scientific Reports 2016 Jun 10;6:27723. doi: 10.1038/srep27723.

739   B.J.J. Menounos et al., 2008, "Western Canadian glaciers advance in concert with climate change circa 4.2 ka." Geophysical Research Letters, 35, L07501, doi:10.1029/2008GL033172.

740   Russell Drysdale et al., "Late Holocene drought responsible for the collapse of Old World civilizations is recorded in an Italian cave flowstone." Geology; 34 (2): 101–104. doi: https://doi.org/10.1130/G22103.1.

741   Lonnie G. Thompson et al., "Kilimanjaro Ice Core Records: Evidence of Holocene Climate Change in Tropical Africa." Science18 Oct 2002: 589-593.

742   M. Davis and L. Thompson, 2006, "An Andean ice-core record of a Middle Holocene mega-drought in North Africa and Asia." Annals of Glaciology, 43, 34-41. doi:10.3189/172756406781812456.

743   Françoise Gasse and Elise Van Campo, 1994, "Abrupt post-glacial climate events in West Asia and North Africa monsoon domains". Earth and Planetary Science Letters 126 (4): 435–456. Bibcode:1994E&PSL.126..435G. doi:10.1016/0012-821X(94)90123-6.

744   J. Ruan et al., 2016, "Evidence of a prolonged drought ca. 4200 year BP correlated with prehistoric settlement abandonment from the Gueldaman GLD1 Cave, Northern Algeria." Climate of the Past, 12(1), 1-4. DOI: 10.5194/cp-12-1-2016.

745   D. Kaniewski et al., "Middle East coastal ecosystem response to middle-to-late Holocene abrupt climate changes." Proceedings of the National Academy of Sciences Sep 2008, 105 (37) 13941-13946; DOI: 10.1073/pnas.0803533105.

746   Fenggui Liu, Zhaodong Feng, "A dramatic climatic transition at ~4000 cal. year BP and its cultural responses in Chinese cultural domains." The Holocene. Volume 22, Issue 10, 1181 – 1197. First Published April 12, 2012. https://doi.org/10.1177/0959683612441839.

747   Jianjun Wang, "The abrupt climate change near 4,400 year BP on the cultural transition in Yuchisi, China and its global linkage." Scientific Reports | 6:27723 | DOI: 10.1038/srep27723. https://www.nature.com/articles/srep27723.pdf.

748   Fenggui Liu and Zhaodong Feng, "A dramatic climatic transition at ~4000 cal. year BP and its cultural responses in Chinese cultural domains." The Holocene 22(10) 1181–1197 © The Author(s) 2012. DOI: 10.1177/0959683612441839. hol.sagepub.com.

749   M. Staubwasser, H. Weiss, 2006, "Holocene Climate and Cultural Evolution in Late Prehistoric–Early Historic West Asia." Quaternary Research, 66(3), 372-387. doi:10.1016/j.yqres.2006.09.001.

750 P. Mayewski et al., 2004, "Holocene climate variability." Quaternary Research, 62(3), 243-255. doi:10.1016/j.yqres.2004.07.001.

751 P. Mayewski et al., 2004, "Holocene climate variability." Quaternary Research, 62(3), 243-255. doi:10.1016/j.yqres.2004.07.001.

752 Bernhard Weninger et al., "The Impact of Rapid Climate Change on prehistoric societies during the Holocene in the Eastern Mediterranean." Documenta Praehistorica XXXVI (2009). UDK 902(4-5)"631\637">551.583.

753 M. Staubwasser, H. Weiss, 2006, "Holocene Climate and Cultural Evolution in Late Prehistoric–Early Historic West Asia." Quaternary Research, 66(3), 372-387. doi:10.1016/j.yqres.2006.09.001.

754 Robert K. Booth et al., "A severe centennial-scale drought in midcontinental North America 4200 years ago and apparent global linkages." The Holocene. Volume 15, Issue 3, 321 – 328. 2005. https://doi.org/10.1191/0959683605hl825ft.

755 J. Stanley et al., 2003, "Nile flow failure at the end of the Old Kingdom, Egypt: Strontium isotopic and petrologic evidence." Geoarchaeology, 18: 395-402. doi:10.1002/gea.10065.

756 Ann Gibbons, "How the Akkadian Empire Was Hung Out to Dry". Science 20 Aug 1993: Volume 261, Issue 5124, DOI: 10.1126/science.261.5124.985.

757 Jianjun Wang, "The abrupt climate change near 4,400 year BP on the cultural transition in Yuchisi, China and its global linkage." Scientific Reports | 6:27723 | DOI: 10.1038/srep27723. https://www.nature.com/articles/srep27723.pdf.

758 C.C. Raible et al., 2016, "Tambora 1815 as a test case for high impact volcanic eruptions: Earth system effects." WIREs Climate Change, 7: 569-589. doi:10.1002/wcc.407.

759 Michael J. Puma et al., "Exploring the potential impacts of historic volcanic eruptions on the contemporary global food system." PAGES Magazine. Science Highlights. Volcanoes and Climate. Volume 23, No 2, December 2015.

760 Anthony J. McMichael, "Insights from past millennia into climatic impacts on human health and survival." Proceedings of the National Academy of Sciences Mar 2012, 109 (13) 4730-4737; DOI: 10.1073/pnas.1120177109. [See page 4735, column 2, paragraph 2].

761 Clive Oppenheimer, Climatic, environmental and human consequences of the largest known historic eruption: Tambora volcano (Indonesia) 1815. Progress in Physical Geography: Earth and Environment (2003). Volume 27, Issue 2, 230 – 259. https://doi.org/10.1191/0309133303pp379ra.

762 C. Newhall et al., 2018, "Anticipating future Volcanic Explosivity Index (VEI) 7 eruptions and their chilling impacts." Geosphere, v. 14, no. 2, p. 1–32, doi:10.1130/GES01513.1.

763 C. Newhall et al., 2018, "Anticipating future Volcanic Explosivity Index (VEI) 7 eruptions and their chilling impacts." Geosphere, v. 14, no. 2, p. 1–32, doi:10.1130/GES01513.1.

764 R.B. Stothers, "Climatic and Demographic Consequences of the Massive Volcanic Eruption of 1258." Climatic Change (2000) 45: 361. https://doi.org/10.1023/A:1005523330643.

765 Michael J. Puma et al., "Exploring the potential impacts of historic volcanic eruptions on the contemporary global food system." PAGES Magazine. Science Highlights. Volcanoes and Climate. Volume 23, No 2, December 2015.

766 C. Oppenheimer, 2003, "Ice core and paleoclimate evidence for the timing and nature of the great mid-13th century volcanic eruption." International Journal of Climatology, 23: 417-426. doi:10.1002/joc.891.

767 C. Newhall et al., 2018, "Anticipating future Volcanic Explosivity Index (VEI) 7 eruptions and their chilling impacts." Geosphere, v. 14, no. 2, p. 1–32, doi:10.1130/GES01513.1.

768 C. Newhall et al., 2018, "Anticipating future Volcanic Explosivity Index (VEI) 7 eruptions and their chilling impacts." Geosphere, v. 14, no. 2, p. 1–32, doi:10.1130/GES01513.1.

769 C.S. Witham, "Volcanic disasters and incidents: A new database." Journal of Volcanology and Geothermal Research. Volume 148, Issues 3–4, 2005. 191-233. https://doi.org/10.1016/j.jvolgeores.2005.04.017.

770 Melanie Rose Auker et al., "A statistical analysis of the global historical volcanic fatalities record." Journal of Applied Volcanology Society and Volcanoes 20132:2. https://doi.org/10.1186/2191-5040-2-2.

771 J.C. Tanguy et al., "Victims from volcanic eruptions: a revised database." Bulletin Volcanology (1998) 60: 137. https://doi.org/10.1007/s004450050222.

772   Michael J. Puma et al., "Exploring the potential impacts of historic volcanic eruptions on the contemporary global food system." PAGES Magazine. Science Highlights. Volcanoes and Climate. Volume 23, No 2, December 2015.

773   C.J.N. Wilson, "The 26.5ka Oruanui eruption, New Zealand: an introduction and overview." Journal of Volcanology and Geothermal Research. Volume 112, Issues 1–4, 2001. 133-174. https://doi.org/10.1016/S0377-0273(01)00239-6.

774   S. Sparks et al., 2005, "Super-eruptions: global effects and future threats." Report of a Geological Society of London Working Group. https://www.geolsoc.org.uk/supereruptions.

775   S. Kutterolf et al., 2008, "Pacific offshore record of plinian arc volcanism in Central America: 2. Tephra volumes and eruptive masses." Geochemistry, Geophysics, Geosystems, 8: Q02S02. https://doi.org/10.1029/2007GC001791.

776   Bethan Harris, "The potential impact of super-volcanic eruptions on the Earth's atmosphere." Weather – August 2008, Volume 63, No. 8. DOI:10.1002/wea.263. https://rmets.onlinelibrary.wiley.com/doi/pdf/10.1002/wea.263.

777   Richard (Bert) Roberts, 2012, "Armageddon and its aftermath: dating the Toba super-eruption." Research Online: http://ro.uow.edu.au/smhpapers/1883.

778   S. Sparks et al., 2005, "Super-eruptions: global effects and future threats." Report of a Geological Society of London Working Group. https://www.geolsoc.org.uk/supereruptions.

779   M.R. Rampino and S. Self, 1992, "Volcanic winter and accelerated glaciation following the Toba super-eruption." Nature, 359, 50-52, doi:10.1038/359050a0.

780   Helen Sian Crosweller et al., "Global database on large magnitude explosive volcanic eruptions (LaMEVE)." Journal of Applied Volcanology Society and Volcanoes 20121:4. https://doi.org/10.1186/2191-5040-1-4. Volcano Global Risk Identification and Analysis Project database (VOGRIPA), British Geological Survey. Data Access: http://www.bgs.ac.uk/vogripa/. Data downloaded 07/05/2018.

781   Helen Sian Crosweller et al., "Global database on large magnitude explosive volcanic eruptions (LaMEVE)." Journal of Applied Volcanology Society and Volcanoes 20121:4. https://doi.org/10.1186/2191-5040-1-4. Volcano Global Risk Identification and Analysis Project database (VOGRIPA), British Geological Survey. Data Access: http://www.bgs.ac.uk/vogripa/. Data downloaded 07/05/2018.

782   S.K. Brown et al., "Characterization of the Quaternary eruption record: analysis of the Large Magnitude Explosive Volcanic Eruptions (LaMEVE) database." J Applied Volcanology. (2014) 3: 5. https://doi.org/10.1186/2191-5040-3-5.

783   C. Newhall et al., 2018, "Anticipating future Volcanic Explosivity Index (VEI) 7 eruptions and their chilling impacts." Geosphere, v. 14, no. 2, p. 1–32, doi:10.1130/GES01513.1. [see table 3].

784   S.C. Loughlin et al., 2015, An introduction to global volcanic hazard and risk. Global Volcanic Hazards and Risk: Cambridge, Cambridge University Press, 1–40, https:// doi .org /10 .1017 / CBO9781361276273.

785   A.S. Richey et al., 2015, "Quantifying renewable groundwater stress with GRACE, Water Resources Research." 51, 5217–5238, doi:10.1002/2015WR017349, and NASA via https://www.nasa.gov/jpl/grace/study-third-of-big-groundwater-basins-in-distress.

786   H Munia et al., "Water stress in global transboundary river basins: significance of upstream water use on downstream stress." Environment Research Letter 11 (2016) 014002. doi:10.1088/1748-9326/11/1/014002 http://iopscience.iop.org/article/10.1088/1748-9326/11/1/014002/pdf.

787   See endnote 75.

788   See endnote 76.

789   Data: Energy Information Administration data was obtained from: International Energy Statistics. These calculations utilized the following data files. Natural gas https://bit.ly/2LC6GBo, Crude Oil https://bit.ly/2IWeEaP, Coal data https://bit.ly/2L6pk3w. [Personal Research: The 50 year reserve timeline estimates are calculated by dividing the 2013 Energy Information Agency's proven global oil, natural gas, and coal reserves by the 2013 levels of production. This calculation tells us there are 50 years of proven oil and gas, and 130 years of coal reserves left. These reserve timeline estimates do not assume any population or economic growth, or a switch to a cold climate regime, which would accelerate energy demand and reduce the reserve timelines.].

790 See endnote 78.

791 See endnote 79.

792 Maurice Smith "Declining Reserve Replacement Ratios Deceiving In Resource Play Environment." November. 28, 2017. JWN Energy. Daily Oil Bulletin. https://www.sproule.com/application/files/2415/1188/2978/Sproule-Declining-Reserve-Replacement-Ratios-Nora-Stewart-Steve-Golko.pdf.

793 Tom Whipple, "Peak Oil Review Dec 26 2017." Originally published by ASPO-US. December 26, 2017. https://www.resilience.org/stories/2017-12-26/peak-oil-review-dec-26-2017/.

794 "All-time low for discovered resources in 2017: Around 7 billion barrels of oil equivalent was discovered." December 21, 2017. https://www.rystadenergy.com/newsevents/news/press-releases/all-time-low-discovered-resources-2017/.

795 Kjell Aleklett and Colin J. Campbell, "The peak and decline of world oil and gas production." Minerals and Energy-Raw Materials Report 18.1 (2003): 5-20.

796 Ian Chapman, 2014, "The end of Peak Oil? Why this topic is still relevant despite recent denials." Energy Policy, 64 . 93-101. http://insight.cumbria.ac.uk/id/eprint/1708/.

797 International Energy Forum. The IOCs And The NOCs In The Modern Energy Context. https://www.ief.org/news/the-iocs-and-the-nocs-in-the-modern-energy-context.

798 International Energy Forum. The IOCs And The NOCs In The Modern Energy Context. https://www.ief.org/news/the-iocs-and-the-nocs-in-the-modern-energy-context.

799 Data: Energy Information Administration. Exports of Crude Oil including Lease Condensate. International Energy Statistics data portal. Based on 2016 data. Petroleum, https://bit.ly/2xfMwHM. Exports of Dry Natural Gas, https://bit.ly/2NhW6nK. Data accessed 16 September 2018.

800 Data (for Figures 8.1.A and B): Oil, gas, and coal data was obtained from the Energy Information Administration from their International Energy Statistics data portal. This graphic utilizes the following data files. Natural gas https://bit.ly/2LC6GBo, Crude Oil https://bit.ly/2IWeEaP, Coal data https://bit.ly/2L6pk3w. Standard fuel units of energy (barrels, cubic feet, and short tons) were converted to British thermal units (Btu) using conversion factors obtained from, https://www.eia.gov/energyexplained/index.php?page=about_btu.

801 R.G. Miller, S.R. Sorrell, 2014, "The future of oil supply." Philosophical Transactions of the Royal Society A 372: 20130179. http://dx.doi.org/10.1098/rsta.2013.0179.

802 Kjell Aleklett and Colin J. Campbell, "The peak and decline of world oil and gas production." Minerals and Energy-Raw Materials Report 18.1 (2003): 5-20. (See the figure on page 6).

803 The US Geological Survey reviewed oil recovery factors for USA oil and gas reservoirs, before applying this to their international oil and gas reserve forecasts (outside the USA). The mean modal recovery was 45% across 67 reservoirs. https://pubs.usgs.gov/sir/2015/5091/sir20155091.pdf.

804 Kjell Aleklett and Colin J. Campbell, "The peak and decline of world oil and gas production." Minerals and Energy-Raw Materials Report 18.1 (2003): 5-20.

805 Ian Chapman, 2014, "The end of Peak Oil? Why this topic is still relevant despite recent denials." Energy Policy, 64 . 93-101. http://insight.cumbria.ac.uk/id/eprint/1708/.

806 Kjell Aleklett and Colin J. Campbell, "The peak and decline of world oil and gas production." Minerals and Energy-Raw Materials Report 18.1 (2003): 5-20.

807 M. Höök, 2014, "Depletion rate analysis of fields and regions: a methodological foundation." Fuel, Volume 121, 1 April 2014, 95–108. http://dx.doi.org/10.1016/j.fuel.2013.12.024.

808 Mahendra K. Verma, "The Reality of Reserve Growth. Reservoir Management." U.S. Geological Survey. GEO ExPro October 2007 35.

809 US Energy Information Administration. "Oil and natural gas resource categories reflect varying degrees of certainty." https://www.eia.gov/todayinenergy/detail.php?id=17151. (See the annotated diagram and descriptions for further explanation).

810 Mahendra K. Verma, "The Reality of Reserve Growth. Reservoir Management." U.S. Geological Survey. GEO ExPro October 2007 35.

811 T.A. Cook, 2013, "Reserve growth of oil and gas fields—Investigations and applications." U.S. Geological Survey Scientific Investigations Report 2013–5063, 29 p., http://pubs.usgs.gov/sir/2013/5063/.

812 https://energy.usgs.gov/OilGas/AssessmentsData/WorldPetroleumAssessment.aspx#3882215-overview. https://pubs.usgs.gov/sir/2015/5091/sir20155091.pdf .

813 David F. Morehouse, "The Intricate Puzzle of Oil and Gas "Reserves Growth." Energy Information Administration. Natural Gas Monthly July 1997. http://large.stanford.edu/courses/2014/ph240/liegl1/docs/morehouse.pdf.

814 Mahendra K. Verma, "The Reality of Reserve Growth. Reservoir Management." U.S. Geological Survey. GEO ExPro October 2007 35.

815 T.R. Klett et al., 2015, U.S. "Geological Survey assessment of reserve growth outside of the United States." U.S. Geological Survey Scientific Investigations Report 2015–5091,13 p., http://dx.doi.org/10.3133/sir20155091.

816 Kjell Aleklett and Colin J. Campbell, "The peak and decline of world oil and gas production." Minerals and Energy-Raw Materials Report 18.1 (2003): 5-20.

817 U.S. Energy Information Administration. "Technically Recoverable Shale Oil and Shale Gas Resources." An Assessment of 137 Shale Formations in 41 Countries Outside the United States. June 2013. [See Table 2, page 3. See pages 15-19, Methodology].

818 U.S. Energy Information Administration. "Technically Recoverable Shale Oil and Shale Gas Resources." An Assessment of 137 Shale Formations in 41 Countries Outside the United States. June 2013. [See pages 15-19, Methodology].

819 U.S. EPA. Hydraulic Fracturing for Oil and Gas: Impacts from the Hydraulic Fracturing Water Cycle on Drinking Water Resources in the United States (Final Report). U.S. Environmental Protection Agency, Washington, DC, EPA/600/R-16/236F, 2016. [See the following hyperlink to the study's cautionary note. https://cfpub.epa.gov/ncea/hfstudy/recordisplay.cfm?deid=332990]

820 US Environmental Protection Agency. Unconventional Oil and Natural Gas Development. https://www.epa.gov/uog.

821 Antonio Troiano et at., "A Coulomb stress model for induced seismicity distribution due to fluid injection and withdrawal in deep boreholes." Geophysical Journal International, Volume 195, Issue 1, 1 October 2013, 504–512, https://doi.org/10.1093/gji/ggt229.

822 R.M.H.E. Van Eijs et al., "Correlation between hydrocarbon reservoir properties and induced seismicity in the Netherlands." Engineering Geology, Volume 84, Issues 3–4, 2006, 99-111, https://doi.org/10.1016/j.enggeo.2006.01.002.

823 Richard Davies et al., "Induced seismicity and hydraulic fracturing for the recovery of hydrocarbons." Marine and Petroleum Geology. Volume 45, 2013, 171-185. https://doi.org/10.1016/j.marpetgeo.2013.03.016.

824 A. Helmstetter et al., 2005, "Importance of small earthquakes for stress transfers and earthquake triggering." Journal of Geophysical Research, 110, B05S08, doi: 10.1029/2004JB003286.

825 W. Ellsworth, 2013, "Injection-Induced Earthquakes." Science 12 Jul 2013: Volume 341, Issue 6142, 1225942, DOI: 10.1126/science.1225942.

826 Jenny Suckale, 2010, "Moderate-to-large seismicity induced by hydrocarbon production." The Leading Edge, 29(3), 310-319. https://doi.org/10.1190/1.3353728.

827 F. Rall Walsh III and Mark D. Zoback, "Oklahoma's recent earthquakes and saltwater disposal." Science Advances, 18 Jun 2015: Volume 1, no. 5, e1500195. DOI: 10.1126/sciadv.1500195.

828 "Report on the Hydrocarbon Exploration and Seismicity in Emilia Region." International Commission on Hydrocarbon Exploration and Seismicity in the Emilia Region. ICHESE February 2014. http://mappegis.regione.emilia-romagna.it/gstatico/documenti/ICHESE/ICHESE_Report.pdf. Linked via, http://ambiente.regione.emilia-romagna.it/geologia-en/notizie/news-2014-1/ichese-commission-on-line-the-final-report.

829 Torild van Eck et al., "Seismic hazard due to small-magnitude, shallow-source, induced earthquakes in The Netherlands." Engineering Geology, Volume 87, Issues 1–2, 2006. 105-121. https://doi.org/10.1016/j.enggeo.2006.06.005.

830 Anti-Fracking Movement Is Large. https://www.americansagainstfracking.org/take-action/epa/, https://www.americansagainstfracking.org/about-the-coalition/members/. https://earthdirectory.net/frack.

831 Conduct a Google Search "what countries have banned fracking?" to see what the tabloid newspapers have to say about this subject.

832 See endnote 79.

833   Kjell Aleklett and Colin J. Campbell, "The peak and decline of world oil and gas production." Minerals and Energy-Raw Materials Report 18.1 (2003): 5-20.

834   Ian Chapman, 2014, "The end of Peak Oil? Why this topic is still relevant despite recent denials." Energy Policy, 64. 93-101. http://insight.cumbria.ac.uk/id/eprint/1708/.

835   Harry J. Longwell, "The future of the oil and gas industry: past approaches, new challenges." World Energy Volume 5 No. 3 2002.

836   Jack Zagar, "The End of Cheap "Conventional" Oil. Independent Petroleum Engineering Consultant. http://www.hubbertpeak.com/Zagar/hawaii/.

837   David F. Morehouse, "The Intricate Puzzle of Oil and Gas "Reserves Growth"." Energy Information Administration. Natural Gas Monthly July 1997. http://large.stanford.edu/courses/2014/ph240/liegl1/docs/morehouse.pdf.

838   Kjell Aleklett and Colin J. Campbell, "The peak and decline of world oil and gas production." Minerals and Energy-Raw Materials Report 18.1 (2003): 5-20.

839   T.A. Cook, 2013, "Reserve growth of oil and gas fields—Investigations and applications." U.S. Geological Survey Scientific Investigations Report 2013–5063, 29 p., http://pubs.usgs.gov/sir/2013/5063/.

840   David F. Morehouse, "The Intricate Puzzle of Oil and Gas "Reserves Growth"." Energy Information Administration. Natural Gas Monthly July 1997. http://large.stanford.edu/courses/2014/ph240/liegl1/docs/morehouse.pdf.

841   R.G. Miller, S.R. Sorrell, 2014, "The future of oil supply." Philosophical Transactions of the Royal Society A 372: 20130179. http://dx.doi.org/10.1098/rsta.2013.0179.

842   Robert L. Hirsch et al., "Peaking of World Oil Production: Impacts, Mitigation, and Risk Management." February 2005. https://www.netl.doe.gov/publications/others/pdf/Oil_Peaking_NETL.pdf.

843   M. Höök, 2014, "Depletion rate analysis of fields and regions: a methodological foundation." Fuel, Volume 121, 1 April 2014, 95–108. http://dx.doi.org/10.1016/j.fuel.2013.12.024.

844   David F. Morehouse, "The Intricate Puzzle of Oil and Gas "Reserves Growth"." Energy Information Administration. Natural Gas Monthly July 1997. http://large.stanford.edu/courses/2014/ph240/liegl1/docs/morehouse.pdf.

845   T.A. Cook, 2013, "Reserve growth of oil and gas fields—Investigations and applications." U.S. Geological Survey Scientific Investigations Report 2013–5063, 29 p., http://pubs.usgs.gov/sir/2013/5063/.

846   "Financial Review of the Global Oil and Natural Gas Industry 2015." Markets and Financial Analysis Team. May 2016. https://www.eia.gov/finance/review/archive/pdf/financial_2015.pdf.

847   U.S. Energy Information Administration. "Technically Recoverable Shale Oil and Shale Gas Resources. An Assessment of 137 Shale Formations in 41 Countries Outside the United States." June 2013. [See Table 2, page 3].

848   Data: Oil, gas, and coal data was obtained from the Energy Information Administration from their International Energy Statistics data portal. This graphic utilizes the following data files. Natural gas https://bit.ly/2LC6GBo, Crude Oil https://bit.ly/2IWeEaP, Coal data https://bit.ly/2L6pk3w. Standard fuel units of energy (barrels, cubic feet, and short tons) were converted to British thermal units (Btu) using conversion factors obtained from, https://www.eia.gov/energyexplained/index.php?page=about_btu.

849   Jean Laherrère, Public domain, via Wikimedia Commons. Image description: Global oil discovery and production, showing peak discovery in the 1960s. March 1, 2015. https://commons.wikimedia.org/wiki/File:World_crude_discovery_production_U-2200Gb_LaherrereMar2015.jpg.

850   Data, U.S. Energy Information Administration. International_data. https://bit.ly/2Rw5hiM. [Worldwide production was 97,977,000 barrels per day in 2017.]

851   "All-time low for discovered resources in 2017: Around 7 billion barrels of oil equivalent was discovered." December 21, 2017. https://www.rystadenergy.com/newsevents/news/press-releases/all-time-low-discovered-resources-2017/.

852   "All-time low for discovered resources in 2017: Around 7 billion barrels of oil equivalent was discovered." December 21, 2017. https://www.rystadenergy.com/newsevents/news/press-releases/all-time-low-discovered-resources-2017/.

853 "Declining Reserve Replacement Ratios Deceiving In Resource Play Environment." November. 28, 2017. View Issue. Maurice Smith. JWN Energy. Daily Oil Bulletin. https://www.sproule. com/application/files/2415/1188/2978/Sproule-Declining-Reserve-Replacement-Ratios-Nora-Stewart-Steve-Golko.pdf.

854 Tom Whipple, "Peak Oil Review" Originally published by ASPO-US. December 26, 2017. https://www.resilience.org/stories/2017-12-26/peak-oil-review-dec-26-2017/

855 Harry J. Longwell, "The future of the oil and gas industry: past approaches, new challenges." World Energy Volume 5 No. 3 2002.

856 James A. Baker III Institute For Public Policy of Rice University In Conjunction With a Program Co-Sponsored By Petroleum Industry Research Foundation No. 14, November 2000 http://large.stanford.edu/publications/power/references/baker/reports/docs/study_14.pdf.

857 Jack Zagar, "The End of Cheap "Conventional" Oil." Independent Petroleum Engineering Consultant. http://www.hubbertpeak.com/Zagar/hawaii/.

858 Rystad Energy Annual Review of World Oil Recoverable Resources: Saudi Arabia adds oil resources ahead of IPO June 20, 2017. https://www.rystadenergy.com/newsevents/news/press-releases/2017-annual-oil-recoverable-resource-review/.

859 Wang Hongjun et al., "Assessment of global unconventional oil and gas resources." Petroleum Exploration and Development. 2016, 43(6): 925–940.

860 Rystad Energy Annual Review of World Oil Recoverable Resources: Saudi Arabia adds oil resources ahead of IPO June 20, 2017. https://www.rystadenergy.com/newsevents/news/press-releases/2017-annual-oil-recoverable-resource-review/.

861 Wang Hongjun et al., "Assessment of global unconventional oil and gas resources." Petroleum Exploration and Development 2016, 43(6): 925–940.

862 Tom Whipple, "Peak Oil Review" Originally published by ASPO-US. December 26, 2017. https://www.resilience.org/stories/2017-12-26/peak-oil-review-dec-26-2017/.

863 Kjell Aleklett et al., "The Peak of the Oil Age: Analyzing the World Oil Production Reference Scenario in World Energy Outlook 2008." Energy Policy, Volume 38, no. 3, Elsevier Ltd, 2010, 1398–1414, doi:10.1016/j.enpol.2009.11.021.

864 Peak oil demand and long-run oil prices. https://www.bp.com/en/global/corporate/energy-economics/spencer-dale-group-chief-economist/peak-oil-demand-and-long-run-oil-prices.html.

865 Data: Energy Information Administration data was obtained from: International Energy Statistics. These calculations utilized the following data files. Natural gas https://bit. ly/2LC6GBo, Crude Oil https://bit.ly/2IWeEaP, Coal data https://bit.ly/2L6pk3w. [Comment: These reserve timeline estimates were calculated by dividing the 2013 Energy Information Agency's proven global oil, natural gas, and coal reserves by the 2013 levels of production. This calculation tells us there are 50 years of proven oil and gas, and 130 years of coal reserves left. These reserve timeline estimates do not assume any population or economic growth, or a switch to a cold climate regime, which would accelerate energy demand and reduce timelines.].

866 See endnote 78.

867 See endnote 75.

868 See endnote 76.

869 Ian Chapman, 2014, "The end of Peak Oil? Why this topic is still relevant despite recent denials." Energy Policy, 64 . 93-101. http://insight.cumbria.ac.uk/id/eprint/1708/

870 U.S. Energy Information Administration. Technically Recoverable Shale Oil and Shale Gas Resources. An Assessment of 137 Shale Formations in 41 Countries Outside the United States. June 2013. [See Table 2, page 3 for this data].

871 Data: Oil, gas, and coal data was obtained from the Energy Information Administration from their International Energy Statistics data portal. This graphic utilizes the following data files: Natural gas https://bit.ly/2LC6GBo, Crude Oil https://bit.ly/2IWeEaP, Coal data https://bit. ly/2L6pk3w. Standard fuel units of energy (barrels, cubic feet, and short tons) were converted to British thermal units (Btu) using conversion factors obtained from, https://www.eia.gov/energyexplained/index.php?page=about_btu.

872 "Uranium 2016: Resources, Production and Demand." NEA No. 7301, OECD 2016. A Joint Report by the Nuclear Energy Agency and the International Atomic Energy Agency.

873   Kjell Aleklett et al., "The Peak of the Oil Age: Analyzing the World Oil Production Reference Scenario in World Energy Outlook 2008." Energy Policy, Volume 38, no. 3, Elsevier Ltd, 2010, 1398–1414, doi:10.1016/j.enpol.2009.11.021.

874   Kjell Aleklett et al., "The Peak of the Oil Age: Analyzing the World Oil Production Reference Scenario in World Energy Outlook 2008." Energy Policy, Volume 38, no. 3, Elsevier Ltd, 2010, 1398–1414, doi:10.1016/j.enpol.2009.11.021.

875   J. Cleland, "World Population Growth; Past, Present and Future." Environ Resource Econ (2013) 55: 543. https://doi.org/10.1007/s10640-013-9675-6.

876   U.S. Energy Information Administration, International Data. World population statistics. https://bit.ly/2LF7Gom.

877   U.S. Energy Information Administration, International Data. World population statistics. https://bit.ly/2LF7Gom.

878   Global Europe 2050. Directorate-General for Research and Innovation. 2012. Socio-economic Sciences and Humanities. EUR 25252.

879   J. Cleland, "World Population Growth; Past, Present and Future." Environ Resource Econ (2013) 55: 543. https://doi.org/10.1007/s10640-013-9675-6.

880   Data: (1) Figure 8.3.A: U.S. Energy Information Administration, International Data. World population statistics. https://bit.ly/2LF7Gom. (2) Figure 8.3.B: Oil, gas, and coal data was obtained from the Energy Information Administration from their International Energy Statistics data portal. This graphic utilizes the following data files. Natural gas https://bit.ly/2LC6GBo, Crude Oil https://bit.ly/2IWeEaP, Coal data https://bit.ly/2L6pk3w. Standard fuel units of energy (barrels, cubic feet, and short tons) were converted to British thermal units (Btu) using conversion factors obtained from, https://www.eia.gov/energyexplained/index.php?page=about_btu.

881   Global Europe 2050. Directorate-General for Research and Innovation. 2012. Socio-economic Sciences and Humanities. EUR 25252.

882   Conduct a Google search, "United Nations, World population, 2050, 2100."

883   J. Cleland, "World Population Growth; Past, Present and Future." Environ Resource Econ (2013) 55: 543. https://doi.org/10.1007/s10640-013-9675-6.

884   P. Gerland et al., "World Population Stabilization Unlikely This Century." Science (New York, NY). 2014;346(6206):234-237. doi:10.1126/science.1257469.

885   Principle of Population, As It Affects The Future Improvement of Society. With Remarks on the Speculations of Mr. Godwin, M. Condorcet and Other Writers (1 ed.). London. Printed for J. Johnson in St Paul's Church-yard. 1798.

886   18th Century Timeline: 1700 – 1799. Technology and Science Ruled "the Age of Enlightenment" https://www.thoughtco.com/18th-century-timeline-1992474.

887   The Brundtland Report, "Our Common Future." To view the report, click the link. http://www.sustainabledevelopment2015.org/AdvocacyToolkit/index.php/earth-summit-history/historical-documents/92-our-common-future.

888   The Brundtland Report, "Our Common Future." To view the report, click the link. http://www.sustainabledevelopment2015.org/AdvocacyToolkit/index.php/earth-summit-history/historical-documents/92-our-common-future.

889   The Brundtland Report, "Our Common Future." To view the report, click the link. http://www.sustainabledevelopment2015.org/AdvocacyToolkit/index.php/earth-summit-history/historical-documents/92-our-common-future.

890   U.S. Energy Information Administration. International Energy Statistics Portal. https://bit.ly/2JahEk1. (2017 World Gross Domestic Product US$113,655.4 billion, at 2010 PPP).

891   Martin L. Weitzman, "A Review of The Stern Review on the Economics of Climate Change." Journal of Economic Literature. Volume XLV (September 2007), 703–724.

892   Richard S. J. Tol and Gary W. Yohe, "A Review of the Stern Review." World Economics. Volume 7, No. 4, October–December 2006.

893   Eric Beinhocker et al., "The carbon productivity challenge: Curbing climate change and sustaining economic growth." McKinsey Global Institute. June 2008.

894   European commission. The Paris Agreement. The Paris climate conference (COP21), December 2015. https://ec.europa.eu/clima/policies/international/negotiations/paris_en.

895 European commission. International climate finance. https://ec.europa.eu/clima/policies/
international/finance_en.
896 OECD, 2018, Insurance spending (indicator). doi: 10.1787/adb73055-en (Accessed on 24 July
2018). [Notes: Insurance spending is defined as the ratio of direct gross premiums to GDP,
which represents the relative importance of the insurance industry in the domestic economy.
This indicator is expressed as a percentage of GDP.].
897 World Life And Nonlife Insurance In 2016. See "Insurance Premiums Per Capita And Percent
Of Gross Domestic Product (GDP)." Insurance Information Institute. Insurance Handbook.
World Insurance Marketplace. https://www.iii.org/publications/insurance-handbook/economic-
and-financial-data/world-insurance-marketplace.
898 Richard G. Newell et al., 2013, "Carbon Markets 15 Years after Kyoto: Lessons Learned, New
Challenges." Journal of Economic Perspectives, 27 (1): 123-46. DOI: 10.1257/jep.27.1.123..
899 "Where Carbon Is Taxed." https://www.carbontax.org/where-carbon-is-taxed/.
900 Carbon Tax Center. "Pricing carbon efficiently and equitably." No U.S. state has a carbon tax.
https://www.carbontax.org/states/.
901 Kevin Kennedy et al., "How the US Can Meet its Emissions Targets with a Carbon Tax." World
Resources Institute. June 21 2018. http://www.wri.org/blog/2018/06/how-us-can-meet-its-
emissions-targets-carbon-tax?
902 EU Emissions Trading System (EU ETS). https://ec.europa.eu/clima/policies/ets_en.
903 Carbon Tax Center. "Pricing carbon efficiently and equitably." Where Carbon Is Taxed. https://
www.carbontax.org/where-carbon-is-taxed/.
904 EU Emissions Trading System (EU ETS). https://ec.europa.eu/clima/policies/ets_en.
905 Richard G. Newell et al., 2013, "Carbon Markets 15 Years after Kyoto: Lessons Learned, New
Challenges." Journal of Economic Perspectives, 27 (1): 123-46. DOI: 10.1257/jep.27.1.123.
906 EU Emissions Trading System (EU ETS). https://ec.europa.eu/clima/policies/ets_en.
907 Carbon Pricing Watch 2016 (May). Washington, DC. Doi: 978-1-4648-0930-9-1 License:
Creative Commons Attribution CC BY 3.0 IGO. This is an adaptation of an original work by
The World Bank. Responsibility for the views and opinions expressed in the adaptation rests
solely with the author or authors of the adaptation and are not endorsed by The World Bank.
https://www.ecofys.com/files/files/world-bank-group_ecofys-carbon-pricing-watch_160525.
pdf. (Accessed 03/04/2018).
908 Richard G. Newell et al., 2013, "Carbon Markets 15 Years after Kyoto: Lessons Learned, New
Challenges." Journal of Economic Perspectives, 27 (1): 123-46. DOI: 10.1257/jep.27.1.123.
909 Sarabjeet Hayer, Policy Department A: Economic and Scientific Policy. "Fossil Fuel Subsidies.
In-Depth Analysis." Directorate General For Internal Policies. IP/A/ENVI 2016-18-REV.
Mars 2017 http://www.europarl.europa.eu/RegData/etudes/IDAN/2017/595372/IPOL_
IDA(2017)595372_EN.pdf.
910 Ambrus Bárány and Dalia Grigonytė, "Measuring Fossil Fuel Subsidies." ECFIN Economic
Briefs 40. March 2015. Brussels. pdf. 13 pages Graph. Bibliogr. ISSN:1831-4473.
911 Tobias S. Schmidt et al., "Renewable energy policy as an enabler of fossil fuel subsidy reform?"
Applying a socio-technical perspective to the cases of South Africa and Tunisia. Global
Environmental Change. Volume 45 (2017) 99–110. DOI: 10.1016/j.gloenvcha.2017.05.004.
912 OECD Survey of Investment Regulation of Pension Funds, 2018. Available at, https://www.
oecd.org/daf/fin/private-pensions/2018-Survey-Investment-Regulation-Pension-Funds.pdf.
913 Data: U.S. Energy Information Administration. International Energy Statistics. Generation of
Electricity Billion Kwh. https://bit.ly/2JtfawJ. (Last Accessed June 1 2018).
914 Energy Information Administration. International Energy Outlook 2017. Release Date:
September 14, 2017, Report Number: DOE/EIA-0484(2017). Data extracted from. Table F1.
Total world delivered energy consumption by end-use sector and fuel, Reference case, 2015-50.
https://www.eia.gov/outlooks/ieo/excel/appf_tables.xlsx. Downloaded 06/04/2018.
915 Data: Energy Information Administration. Primary energy and electricity consumption data
were obtained from the International Energy Statistics data portal. World primary energy
consumption by country (Quadrillion British thermal units, 1015 Btu), https://bit.ly/2xAB4sR.
World electricity consumption by country (Billion Kwh) https://bit.ly/2J8TJ0t. Kilowatt hours
(Kwh) were converted to British thermal units (Btu) using the 2015 conversion factor obtained

from, https://www.eia.gov/energyexplained/index.php?page=about_btu. 1 kilowatt-hour = 3,412 Btu.

916   U.S. Energy Information Administration (EIA). International Energy Outlook 2017. Release Date: September 14, 2017. Report Number: DOE/EIA-0484(2017). https://www.eia.gov/outlooks/ieo/pdf/industrial.pdf.

917   U.S. Energy Information Administration (EIA), Annual Energy Outlook 2017, DOE/EIA-0383(2017) (Washington, DC: January 2017). Data used; https://www.eia.gov/outlooks/ieo/excel/appf_tables.xlsx.

918   Data: International Energy Outlook 2017. Release Date: September 14, 2017, Report Number: DOE/EIA-0484(2017). Data extracted from. Table F1. Total world delivered energy consumption by end-use sector and fuel, Reference case, 2015-50. https://www.eia.gov/outlooks/ieo/excel/appf_tables.xlsx. Downloaded 06/04/2018.

919   U.S. Energy Information Administration (EIA). International Energy Outlook 2017. Release Date: September 14, 2017. Report Number: DOE/EIA-0484(2017). https://www.eia.gov/outlooks/ieo/pdf/industrial.pdf.

920   Data: Report: International Energy Outlook 2017. Release Date: September 14, 2017, Report Number: DOE/EIA-0484(2017). (1) Figure 9.3.A: Data extracted from. Table L1. Transportation sector energy consumption by region and fuel, Reference case, 2015-50. https://www.eia.gov/outlooks/ieo/excel/appl_tables.xlsx. Downloaded 06/04/2018. (2) Figure 9.3.B: Data extracted from. Table K1. Industrial sector energy consumption by region and sector, Reference case, 2015-50. https://www.eia.gov/outlooks/ieo/excel/appk_tables.xlsx. Downloaded 06/04/2018.

921   List of OECD Member countries - Ratification of the Convention on the OECD. http://www.oecd.org/about/membersandpartners/list-oecd-member-countries.htm.

922   U.S. Energy Information Administration (EIA). International Energy Outlook 2017. Release Date: September 14, 2017, Report Number: DOE/EIA-0484(2017). Data adapted from https://www.eia.gov/outlooks/ieo/excel/appl_tables.xlsx. Accessed 06/04/2018.

923   U.S. Energy Information Administration. Transportation sector energy consumption. Overview. International Energy Outlook 2016. https://www.eia.gov/outlooks/ieo/pdf/transportation.pdf. [See Overview on page 1 of 11 of Chapter 8].

924   U.S. Energy Information Administration. Transportation sector energy consumption. Overview. International Energy Outlook 2016. https://www.eia.gov/outlooks/ieo/pdf/transportation.pdf. [See Overview on page 4 and 5 of 11 of Chapter 8].

925   International Energy Outlook 2017. Release Date: September 14, 2017, Report Number: DOE/EIA-0484(2017). Data extracted from. Table F1. Total world delivered energy consumption by end-use sector and fuel. https://bit.ly/2NM2GPG. (Accessed 06/04/2018).

926   U.S. Energy Information Administration. International Energy Statistics. Generation of Electricity Billion Kwh. https://bit.ly/2JtfawJ. (Last Accessed Jun 01 2018).

927   International Energy Outlook 2017. Release Date: September 14, 2017, Report Number: DOE/EIA-0484(2017). Data extracted from. Table F1. Total world delivered energy consumption by end-use sector and fuel. https://bit.ly/2NM2GPG. (Accessed 06/04/2018).

928   Data: U.S. Energy Information Administration. International Energy Statistics. Generation of Electricity Billion Kwh. https://bit.ly/2JtfawJ. (Last Accessed June 1 2018).

929   Data: US Energy Information Administration: World electricity generation data by category and country were obtained from the International Energy Statistics data portal. https://bit.ly/2J8TJ0t.

930   US Department of Energy. Office of Energy Efficiency & Renewable Energy. Hydropower Vision. A New Chapter for America's. Renewable Electricity Source. https://www.energy.gov/sites/prod/files/2018/02/f49/Hydropower-Vision-021518.pdf.

931   Microhydropower Systems. https://www.energy.gov/energysaver/buying-and-making-electricity/microhydropower-systems.

932   Run of River Power. http://www.energybc.ca/runofriver.html.

933   U.S. Energy Information Administration. International Energy Statistics. Generation of Electricity Billion Kwh. https://bit.ly/2JtfawJ. Last Accessed June 01 2018.

934   US Energy Information Administration: World electricity generation data by category and country were obtained from the International Energy Statistics data portal. https://bit.ly/2J8TJ0t.

935 Camila Stark et al., "Renewable Electricity: Insights for the Coming Decade." Joint Institute for Strategic Energy Analysis. Technical Report NREL/TP-6A50-63604. February 2015.

936 Camila Stark et al., "Renewable Electricity: Insights for the Coming Decade." Joint Institute for Strategic Energy Analysis. Technical Report NREL/TP-6A50-63604. February 2015.

937 Camila Stark et al., "Renewable Electricity: Insights for the Coming Decade." Joint Institute for Strategic Energy Analysis. Technical Report NREL/TP-6A50-63604. February 2015.

938 Camila Stark et al., "Renewable Electricity: Insights for the Coming Decade." Joint Institute for Strategic Energy Analysis. Technical Report NREL/TP-6A50-63604. February 2015.

939 US Department of energy. Solar Energy Technologies Office. Photovoltaics. Photovoltaics fact sheet. https://www.energy.gov/sites/prod/files/2016/02/f29/PV%20Fact%20Sheet-508web.pdf.

940 US Energy Information Administration: World electricity generation data by category and country were obtained from the International Energy Statistics data portal. https://bit.ly/2J8TJ0t.

941 Camila Stark et al., "Renewable Electricity: Insights for the Coming Decade." Joint Institute for Strategic Energy Analysis. Technical Report NREL/TP-6A50-63604. February 2015.

942 U.S. Department of Energy Solar Energy Technologies Office. Concentrating Solar Power. https://www.energy.gov/eere/solar/concentrating-solar-power.

943   Geothermal energy and Engineered Geothermal System. https://setis.ec.europa.eu/technologies/geothermal-energy.

944 EGEC Geothermal. The Voice of Geothermal in Europe. https://www.egec.org/about/#aboutgeot.

945 U.S. Department of Energy. Office of Energy Efficiency & Renewable Energy. Geothermal Technologies Office. Geothermal. https://www.energy.gov/eere/geothermal/geothermal-energy-us-department-energy.

946 US Energy Information Administration: World electricity generation data by category and country were obtained from the International Energy Statistics data portal. https://bit.ly/2J8TJ0t.

947 EGEC Geothermal. The Voice of Geothermal in Europe. https://www.egec.org/about/#aboutgeot.

948 https://www.energy.gov/eere/buildings/building-technologies-office, https://www.energy.gov/energysaver/choosing-and-installing-geothermal-heat-pumps.

949 European Commission Study Report. Optimal use of biogas from waste streams. An assessment of the potential of biogas from digestion in the EU beyond 2020. Study Authors. CE Delft: Bettina Kampman, et al. December 2016.

950 United States Environmental Protection Agency. Learn About Biogas Recovery. https://www.epa.gov/agstar/learn-about-biogas-recovery.

951 United States Environmental Protection Agency. Recovering Value from Waste. Anaerobic Digester System Basics. December 2011. https://www.epa.gov/sites/production/files/2014-12/documents/recovering_value_from_waste.pdf.

952 United States Environmental Protection Agency. Learn About Biogas Recovery. https://www.epa.gov/agstar/learn-about-biogas-recovery.

953 United States Environmental Protection Agency. Recovering Value from Waste. Anaerobic Digester System Basics. December 2011. https://www.epa.gov/sites/production/files/2014-12/documents/recovering_value_from_waste.pdf.

954 European Commission Study Report. Optimal use of biogas from waste streams. An assessment of the potential of biogas from digestion in the EU beyond 2020. Study Authors. CE Delft: Bettina Kampman, et al. December 2016.

955 M.E. Brown et al., 2015, "Climate Change, Global Food Security, and the U.S. Food System." http://www.usda.gov/oce/climate_change/FoodSecurity2015Assessment/FullAssessment.pdf.

956 European Commission. Biomass. https://ec.europa.eu/energy/en/topics/renewable-energy/biomass.

957 https://www.epa.gov/agstar/agstar-data-and-trend. This page provides national market data and trends related to these biogas recovery systems, exemplified by anaerobic digesters operating on livestock farms across the United States.

958 United States Environmental Protection Agency. AgSTAR Stories from the Farm. https://www.epa.gov/agstar/agstar-stories-farm. https://www.epa.gov/agstar/agstar-data-and-trends#adfacts.

959 Renewable Energy Production on Farms. https://ag.umass.edu/crops-dairy-livestock-equine/fact-sheets/renewable-energy-production-on-farms.

960 Opportunities for Combined Heat and Power at Wastewater Treatment Facilities: Market Analysis and Lessons from the Field. October 2011. Report prepared by: Eastern Research Group, Inc. (ERG) and Resource Dynamics Corporation (RDC) for the U.S. Environmental Protection Agency, and Combined Heat and Power Partnership, October 2011.

961 D. Panepinto et al., "Energy from Biomass for Sustainable Cities." IOP Conf. Series: Earth and Environmental Science 72 (2017) 012021 doi:10.1088/1755-1315/72/1/012021.

962 Turning Food Waste into Energy to Power Homes. https://www.biogasworld.com/news/turning-food-waste-into-energy-to-power-homes/.

963 Smart City Sweden. http://smartcitysweden.com/focus-areas/bio-energy/

964 Asia Biomass Office. https://www.asiabiomass.jp/english/topics/1612_02.html.

965 Energy from municipal solid waste. https://www.eia.gov/energyexplained/?page=biomass_waste_to_energy.

966 Using biogas technologies to reduce methane emissions from waste https://www.biogas.org.nz/.

967 Forestry New Zealand. Planting one billion trees. https://www.mpi.govt.nz/funding-and-programmes/forestry/planting-one-billion-trees/.

968 Australian Government. Department of the Environment and Energy. 20 million Trees by 2020. http://www.environment.gov.au/land/20-million-trees.

969 The Billion Tree Campaign. https://www.plant-for-the-planet.org/en/treecounter/billion-tree-campaign-2.

970 The Great Green Wall in Africa (Sahel). http://www.greatgreenwall.org/great-green-wall/#great-green-wall-internal.

971 Andrés Viña et al., "Effects of conservation policy on China's forest recovery." Science Advances 18 Mar 2016: Volume 2, no. 3, e1500965. DOI: 10.1126/sciadv.1500965.

972 The Power of Trees. http://trees.org/.

973 National Grid. Electricity in New Zealand - Electricity Authority. https://www.ea.govt.nz/dmsdocument/20410.

974 European Network of Transmission System Operators. ENTSO-E Transmission System Map. https://www.entsoe.eu/data/map/.

975 Department of Energy. Office of Electricity. Grid Modernization and the Smart Grid. https://www.energy.gov/oe/activities/technology-development/grid-modernization-and-smart-grid.

976 European Network of Transmission System Operators for Electricity. E-Highway2050. Unveiling the Electricity Highways Project Results: "Europe's Future Secure and Sustainable Electricity Infrastructure". https://www.entsoe.eu/outlooks/ehighways-2050/.

977 European Commission. Smart-grid: from innovation to deployment. Communication from the commission to the European Parliament, The council, The European Economic and Social committee and the Committee of the Regions. Smart-grid: from innovation to deployment /* COM/2011/0202 final */ http://eur-lex.europa.eu/legal-content/EN/TXT/?qid=1409145686999&uri=CELEX:52011DC0202.

978 Department of Energy. Office of Electricity. Grid Modernization and the Smart Grid. https://www.energy.gov/oe/activities/technology-development/grid-modernization-and-smart-grid.

979 Camila Stark et al., "Renewable Electricity: Insights for the Coming Decade." Joint Institute for Strategic Energy Analysis. Technical Report NREL/TP-6A50-63604. 2015. https://www.nrel.gov/docs/fy15osti/63604.pdf.

980 Katrin Schaber et al., "Parametric study of variable renewable energy integration in Europe: Advantages and costs of transmission grid extensions." Energy Policy, Volume 42, 2012, 498-508, https://doi.org/10.1016/j.enpol.2011.12.016.

981 Department of Energy's Office of Electricity Delivery and Energy Reliability. What is the Smart Grid? https://www.smartgrid.gov/the_smart_grid/index.html.

982 European Commission. Smart-grid: from innovation to deployment. Communication from the commission to the European Parliament, The council, The European Economic and Social committee and the Committee of the Regions. http://eur-lex.europa.eu/legal-content/EN/TXT/?qid=1409145686999&uri=CELEX:52011DC0202.

983 European Commission. Smart-grid and meters. https://ec.europa.eu/energy/en/topics/markets-and-consumers/smart-grids-and-meters.

984 Department of Electricity. Office of Electricity. https://www.energy.gov/oe/activities/
technology-development/grid-modernization-and-smart-grid.
985 European Commission. Energy storage. https://ec.europa.eu/energy/en/topics/technology-and-
innovation/energy-storage.
986 Department of Energy's Office of Electricity Delivery and Energy Reliability. Pumped-storage
hydropower. https://www.energy.gov/eere/articles/5-promising-water-power-technologies.
987 Energy Storage Association. White Paper, 35x25: A Vision for Energy Storage. http://
energystorage.org/vision2025.
988 Department of Energy. Energy Storage. https://www.energy.gov/science-innovation/electric-
power/storage.
989 National Renewable Energy Laboratory. Valuing the Resilience Provided by Solar and Battery
Energy Storage Systems. https://www.energy.gov/sites/prod/files/2018/03/f49/Valuing-
Resilience.pdf.
990 Energy Storage Association. Energy Storage Technologies. http://energystorage.org/energy-
storage/energy-storage-technologies.
991 W. Breuer et al., Siemens AG, Energy Sector, Power Transmission Division, Germany. "Highly
Efficient Solutions for Smart and Bulk Power Transmission of "Green Energy"." Presented at
21TH World Energy Congress, Montreal, Canada. September 12–16, 2010. Updated Version,
July 2011.
992 Alexandre Oudalov et al., "A Method for a Comparison of Bulk Energy Transport Systems."
Environ. Sci. Technol., 2009, 43 (20), 7619–7625. DOI: 10.1021/es900687e.
993 European Commission. EU Science Hub. Report. The European Commission's science
and knowledge service. "HVDC Submarine Power Cables in the World: State-of-the-Art
Knowledge." https://ec.europa.eu/jrc/en/publication/hvdc-submarine-power-cables-world-state-
art-knowledge.
994 P. Kuhn et al., "Challenges and opportunities of power systems from smart homes to super-
grids." Ambio (2016) 45(Supplement 1): 50. https://doi.org/10.1007/s13280-015-0733-x.
995 European Commission. Smart-grid Task Force. https://ec.europa.eu/energy/en/topics/markets-
and-consumers/smart-grids-and-meters/smart-grids-task-force.
996 US Department of Energy. Office of Electricity Delivery and Energy Reliability. What is the
Smart Grid? https://www.smartgrid.gov/the_smart_grid/smart_grid.html.
997 European Commission. European Parliament - Europa EU. Directorate General for Internal
Policies. Policy Department A: Economic and Scientific Policy. Industry Research and
Energy. Decentralized Energy Systems. Decentralized Energy Production. IP/A/ITRE/
ST/2009-16 JUNE 2010. PE 440.280. www.europarl.europa.eu/document/activities/
cont/.../20110629ATT22897EN.pdf.
998 US Department of Energy. Office of Energy Efficiency & Renewable Energy. How Distributed
Wind Works. https://www.energy.gov/eere/wind/how-distributed-wind-works.
999 World Alliance for Decentralized Energy. http://www.localpower.org/abt_mission.html.
1000 Peter Alstone et al., "Decentralized energy systems for clean electricity access." Nature
Climate Change Volume 5, 305–314 (2015).
1001 European Commission. Energy Strategy and Energy Union. Secure, competitive, and
sustainable energy. https://ec.europa.eu/energy/en/topics/energy-strategy-and-energy-union,
https://ec.europa.eu/energy/en/topics/enforcement-laws.
1002 Federal Energy Regulatory Commission. Office of Energy Market Regulation (OEMR). https://
www.ferc.gov/about/offices/oemr.asp.
1003 National Renewable Energy Laboratory. State, Local, & Tribal Governments. Feed-In Tariffs.
https://www.nrel.gov/technical-assistance/basics-tariffs.html.
1004 US Department of Energy. Office of Energy Efficiency & Renewable Energy Feed-in Tariff
Resources. https://www.energy.gov/eere/slsc/feed-tariff-resources.
1005 US Energy Information Agency. Feed-in tariff: A policy tool encouraging deployment of
renewable electricity technologies. https://www.eia.gov/todayinenergy/detail.php?id=11471.
1006 https://web.archive.org/web/20090509184329/http://ec.europa.eu/energy/climate_actions/
doc/2008_res_working_document_en.pdf. Page 3 of 38.

1007 National Renewable Energy Laboratory. City-Level Energy Decision Making: Examples from 20 Cities. https://www.nrel.gov/technical-assistance/blog/posts/city-level-energy-decision-making-examples-from-20-cities.html.

1008 Renewable Energy in Cities: State of the Movement. https://www.renewablecities.ca/articles/renewable-energy-in-cities-state-of-the-movement.

1009 City of Vancouver. Greenest City 2020 Action Plan. 2016-2017 Implementation Update. vancouver.ca/green-vancouver/39764.aspx.

1010 Tania Urmee et al., "Green Growth in cities: two Australian cases." Renew. Energy Environ. Sustain. 2, 43 (2017). T. Urmee et al., published by EDP Sciences, 2017. DOI: 10.1051/rees/2017007.

1011 100% Renewable Energy Cities & Regions Network. http://www.iclei.org/activities/agendas/low-carbon-city/iclei-100re-cities-regions-network.html.

1012 Top 5 Green Public Transport Projects Globally. https://impact4all.org/revealed-top-5-renewable-public-transport-systems/.

1013 The International Council on Clean Transportation (ICCT). 2017 Global Update. Light Duty Vehicle. Greenhouse Gas and Fuel Economy Standards. Zifei Yang and Anup Bandivadekar. https://www.theicct.org/sites/default/files/publications/2017-Global-LDV-Standards-Update_ICCT-Report_23062017_vF.pdf.

1014 Optimal use of biogas from waste streams. An assessment of the potential of biogas from digestion in the EU beyond 2020. https://ec.europa.eu/energy/sites/ener/files/documents/ce_delft_3g84_biogas_beyond_2020_final_report.pdf. [See transport sector page 39].

1015 Clean transport, Urban transport. Green propulsion in transport. https://ec.europa.eu/transport/themes/urban/vehicles/road_en.

1016 European Biofuels Technology Platform. Liquid, synthetic hydrocarbons. http://www.etipbioenergy.eu/images/synthetic-hydrocarbons-fact-sheet.pdf.

1017 IRENA, 2017, Biogas for road vehicles: Technology brief. International Renewable Energy Agency, Abu Dhabi.

1018 J. Shen et al., 2002, "Opportunities for the Early Production of Fischer-Tropsch (F-T) Fuels in the U.S. An Overview." US Department of Energy, 8th Diesel Emissions Reduction Conference (DEER). August 2002. https://www.eere.energy.gov/vehiclesandfuels/pdfs/deer_2002/session4/2002_deer_shen.pdf.

1019 Synthetic Diesel Fuel. https://www.dieselnet.com/tech/fuel_syn.php.

1020 Synthetic Fuels. http://www.futurecars.com/futurefuels/synthetic_fuels.html.

1021 Paulina Jaramillo et al., "Comparative Analysis of the Production Costs and Life-Cycle GHG Emissions of FT Liquid Fuels from Coal and Natural Gas." Environmental Science & Technology 2008 42 (20), 7559-7565. DOI: 10.1021/es8002074.

1022 NGV Global's Natural Gas Vehicle Knowledge Base. http://www.iangv.org/.

1023 Alternative Fuels Data Center. Natural Gas Vehicles. https://www.afdc.energy.gov/vehicles/natural_gas.html.

1024 Are natural gas cars a real alternative? https://www.energuide.be/en/questions-answers/are-natural-gas-cars-a-real-alternative/198/.

1025 Global Water Security. Intelligence Community Assessment, ICA 2012-08, 2 February 2012. https://www.dni.gov/files/documents/Special%20Report_ICA%20Global%20Water%20Security.pdf.

1026 European Roadmap Electrification of Road Transport. 3rd Edition, Version: 10. June 2017.

1027 European Commission. Electrification of the Transport System Studies and reports. Directorate-General for Research and Innovation. 2017 Smart Green and Integrated Transport. https://bit.ly/2syrC3y.

1028 Trafikverket. First electric road in Sweden inaugurated. News release 22 June 2016. http://www.trafikverket.se/en/startpage/about-us/news/2016/2016-06/first-electric-road-in-sweden-inaugurated.

1029 U.S. Department of Energy, Office of Energy Efficiency and Renewable Energy. Fuel Cell Technologies Market Report 2016. Prepared by Sandra Curtin and Jennifer Gangi of the Fuel Cell and Hydrogen Energy Association, in Washington, D.C. https://energy.gov/sites/prod/files/2017/10/f37/fcto_2016_market_report.pdf.

1030 Hydrogen Production: Photoelectrochemical Water Splitting. https://energy.gov/eere/fuelcells/
hydrogen-production-photoelectrochemical-water-splitting.
1031 Robert K. Booth et al., "A severe centennial-scale drought in midcontinental North America
4200 years ago and apparent global linkages." The Holocene. Volume15, Issue 3, 321 – 328.
2005. https://doi.org/10.1191/0959683605hl825ft.
1032 J. Stanley et al., 2003, "Nile flow failure at the end of the Old Kingdom, Egypt: Strontium
isotopic and petrologic evidence." Geoarchaeology, 18: 395-402. doi:10.1002/gea.10065.
1033 Ann Gibbons, "How the Akkadian Empire Was Hung Out to Dry." Science 20 Aug 1993:
Volume 261, Issue 5124, 985. DOI: 10.1126/science.261.5124.985.
1034 Jianjun Wang, "The abrupt climate change near 4,400 year BP on the cultural transition in
Yuchisi, China and its global linkage." Scientific Reports | 6:27723 | DOI: 10.1038/srep27723.
https://www.nature.com/articles/srep27723.pdf.
1035 A.S. Richey et al., 2015, "Quantifying renewable groundwater stress with GRACE." Water
Resources Research, 51, 5217–5238, doi:10.1002/2015WR017349, and NASA via https://www.
nasa.gov/jpl/grace/study-third-of-big-groundwater-basins-in-distress.
1036 Christer Nilsson et al., "Fragmentation and Flow Regulation of the World's Large
River Systems." Science 15 Apr 2005: Volume 308, Issue 5720, 405-408. DOI: 10.1126/
science.1107887.
1037 Muhammad Mizanur Rahaman and Olli Varis, 2005, "Integrated water resources management:
evolution, prospects and future challenges, Sustainability." Science, Practice and Policy, 1:1,
15-21, DOI: 10.1080/15487733.2005.11907961.
1038 Arjen Y. Hoekstra, Mesfin M. Mekonnen, "The water footprint of humanity." Proceedings of
the National Academy of Sciences Feb 2012, 109 (9) 3232-3237; DOI: 10.1073/pnas.1109936109.
1039 European Commission. Water use in industry. Cooling for electricity production dominates
water use in industry. http://ec.europa.eu/eurostat/statistics-explained/index.php/
Archive:Water_use_in_industry.
1040 R. Quentin Grafton et al., "Global insights into water resources, climate change
and governance." Nature Climate Change Volume 3, 315–321 (2013). DOI: 10.1038/
NCLIMATE1746.
1041 Global Water Security. Intelligence Community Assessment, ICA 2012-08, 2 February 2012.
https://www.dni.gov/files/documents/Special%20Report_ICA%20Global%20Water%20
Security.pdf.
1042 Mark W. Rosegrant et al., "Water for Agriculture: Maintaining Food Security Under Growing
Scarcity (November 2009)." Annual Review of Environment and Resources, Volume 34, 205-
222, 2009. Available at SSRN: https://ssrn.com/abstract=1599085 or http://dx.doi.org/10.1146/
annurev.environ.030308.090351.
1043 Arjen Y. Hoekstra, Mesfin M. Mekonnen, "The water footprint of humanity." Proceedings of
the National Academy of Sciences Feb 2012, 109 (9) 3232-3237; DOI: 10.1073/pnas.1109936109.
1044 Y. Wada et al., 2010, "Global depletion of groundwater resources." Geophysical Research
Letters, 37, L20402, doi: 10.1029/2010GL044571.
1045 A. S. Richey et al., 2015, "Quantifying renewable groundwater stress with GRACE." Water
Resources Research, 51, 5217–5238, doi:10.1002/2015WR017349, and NASA via https://www.
nasa.gov/jpl/grace/study-third-of-big-groundwater-basins-in-distress.
1046 Tom Gleeson et al., "Water balance of global aquifers revealed by groundwater footprint."
Nature Volume 488, 197–200 (09 August 2012). doi:10.1038/nature11295.
1047 H. Munia et al. "Water stress in global transboundary river basins: significance
of upstream water use on downstream stress." Environment Research Letter
11 (2016) 014002. doi:10.1088/1748-9326/11/1/014002 http://iopscience.iop.org/
article/10.1088/1748-9326/11/1/014002/pdf.
1048 V.M. Tiwari et al., 2009, "Dwindling groundwater resources in northern India, from satellite
gravity observations." Geophysical Research Letters, 36, L18401, doi: 10.1029/2009GL039401.
1049 R.B. Jackson et al., 2001, "Water in a Changing World." Ecological Applications, 11: 1027-1045.
doi:10.1890/1051-0761(2001)011[1027:WIACW]2.0.CO;2.
1050 Prasad S. Thenkabail et al., "Global irrigated area map (GIAM), derived from remote sensing,
for the end of the last millennium." International Journal of Remote Sensing. Volume 30, 2009 -
Issue 14. https://doi.org/10.1080/01431160802698919.

1051 M.M. Mekonnen and A.Y. Hoekstra, "The green, blue and grey water footprint of crops and derived crop products, Hydrology and Earth System Sciences." 15, 1577-1600, https://doi.org/10.5194/hess-15-1577-2011, 2011.

1052 US Energy Information Administration. International Energy Statistics data portal. World population statistics by region. http://bit.ly/2JGsXju.

1053 David Satterthwaite et al., "Adapting to Climate Change in Urban Areas. The possibilities and constraints in low- and middle-income nations." http://pubs.iied.org/pdfs/10549IIED.pdf.

1054 K.C. Seto et al., 2011, "A Meta-Analysis of Global Urban Land Expansion." PLoS ONE 6(8): e23777. https://doi.org/10.1371/journal.pone.0023777.

1055 Megacities: Bangalore, Bangkok, Beijing, Bogota, Bombay, Buenos Aires, Cairo, Calcutta, Changzhou, Chengdu, Chennai, Chongqing, Dehli, Dhaka, Guangzhou-Foshan, Harbin, Hangzhou, Istanbul, Jakarta, Jinan, Karachi, Kinshasa, Lagos, Lahore, Lima, London, Los Angeles, Madras, Manila, Mexico City, Moscow, Mumbai, Nagoya, Nanjing, New York, Paris, Rio de Janeiro, Sao Paulo, Seoul, Shanghai, Shantou, Tehran, Tianjin, Tokyo, Wuhan, Xi'an.

1056 Mesfin M. Mekonnen, Arjen Y. Hoekstra, "Four billion people facing severe water scarcity." Science Advances. 12 Feb 2016: e1500323. DOI: 10.1126/sciadv.1500323.

1057 M. Kummu et al., "The world's road to water scarcity: shortage and stress in the 20th century and pathways towards sustainability." Scientific Reports Volume 6, Article number: 38495 (2016). DOI: 10.1038/srep38495.

1058 USAID. Water and Sanitation. https://www.usaid.gov/what-we-do/water-and-sanitation.

1059 The US Geological Survey. The USGS Water Science School. The World's Water. https://water.usgs.gov/edu/earthwherewater.html.

1060 The US Geological Survey. The water cycle. https://water.usgs.gov/edu/watercycle.html.

1061 The US Geological Survey. https://water.usgs.gov/edu/watercyclegwdischarge.html.

1062 A.S. Richey at al., 2015, "Quantifying renewable groundwater stress with GRACE." Water Resources Research. 51, 5217–5238, doi:10.1002/2015WR017349, and NASA via https://www.nasa.gov/jpl/grace/study-third-of-big-groundwater-basins-in-distress.

1063 H. Munia et al., "Water stress in global transboundary river basins: significance of upstream water use on downstream stress." Environment Research Letter 11 (2016) 014002. doi:10.1088/1748-9326/11/1/014002 http://iopscience.iop.org/article/10.1088/1748-9326/11/1/014002/pdf.

1064 Tom Gleeson et al., "Water balance of global aquifers revealed by groundwater footprint." Nature Volume 488, 197–200 (09 August 2012). doi:10.1038/nature11295.

1065 C.A. Schlosser et al., 2014, "The future of global water stress: An integrated assessment." Earth's Future, 2: 341-361. doi:10.1002/2014EF000238.

1066 R.B. Jackson et al., 2001, "Water in a Changing World." Ecological Applications, 11: 1027-1045. doi:10.1890/1051-0761(2001)011[1027:WIACW]2.0.CO;2.

1067 R. Quentin Grafton et al., "Global insights into water resources, climate change and governance." Nature Climate Change Volume 3, 315–321 (2013). DOI: 10.1038/NCLIMATE1746.

1068 USAID. Water and Sanitation. https://www.usaid.gov/what-we-do/water-and-sanitation.

1069 Mesfin M. Mekonnen, Arjen Y. Hoekstra, "Four billion people facing severe water scarcity." Science Advances. 12 Feb 2016: e1500323. DOI: 10.1126/sciadv.1500323.

1070 M. Kummu et al., "The world's road to water scarcity: shortage and stress in the 20th century and pathways towards sustainability." Scientific Reports Volume 6, Article number: 38495 (2016). DOI: 10.1038/srep38495.

1071 The United States Geological Survey. Ground water. https://pubs.usgs.gov/gip/gw/gw_a.html, https://pubs.usgs.gov/gip/gw/how_a.html.

1072 A.S. Richey et al., 2015, "Quantifying renewable groundwater stress with GRACE." Water Resources Research, 51, 5217–5238, doi:10.1002/2015WR017349, and NASA via https://www.nasa.gov/jpl/grace/study-third-of-big-groundwater-basins-in-distress.

1073 Y. Wada et al., 2010, "Global depletion of groundwater resources." Geophysical Research Letters, 37, L20402, doi: 10.1029/2010GL044571.

1074 Tom Gleeson et al., "Water balance of global aquifers revealed by groundwater footprint." Nature Volume 488, 197–200 (09 August 2012). doi:10.1038/nature11295.

1075 R.B. Jackson et al., 2001, "Water in a Changing World." Ecological Applications, 11: 1027-1045. doi:10.1890/1051-0761(2001)011[1027:WIACW]2.0.CO;2

1076 V.M. Tiwari et al., 2009, "Dwindling groundwater resources in northern India, from satellite gravity observations." Geophysical Research Letters, 36, L18401, doi: 10.1029/2009GL039401.

1077 Christer Nilsson et al., "Fragmentation and Flow Regulation of the World's Large River Systems." Science 15 Apr 2005: Volume 308, Issue 5720, 405-408. DOI: 10.1126/science.1107887.

1078 Muhammad Mizanur Rahaman and Olli Varis, 2005, "Integrated water resources management: evolution, prospects and future challenges." Sustainability: Science, Practice and Policy, 1:1, 15-21, DOI: 10.1080/15487733.2005.11907961.

1079 R.B. Jackson et al., 2001, "Water in a Changing World." Ecological Applications, 11: 1027-1045. doi:10.1890/1051-0761(2001)011[1027:WIACW]2.0.CO;2

1080 C. Zarfl et al., "A global boom in hydropower dam construction." Aquatic Sciences (2015) 77: 161. https://doi.org/10.1007/s00027-014-0377-0.

1081 R. Quentin Grafton et al., "Global insights into water resources, climate change and governance." Nature Climate Change Volume 3, 315–321 (2013). DOI: 10.1038/NCLIMATE1746.

1082 Prasad S. Thenkabail et al., "Global irrigated area map (GIAM), derived from remote sensing, for the end of the last millennium." International Journal of Remote Sensing. Volume 30, 2009 - Issue 14. https://doi.org/10.1080/01431160802698919.

1083 M.M. Mekonnen and A.Y. Hoekstra, "The green, blue and grey water footprint of crops and derived crop products." Hydrology and Earth System Sciences, 15, 1577-1600, https://doi.org/10.5194/hess-15-1577-2011, 2011.

1084 Aaron T. Wolf et al., "International River Basins of the World." 387-427 21 Jul 2010. https://doi.org/10.1080/07900629948682.

1085 H Munia et al., "Water stress in global transboundary river basins: significance of upstream water use on downstream stress." Environment Research Letter 11 (2016) 014002. doi:10.1088/1748-9326/11/1/014002 http://iopscience.iop.org/article/10.1088/1748-9326/11/1/014002/pdf.

1086 Global Water Security. Intelligence Community Assessment, ICA 2012-08, 2 February 2012. https://www.dni.gov/files/documents/Special%20Report_ICA%20Global%20Water%20Security.pdf

1087 Muhammad Mizanur Rahaman and Olli Varis, 2005, Integrated water resources management: evolution, prospects and future challenges, Sustainability: Science, Practice and Policy, 1:1, 15-21, DOI: 10.1080/15487733.2005.11907961.

1088 U.S. Government Global Water Strategy, 2017. https://www.usaid.gov/sites/default/files/documents/1865/Global_Water_Strategy_2017_final_508v2.pdf.

1089 Global Water Security. Intelligence Community Assessment, ICA 2012-08, 2 February 2012. https://www.dni.gov/files/documents/Special%20Report_ICA%20Global%20Water%20Security.pdf

1090 S. Mark Howden et al., "Adapting agriculture to climate change." Proceedings of the National Academy of Sciences Dec 2007, 104 (50) 19691-19696; DOI: 10.1073/pnas.0701890104.

1091 Rosegrant, Mark W et al., "Water for Agriculture: Maintaining Food Security Under Growing Scarcity (November 2009)." Annual Review of Environment and Resources, Volume 34, 205-222, 2009. Available at SSRN: https://ssrn.com/abstract=1599085 or http://dx.doi.org/10.1146/annurev.environ.030308.090351.

1092 Arjen Y. Hoekstra, Mesfin M. Mekonnen, "The water footprint of humanity." Proceedings of the National Academy of Sciences Feb 2012, 109 (9) 3232-3237; DOI: 10.1073/pnas.1109936109.

1093 Global Water Security. Intelligence Community Assessment, ICA 2012-08, 2 February 2012. https://www.dni.gov/files/documents/Special%20Report_ICA%20Global%20Water%20Security.pdf

1094 M.M. Mekonnen, A.Y. Hoekstra, 2012, "A global assessment of the Water Footprint of Farm Animal Products." Ecosystems, 15(3), 401-415. DOI: 10.1007/s10021-011-9517-8.

1095 M.M. Mekonnen, A.Y. Hoekstra, "The green, blue and grey water footprint of crops and derived crop products." Hydrology and Earth System Sciences, 15, 1577-1600, https://doi.org/10.5194/hess-15-1577-2011, 2011.

1096 Kate A. Brauman, et al. "Improvements in crop water productivity increase water sustainability and food security—a global analysis." Environmental Research Letters. 8 (2013) 024030 (7pp). doi:10.1088/1748-9326/8/2/024030.

1097 M.M. Mekonnen, A.Y. Hoekstra, "The green, blue and grey water footprint of crops and derived crop products." Hydrology and Earth System Sciences, 15, 1577-1600, https://doi.org/10.5194/hess-15-1577-2011, 2011.

1098 Graham K. MacDonald et al., "Rethinking Agricultural Trade Relationships in an Era of Globalization." BioScience, Volume 65, Issue 3, 1 March 2015, 275–289, https://doi.org/10.1093/biosci/biu225. (See Figure 4, irrigation water usage, especially for the USA and how much water it exports with food exports).

1099 David Pimentel et al., "Water Resources: Agricultural and Environmental Issues." BioScience, Volume 54, Issue 10, 1 October 2004, 909–918, https://doi.org/10.1641/0006-3568(2004)054[0909:WRAAEI]2.0.CO;2.

1100 Lisa Pfeiffer and C.Y. Cynthia Lin, "Does Efficient Irrigation Technology Lead to Reduced Groundwater Extraction? Empirical Evidence." Journal of Environmental Economics and Management. Volume 67, Issue 2, March 2014, 189-208. https://doi.org/10.1016/j.jeem.2013.12.002.

1101 Frank A. Ward, Manuel Pulido-Velazquez, "Water conservation in irrigation can increase water use." Proceedings of the National Academy of Sciences Nov 2008, 105 (47) 18215-18220; DOI: 10.1073/pnas.0805554105.

1102 David Pimentel et al., "Water Resources: Agricultural and Environmental Issues." BioScience, Volume 54, Issue 10, 1 October 2004, 909–918, https://doi.org/10.1641/0006-3568(2004)054[0909:WRAAEI]2.0.CO;2.

1103 Arjen Y. Hoekstra, Mesfin M. Mekonnen, "The water footprint of humanity." Proceedings of the National Academy of Sciences Feb 2012, 109 (9) 3232-3237; DOI: 10.1073/pnas.1109936109.

1104 European Commission. "Water use in industry. Cooling for electricity production dominates water use in industry." http://ec.europa.eu/eurostat/statistics-explained/index.php/Archive:Water_use_in_industry.

1105 European Commission. Enterprise and Industry. Sustainable Industry – Going for Growth and Resource Efficiency. Report Authors. Authors: Koen Rademaekers, Sahar Samir Zaki, Matthew Smith. Ref. Ares(2014)1209330 - 16/04/2014. https://ec.europa.eu/docsroom/documents/5188/attachments/1/translations/en/renditions/pdf.

1106 United Nations Industrial Development Organization. See Flagship publications. Industrial Development Report 2018. Demand for Manufacturing: Driving Inclusive and Sustainable Industrial Development.

1107 European Commission. Water use in industry. Cooling for electricity production dominates water use in industry. http://ec.europa.eu/eurostat/statistics-explained/index.php/Archive:Water_use_in_industry.

1108 European Commission. Enterprise and Industry. Sustainable Industry – Going for Growth & Resource Efficiency. Report Authors. Authors: Koen Rademaekers, Sahar Samir Zaki, Matthew Smith. Ref. Ares(2014)1209330 - 16/04/2014. https://ec.europa.eu/docsroom/documents/5188/attachments/1/translations/en/renditions/pdf.

1109 European Commission. Enterprise and Industry. Sustainable Industry – Going for Growth & Resource Efficiency. Report Authors. Authors: Koen Rademaekers, Sahar Samir Zaki, Matthew Smith. Ref. Ares(2014)1209330 - 16/04/2014. https://ec.europa.eu/docsroom/documents/5188/attachments/1/translations/en/renditions/pdf.

1110 European Commission. Water use in industry. Cooling for electricity production dominates water use in industry. http://ec.europa.eu/eurostat/statistics-explained/index.php/Archive:Water_use_in_industry.

1111 E S Spang et al., "The water consumption of energy production: an international comparison." Environmental Research Letters, 9 (2014) 105002 (14pp) doi:10.1088/1748-9326/9/10/105002.

1112 Erik Mielke et al., "Water Consumption of Energy Resource Extraction, Processing, and Conversion, A review of the literature for estimates of water intensity of energy-resource extraction, processing to fuels, and conversion to electricity." Energy Technology Innovation Policy Discussion Paper No. 2010-15, Belfer Center for Science and International Affairs, Harvard Kennedy School, Harvard University, October 2010.

1113 Arjen Y. Hoekstra, Mesfin M. Mekonnen, "The water footprint of humanity." Proceedings of the National Academy of Sciences Feb 2012, 109 (9) 3232-3237; DOI: 10.1073/pnas.1109936109.

1114 K.C. Seto et al., 2011, "A Meta-Analysis of Global Urban Land Expansion." PLoS ONE 6(8): e23777. doi:10.1371/journal.pone.0023777.

1115 David Satterthwaite et al., "Adapting to Climate Change in Urban Areas. The possibilities and constraints in low- and middle-income nations." http://pubs.iied.org/pdfs/10549IIED.pdf.

1116 M. Kummu et al., 2011, "How Close Do We Live to Water? A Global Analysis of Population Distance to Freshwater Bodies." PLoS ONE 6(6): e20578. doi:10.1371/journal.pone.0020578.

1117 David Satterthwaite et al., "Adapting to Climate Change in Urban Areas. The possibilities and constraints in low- and middle-income nations." http://pubs.iied.org/pdfs/10549IIED.pdf.

1118 European Commission: Cities of Tomorrow. Challenges, visions, ways forward. http://ec.europa.eu/regional_policy/sources/docgener/studies/pdf/citiesoftomorrow/citiesoftomorrow_final.pdf.

1119 K.C. Seto et al., 2011, "A Meta-Analysis of Global Urban Land Expansion." PLoS ONE 6(8): e23777. https://doi.org/10.1371/journal.pone.0023777.

1120 US Census: Coastal Areas. https://www.census.gov/topics/preparedness/about/coastal-areas.html

1121 The World Bank (2010). The International Bank for Reconstruction and Development. Climate Risks and Adaptation In Asian Coastal Megacities. A Synthesis Report.

1122 K.C. Seto et al., 2011, "A Meta-Analysis of Global Urban Land Expansion." PLoS ONE 6(8): e23777. doi:10.1371/journal.pone.0023777.

1123 B. Neumann et al., "Future Coastal Population Growth and Exposure to Sea-Level Rise and Coastal Flooding - A Global Assessment." Kumar L, ed. PLoS ONE. 2015;10(3):e0118571. doi:10.1371/journal.pone.0118571.

1124 R.J. Nicholls, "Coastal megacities and climate change." GeoJournal (1995) 37: 369. https://doi.org/10.1007/BF00814018.

1125 Gordon McGranahan, et al. "The rising tide: assessing the risks of climate change and human settlements in low elevation coastal zones." Environment and Urbanization. Volume19, Issue 1, 17 – 37. April 1, 2007. https://doi.org/10.1177/0956247807076960.

1126 Muhammad Mizanur Rahaman and Olli Varis, 2005, "Integrated water resources management: evolution, prospects and future challenges." Sustainability: Science, Practice and Policy, 1:1, 15-21, DOI: 10.1080/15487733.2005.11907961.

1127 Muhammad Mizanur Rahaman and Olli Varis, 2005, "Integrated water resources management: evolution, prospects and future challenges." Sustainability: Science, Practice and Policy, 1:1, 15-21, DOI: 10.1080/15487733.2005.11907961.

1128 U.S. Government Global Water Strategy, 2017. https://www.usaid.gov/sites/default/files/documents/1865/Global_Water_Strategy_2017_final_508v2.pdf.

1129 Muhammad Mizanur Rahaman et al., "EU Water Framework Directive vs. Integrated Water Resources Management: The Seven Mismatches." 565-575. Published online: 22 Jan 2007. https://doi.org/10.1080/07900620412331319199.

1130 European Commission. River Basin Management Plans available in each European River Basin District. http://ec.europa.eu/environment/water/participation/map_mc/map.htm. (Click on the country maps to review the River Basin Management Plans for each major European Water Basin).

1131 Mark Giordano and Tushaar Shah, 2014, From IWRM back to integrated water resources management, International Journal of Water Resources Development, 30:3, 364-376. DOI: 10.1080/07900627.2013.851521.

1132 Chay Asdak and Munawir, "Integrated Water Resources Conservation Management for a Sustainable Food Security." Second International Conference on Sustainable Agriculture and Food Security: A Comprehensive Approach, KnE Life Sciences, 238–270. DOI 10.18502/kls.v2i6.1045.

1133 Mark Giordano and Tushaar Shah, 2014, "From IWRM back to integrated water resources management." International Journal of Water Resources Development, 30:3, 364-376. DOI: 10.1080/07900627.2013.851521.

1134 Muhammad Mizanur Rahaman et al., "EU Water Framework Directive vs. Integrated Water Resources Management: The Seven Mismatches." 565-575. Published online: 22 Jan 2007. https://doi.org/10.1080/07900620412331319199.

1135 Muhammad Mizanur Rahaman and Olli Varis, 2005, "Integrated water resources management: evolution, prospects and future challenges." Sustainability: Science, Practice and Policy, 1:1, 15-21, DOI: 10.1080/15487733.2005.11907961.

1136 S. Hohensinner et al., 2018, "River Morphology, Channelization, and Habitat Restoration." In: Schmutz S., Sendzimir J. (eds) Riverine Ecosystem Management. Aquatic Ecology Series, Volume 8. Springer, Cham. DOI https://doi.org/10.1007/978-3-319-73250-3_3.

1137 E. Kiedrzyńska et al., "Sustainable floodplain management for flood prevention and water quality improvement." M. Nat Hazards (2015) 76: 955. https://doi.org/10.1007/s11069-014-1529-1.

1138 S. Hohensinner et al., 2018, "River Morphology, Channelization, and Habitat Restoration." In: Schmutz S., Sendzimir J. (eds) "Riverine Ecosystem Management." Aquatic Ecology Series, Volume 8. Springer, Cham. DOI https://doi.org/10.1007/978-3-319-73250-3_3.

1139 A. Nicol et al., (Eds.) 2015, "Water-smart agriculture in East Africa. Colombo, Sri Lanka." International Water Management Institute (IWMI). CGIAR Research Program on Water, Land and Ecosystems (WLE); Kampala,Uganda: Cooperative for Assistance and Relief Everywhere (CARE). 352p. doi: 10.5337/2015.203.

1140 Chay Asdak and Munawir, "Integrated Water Resources Conservation Management for a Sustainable Food Security." Second International Conference on Sustainable Agriculture and Food Security: A Comprehensive Approach, KnE Life Sciences, 238–270. DOI 10.18502/kls.v2i6.1045

1141 Bancy M. Mati, Working Paper 13. "100 Ways to Manage Water for Smallholder Agriculture in Eastern and Southern Africa A Compendium of Technologies and Practices." March 2007. IMAWESA.

1142 Mahdavi A et al., "Identification of groundwater artificial recharge sites using Fuzzy logic: A case study of Shahrekord plain, Iran." International magazine of geo-science publication, 2012; 1-14.

1143 Prasenjit Bhowmick et al., "A review on GIS based Fuzzy and Boolean logic modeling approach to identify the suitable sites for Artificial Recharge of Groundwater." Sch. J. Eng. Tech., 2014; 2(3A):316-319.

1144 H. Nasiri et al., "Determining the most suitable areas for artificial groundwater recharge via an integrated PROMETHEE II-AHP method in GIS environment (case study: Garabaygan Basin, Iran)." Environ Monit Assess (2013) 185: 707. https://doi.org/10.1007/s10661-012-2586-0.

1145 Prasenjit Bhowmick et al., "A review on GIS based Fuzzy and Boolean logic modeling approach to identify the suitable sites for Artificial Recharge of Groundwater." Sch. J. Eng. Tech., 2014; 2(3A):316-319.

1146 A. Nicol et al., (Eds.) 2015, "Water-smart agriculture in East Africa. Colombo, Sri Lanka." International Water Management Institute (IWMI). CGIAR Research Program on Water, Land and Ecosystems (WLE); Kampala,Uganda: Cooperative for Assistance and Relief Everywhere (CARE). 352p. doi: 10.5337/2015.203.

1147 Bancy M. Mati, "100 Ways to Manage Water for Smallholder Agriculture in Eastern and Southern Africa. A Compendium of Technologies and Practices." March 2007. SWMnet Working Paper 13. IMAWESA.

1148 The US Geological Survey. Groundwater Storage - The Water Cycle. https://water.usgs.gov/edu/watercyclegwstorage.html.

1149 H. Bouwer, "Artificial recharge of groundwater: hydrogeology and engineering." Hydrogeology Journal (2002) Volume 10, Issue 1, 121–142. https://doi.org/10.1007/s10040-001-0182-4.

1150 Takashi Asano, "Water Reuse via Groundwater Recharge." International Review for Environmental Strategies (IRES) Volume 6, No. 2, 205 – 216, 2006.

1151 "How a Watershed Flood Control Dam Works." Oklahoma Conservation Commission. The USDA Watershed Program. https://www.lowellma.gov/DocumentCenter/View/1057

1152 The US Environmental Protection Agency. Green Infrastructure. Manage Flood Risk. https://www.epa.gov/green-infrastructure/manage-flood-risk.

1153 Hua-peng Qin et al., "The effects of low impact development on urban flooding under different rainfall characteristics." Journal of Environmental Management (2013), http://dx.doi.org/10.1016/j.jenvman.2013.08.026.

1154 The US Environmental Protection Agency. Green Infrastructure. What is Green Infrastructure? https://www.epa.gov/green-infrastructure/what-green-infrastructure.

1155 S. Lattemann et al., "Global Desalination Situation." Sustainability Science and Engineering, Volume 2, 2010, 7-39. https://doi.org/10.1016/S1871-2711(09)00202-5.

1156 Noreddine Ghaffour, et al. "Technical review and evaluation of the economics of water desalination: Current and future challenges for better water supply sustainability." http://dx.doi.org/10.1016/j.desal.2012.10.015.

1157 Ali A. Al-Karaghouli and L.L. Kazmerski, 2011, "Renewable Energy Opportunities in Water Desalination, Desalination." Michael Schorr, IntechOpen, DOI: 10.5772/14779. Available from: https://www.intechopen.com/books/desalination-trends-and-technologies/renewable-energy-opportunities-in-water-desalination.

1158 Ali A. Al-Karaghouli and L.L. Kazmerski, 2011, "Renewable Energy Opportunities in Water Desalination, Desalination." Michael Schorr, IntechOpen, DOI: 10.5772/14779. Available from: https://www.intechopen.com/books/desalination-trends-and-technologies/renewable-energy-opportunities-in-water-desalination.

1159 Mike Mickley, 2013, "US Municipal Desalination Plants: Number, Types, Locations, Sizes, and Concentrate Management Practices." IDA Journal of Desalination and Water Reuse, 4:1, 44-51, DOI: 10.1179/ida.2012.4.1.44.

1160 H. March et al., "The End of Scarcity? Water Desalination as The New Cornucopia for Mediterranean Spain." Journal of Hydrology (2014). http://dx.doi.org/10.1016/j.jhydrol.2014.04.023.

1161 J. Aparicio et al., "Economic evaluation of small desalination plants from brackish aquifers." Application to Campo de Cartagena (SE Spain). Desalination 411 (2017) 38–44. http://dx.doi.org/10.1016/j.desal.2017.02.004.

1162 Xiang Zheng et al., "Seawater desalination in China: Retrospect and prospect." April 2014. The Chemical Engineering Journal 242:404–413. DOI: 10.1016/j.cej.2013.12.

1163 Ras Al Khair Desalination Plant. https://www.water-technology.net/projects/ras-al-khair-desalination-plant/.

1164 Magtaa Reverse Osmosis Desalination Plant, Algeria. https://www.water-technology.net/projects/magtaa-desalination/.

1165 Adelaide Desalination Plant. https://www.sawater.com.au/community-and-environment/our-water-and-sewerage-systems/water-treatment/desalination/adelaide-desalination-plant-adp.

1166 Sorek, The World's Largest and Most Advanced SWRO Desalination Plant. http://www.ide-tech.com/blog/b_case_study/sorek-project/.

1167 Ali A. Al-Karaghouli and L.L. Kazmerski, 2011, "Renewable Energy Opportunities in Water Desalination, Desalination." Michael Schorr, IntechOpen, DOI: 10.5772/14779. Available from: https://www.intechopen.com/books/desalination-trends-and-technologies/renewable-energy-opportunities-in-water-desalination.

1168 V.G. Gude, "Energy storage for desalination processes powered by renewable energy and waste heat sources." Appl Energy (2014), http://dx.doi.org/10.1016/j.apenergy.2014.06.061.

1169 Noreddine Ghaffour et al., "Technical review and evaluation of the economics of water desalination: Current and future challenges for better water supply sustainability." http://dx.doi.org/10.1016/j.desal.2012.10.015.

1170 Ali A. Al-Karaghouli and L.L. Kazmerski, 2011, "Renewable Energy Opportunities in Water Desalination, Desalination." Michael Schorr, IntechOpen, DOI: 10.5772/14779. Available from: https://www.intechopen.com/books/desalination-trends-and-technologies/renewable-energy-opportunities-in-water-desalination.

1171 V.G. Gude, "Energy storage for desalination processes powered by renewable energy and waste heat sources." Appl Energy (2014), http://dx.doi.org/10.1016/j.apenergy.2014.06.061.

1172 A. Mohamed et al., "Renewable Energy Powered Desalination Systems: Technologies and Economics – State of the Art." Twelfth International Water Technology Conference, IWTC12 2008, Alexandria, Egypt 1099.

1173 Ali A. Al-Karaghouli and L.L. Kazmerski, 2011, "Renewable Energy Opportunities in Water Desalination, Desalination." Michael Schorr, IntechOpen, DOI: 10.5772/14779. Available from: https://www.intechopen.com/books/desalination-trends-and-technologies/renewable-energy-opportunities-in-water-desalination.

1174 E.R. Shouman et al., 2015, "Economics of Renewable Energy for Water Desalination in Developing Countries." International Journal of Economics and Management Sciences 5:305. doi:10.4172/21626359.1000305.

1175 Chennan Li et al., "Solar assisted sea water desalination: A review." Renewable and Sustainable Energy Reviews 19(2013)136–163. http://dx.doi.org/10.1016/j.rser.2012.04.059.

1176 Ali A. Al-Karaghouli and L.L. Kazmerski, 2011, "Renewable Energy Opportunities in Water Desalination, Desalination." Michael Schorr, IntechOpen, DOI: 10.5772/14779. Available from: https://www.intechopen.com/books/desalination-trends-and-technologies/renewable-energy-opportunities-in-water-desalination.

1177 Manchanda and Kumar, "A comprehensive decade review and analysis on designs and performance parameters of passive solar still." Renewables (2015) 2:17. DOI 10.1186/s40807-015-0019-8.

1178 Rasika R Dahake et al., "A Review on Solar Still Water Purification." International Journal for Innovative Research in Science & Technology Volume 3 Issue 9 2017 59-63.

1179 Hikmet Ş. Aybar, "A review of desalination by solar still." May 2007. NATO Security through Science Series C: Environmental Security. In book: Solar Desalination for the 21st Century. DOI: 10.1007/978-1-4020-5508-9_15.

1180 A.Z.A. Saifullaha et al., "Solar pond and its application to desalination." Asian Transactions on Science & Technology (ATST ISSN: 2221-4283) Volume 02 Issue 03.

1181 Ange Abena Mbarga et al., "Integration of Renewable Energy Technologies With Desalination." Current Sustainable Renewable Energy Rep (2014) 1:11–18. DOI 10.1007/s40518-013-0002-1.

1182 China South–North Water Transfer Project Official Website. http://www.nsbd.gov.cn/zx/english/

1183 The Trans Africa Pipeline. http://transafricapipeline.org.

1184 Michael J Puma et al., "Assessing the evolving fragility of the global food system." Environment Research Letter 10 (2015) 024007 doi:10.1088/1748-9326/10/2/024007.

1185 C. Sage, "The interconnected challenges for food security from a food regimes perspective: Energy, climate and malconsumption." Journal of Rural Studies (2012), doi:10.1016/j.jrurstud.2012.02.005.

1186 David Satterthwaite et al., "Review. Urbanization and its implications for food and farming." Phil. Trans. R. Soc. B (2010) 365, 2809–2820. doi:10.1098/rstb.2010.0136.

1187 Global Europe 2050. Directorate-General for Research and Innovation. 2012. Socio-economic Sciences and Humanities. EUR 25252.

1188 Conduct a Google search, "United Nations, World population, 2050, 2100."

1189 J. Cleland, "World Population Growth; Past, Present and Future." Environ Resource Econ (2013) 55: 543. https://doi.org/10.1007/s10640-013-9675-6.

1190 P. Gerland et al., "World Population Stabilization Unlikely This Century." Science (New York, NY). 2014;346(6206):234-237. doi:10.1126/science.1257469.

1191 Tony Fischer et al., "Crop yields and global food security. Will yield increase continue to feed the world?" Australia Centre for International Agricultural Research. Grains Research and Development Corporation. https://www.aciar.gov.au/node/12101.

1192 M.E. Brown et al., 2015, "Climate Change, Global Food Security, and the U.S. Food System." http://www.usda.gov/oce/climate_change/FoodSecurity2015Assessment/FullAssessment.pdf.

1193 David Satterthwaite et al., Review. "Urbanization and its implications for food and farming." Phil. Trans. R. Soc. B (2010) 365, 2809–2820. doi:10.1098/rstb.2010.0136.

1194 C. Sage, "The interconnected challenges for food security from a food regimes perspective: Energy, climate and malconsumption." Journal of Rural Studies (2012), doi:10.1016/j.jrurstud.2012.02.005.

1195 Alex Evans, "The Feeding of the Nine Billion Global Food Security for the 21st Century." A Chatham House Report.

1196 Philippe Marchand et al., 2016, "Reserves and trade jointly determine exposure to food supply shocks." Environment Research Letter 11 095009. Environment Research Letter 11 (2016) 095009 doi:10.1088/1748-9326/11/9/095009.

1197 Graham K. MacDonald et al., "West; Rethinking Agricultural Trade Relationships in an Era of Globalization." BioScience, Volume 65, Issue 3, 1 March 2015, 275–289, https://doi.org/10.1093/biosci/biu225.

1198 M. Porkka et al., 2013, "From Food Insufficiency towards Trade Dependency: A Historical Analysis of Global Food Availability." PLoS ONE 8(12): e82714. doi:10.1371/journal.pone.0082714.

1199 European Commission. Agriculture and Rural Development. Agricultural trade in 2013: EU gains in commodity exports. Monitoring Agri-trade Policy. https://ec.europa.eu/agriculture/sites/agriculture/files/trade-analysis/.../2014-1_en.pdf.

1200 European Commission. Agricultural and food trade. https://ec.europa.eu/agriculture/sites/agriculture/files/.../agricultural-food-trade.pdf.

1201 Graham K. MacDonald et al., "Rethinking Agricultural Trade Relationships in an Era of Globalization." BioScience, Volume 65, Issue 3, 1 March 2015, 275–289, https://doi.org/10.1093/biosci/biu225.

1202 Graham K. MacDonald et al., "Rethinking Agricultural Trade Relationships in an Era of Globalization." BioScience, Volume 65, Issue 3, 1 March 2015, 275–289, https://doi.org/10.1093/biosci/biu225. (See Figures 1-4 and 8 for food trade and country interdependencies by crop or food type, and by calories, value, irrigation water, and land area).

1203 Christopher Bren d'Amour et al., "Teleconnected food supply shocks." Environment Research Letter 11 (2016) 035007 doi:10.1088/1748-9326/11/3/035007. (see table 1, page 2 for the list of countries for each main crop).

1204 Bo Chen and Sayed H. Saghaian, 2016, "Market Integration and Price Transmission in the World Rice Export Markets." Journal of Agricultural and Resource Economics 41(3):444–457. https://uknowledge.uky.edu/agecon_facpub/8.

1205 S. Muthayya et al., 2014, "An overview of global rice production, supply, trade, and consumption." Ann. N.Y. Acad. Sci., 1324: 7-14. doi:10.1111/nyas.12540.

1206 M.E. Brown et al., 2015, "Climate Change, Global Food Security, and the U.S. Food System." http://www.usda.gov/oce/climate_change/FoodSecurity2015Assessment/FullAssessment.pdf.

1207 Tony Fischer et al., "Crop yields and global food security. Will yield increase continue to feed the world?" Australia Centre for International Agricultural Research. Grains Research and Development Corporation. https://www.aciar.gov.au/node/12101.

1208 Michael J Puma et al., "Assessing the evolving fragility of the global food system." Environment Research Letter 10 (2015) 024007 doi:10.1088/1748-9326/10/2/024007.

1209 Philippe Marchand et al., 2016, "Reserves and trade jointly determine exposure to food supply shocks." Environment Research Letter 11 095009. Environment Research Letter 11 (2016) 095009 doi:10.1088/1748-9326/11/9/095009.

1210 Christopher Bren d'Amour et al., "Teleconnected food supply shocks." Environment Research Letter 11 (2016) 035007 doi:10.1088/1748-9326/11/3/035007.

1211   Crop Stakeholder Organizations: 1) International Maize and Wheat Improvement Center (CIMMYT), https://www.cimmyt.org/, 2) International Rice Research Institute (IRRI), http://irri.org/. 3) The Pan-African Bean Research Alliance (PABRA), http://www.pabra-africa.org/seeds-systems. 4) International Crops Research Institute for the Semi-Arid Tropics (ICRISAT), http://www.icrisat.org. 5) International Potato Centre, https://cipotato.org/.

1212 Christopher Bren d'Amour et al., "Teleconnected food supply shocks." Environment Research Letter 11 (2016) 035007 doi:10.1088/1748-9326/11/3/035007.

1213 Christopher Bren d'Amour et al., "Teleconnected food supply shocks." Environment Research Letter 11 (2016) 035007 doi:10.1088/1748-9326/11/3/035007.

1214 Alex Evans, "The Feeding of the Nine Billion Global Food Security for the 21st Century." A Chatham House Report.

1215 M.M. Mekonnen and A. Y. Hoekstra, "The green, blue and grey water footprint of crops and derived crop products." Hydrology and Earth System Sciences, 15, 1577-1600, https://doi.org/10.5194/hess-15-1577-2011, 2011.

1216 Kate A Brauman et al., "Improvements in crop water productivity increase water sustainability and food security—a global analysis." Environmental Research Letters. 8 (2013) 024030 (7pp). doi:10.1088/1748-9326/8/2/024030. (See Figures 1, 5, and 6).

1217 Tony Fischer et al., "Crop yields and global food security. Will yield increase continue to feed the world?" Australia Centre for International Agricultural Research. Grains Research and Development Corporation. (See Table 1.2, page 11).

1218   Crop Stakeholder Organizations: 1) International Maize and Wheat Improvement Center (CIMMYT), https://www.cimmyt.org/, 2) International Rice Research Institute (IRRI), http://irri.org/. 3) The Pan-African Bean Research Alliance (PABRA), http://www.pabra-africa.org/seeds-systems. 4) International Crops Research Institute for the Semi-Arid Tropics (ICRISAT), http://www.icrisat.org. 5) International Potato Centre, https://cipotato.org/.

1219 E. Jean Finnegan et al., 2010, Vernalization. In eLS, (Ed.). doi:10.1002/9780470015902.a0002048.pub3.

1220 P Chouard, "Vernalization and its Relations to Dormancy." Annual Review of Plant Physiology. Volume 11:191-238 (Volume publication date June 1960). https://doi.org/10.1146/annurev.pages11.060160.001203.

1221 Tony Fischer et al., "Crop yields and global food security. Will yield increase continue to feed the world?" Australia Centre for International Agricultural Research. Grains Research and Development Corporation. (See Table 3.1, page 67).

1222 United States Department of Agriculture. Natural Resources Conservation Service. Fact Sheet. Cover Crop (340) Tennessee. 2015. https://efotg.sc.egov.usda.gov/references/public/TN/CoverCrop_340_FactSheet_Final_July2015.pdf. [See Table 1 Comparison of Maturity and Cold Tolerance of Small Grains and Annual Ryegrass, page 2.].

1223 N.N. Săulescu and H.J. Braun, Chapter 9, "Cold Tolerance." Extracted from: Reynolds, M.P., J.I. Ortiz-Monasterio, and A. McNab (eds.). 2001. Application of Physiology in Wheat Breeding. Mexico, D.F.: CIMMYT. http://www.plantstress.com/articles/up_cold_files/cold_chapter.pdf.

1224 Vernalization of Winter Wheat. http://igrow.org/agronomy/wheat/vernalization-of-winter-wheat/.

1225 US Department of Agriculture (USDA). Commodity Intelligence Report, 2016. Russia: Sown Area for 2016/17 Winter Grains Falls Short of Ministry Forecast. https://pecad.fas.usda.gov/highlights/2016/01/rs_13jan2016/index.htm.

1226 United States Department of Agriculture. Natural Resources Conservation Service. Fact Sheet. Cover Crop (340) Tennessee. 2015. https://efotg.sc.egov.usda.gov/references/public/TN/CoverCrop_340_FactSheet_Final_July2015.pdf. [See Table 1 Comparison of Maturity and Cold Tolerance of Small Grains and Annual Ryegrass, page 2.].

1227 L.V. Gusta and B.J. O'Connor, "Frost Tolerance of Wheat, Oats, Barley, Canola and Mustard and the Role of Ice-Nucleating Bacteria." Canadian Journal of Plant Science, 1987, 67(4): 1155-1165, https://doi.org/10.4141/cjps87-155.

1228 N.N. Săulescu and H.J. Braun, Chapter 9, "Cold Tolerance." Extracted from: Reynolds, M.P., J.I. Ortiz-Monasterio, and A. McNab (eds.). 2001. Application of Physiology in Wheat Breeding. Mexico, D.F.: CIMMYT. http://www.plantstress.com/articles/up_cold_files/cold_chapter.pdf.

1229 Tony Fischer et al., "Crop yields and global food security. Will yield increase continue to feed the world?" Australia Centre for International Agricultural Research. Grains Research and Development Corporation. (See page 137).

1230 Renata Pereira da Cruz et al., 2013, "Avoiding damage and achieving cold tolerance in rice plants." https://doi.org/10.1002/fes3.25.

1231 E. Shakiba E et al., 2017, "Genetic architecture of cold tolerance in rice (Oryza sativa) determined through high resolution genome-wide analysis." PLoS ONE 12(3): e0172133. doi:10.1371/journal.pone.0172133.

1232 Moon-Hee Lee, "Low Temperature Tolerance in Rice: The Korean Experience. Increased Lowland Rice Production in the Mekong Region." Edited by Shu Fukai and Jaya Basnayake. ACIAR Proceedings 101. (printed version published in 2001).

1233 International Rice Research Institute. "Stress-tolerant Rice for Africa and South Asia." http://strasa.irri.org/stresses/cold-tolerant.

1234 V.M. Rodríguez et al., "Combining maize base germplasm for cold tolerance breeding." http://digital.csic.es/bitstream/10261/9598/3/Combining_maize_base_germplasm.pdf.

1235 H. A. Eagles, "Cold tolerance and its relevance to maize breeding in New Zealand." Proceedings Agronomy Society of New Zealand 9; 1979.

1236 Pedro Revilla et al., "Association mapping for cold tolerance in two large maize inbred panels." BMC Plant Biology (2016) 16:127. DOI 10.1186/s12870-016-0816-2.

1237 Alicja Sobkowiak et al., "Molecular foundations of chilling-tolerance of modern maize." Sobkowiak et al. BMC Genomics (2016) 17:125. DOI 10.1186/s12864-016-2453-4.

1238 H. Campos et al., "Improving drought tolerance in maize: a view from industry." Field Crops Research, Volume 90, Issue 1, 2004, 19-34, https://doi.org/10.1016/j.fcr.2004.07.003.

1239 H. Campos et al., "Changes in drought tolerance in maize associated with fifty years of breeding for yield in the US Corn Belt." January 2006Maydica 51(2):369-381.

1240 CGIAR. Research Program on Maize. http://maize.org/maize-impacts/

1241 Monica Fisher et al., "Drought tolerant maize for farmer adaptation to drought in sub-Saharan Africa: Determinants of adoption in eastern and southern Africa." Climatic Change (2015) 133: 283. https://doi.org/10.1007/s10584-015-1459-2.

1242 Examples of Genetically Modified Drought-Tolerant Crops: 1) DroughtGard hybrids. https://www.genuity.com/corn/Pages/DroughtGard-Hybrids.aspx. 2) Syngenta US. Agrisure Artesian. http://www.syngenta-us.com/corn/agrisure/agrisure-artesian. 3) DuPont Pioneer Optimum AQUAmax. https://www.pioneer.com/home/site/us/products/corn/seed-traits-technologies-corn/optimum-aquamax-hybrids.

1243 CIMMYT Global Wheat Program http://www.cimmyt.org/global-wheat-research/.

1244 Delphine Fleury et al., "Genetic and genomic tools to improve drought tolerance in wheat." Journal of Experimental Botany, Volume 61, Issue 12, 1 July 2010, 3211–3222, https://doi.org/10.1093/jxb/erq152.

1245 Learnmore Mwadzingeni et al., "Breeding wheat for drought tolerance: Progress and technologies." Journal of Integrative Agriculture. Volume 15, Issue 5, 2016. 935-943. https://doi.org/10.1016/S2095-3119(15)61102-9. [See table 1, page 936 for progress made with CIMMYT programs].

1246 Learnmore Mwadzingeni et al., "Breeding wheat for drought tolerance: Progress and technologies." Journal of Integrative Agriculture. Volume 15, Issue 5, 2016. 935-943. https://doi.org/10.1016/S2095-3119(15)61102-9.

1247 Delphine Fleury et al., "Genetic and genomic tools to improve drought tolerance in wheat." Journal of Experimental Botany, Volume 61, Issue 12, 1 July 2010, 3211–3222, https://doi.org/10.1093/jxb/erq152.

1248 Saeed Rauf et al., "Breeding Strategies to Enhance Drought Tolerance in Crops." January 2015. DOI: 10.13140/2.1.2343.9682. In book: Advances in Plant Breeding Strategies; Agronomic, Abiotic and Biotic Stress Traits Edition: 2 Chapter: 11 Publisher: SpringerEditors: J.M. Al-Khayri et al. (eds)

1249 Matthew Reynolds et al., "Raising yield potential of wheat. Overview of a consortium approach and breeding strategies." Journal of Experimental Botany, Volume 62, Issue 2, 1 January 2011, 439–452, https://doi.org/10.1093/jxb/erq311.

1250 Saeed Rauf et al., "Breeding Strategies to Enhance Drought Tolerance in Crops." January 2015. DOI: 10.13140/2.1.2343.9682. In book: Advances in Plant Breeding Strategies; Agronomic, Abiotic and Biotic Stress Traits Edition: 2 Chapter: 11 Publisher: SpringerEditors: J.M. Al-Khayri et al. (eds)

1251 Delphine Fleury et al., "Genetic and genomic tools to improve drought tolerance in wheat." Journal of Experimental Botany, Volume 61, Issue 12, 1 July 2010, 3211–3222, https://doi.org/10.1093/jxb/erq152.

1252 Todaka Daisuke et al., "Recent advances in the dissection of drought-stress regulatory networks and strategies for development of drought-tolerant transgenic rice plants." Review Article. Frontiers in Plant Science. Volume 6 18 February 2015. doi: 10.3389/fpls.2015.00084.

1253 Arvind Kumar et al., "Breeding high-yielding drought-tolerant rice: genetic variations and conventional and molecular approaches." Journal of Experimental Botany, Volume 65, Issue 21, 1 November 2014, 6265–6278, https://doi.org/10.1093/jxb/eru363.

1254 Arvind Kumar et al., "Breeding high-yielding drought-tolerant rice: genetic variations and conventional and molecular approaches." Journal of Experimental Botany, Volume 65, Issue 21, 1 November 2014, 6265–6278, https://doi.org/10.1093/jxb/eru363.

1255 Todaka Daisuke et al., "Recent advances in the dissection of drought-stress regulatory networks and strategies for development of drought-tolerant transgenic rice plants." Review Article. Frontiers in Plant Science. Volume 6 18 February 2015. doi: 10.3389/fpls.2015.00084.

1256 United States Department of Agriculture. Natural Resources Conservation Service. Fact Sheet. Cover Crop (340) Tennessee. 2015. https://efotg.sc.egov.usda.gov/references/public/TN/CoverCrop_340_FactSheet_Final_July2015.pdf. [See Table 1 Comparison of Maturity and Cold Tolerance of Small Grains and Annual Ryegrass, page 2.].

1257 United States Department of Agriculture. Natural Resources Conservation Service. Plant Guide. SORGHUM. Sorghum Bicolor (L.) Moench Plant Symbol = SOBI2.

1258 Belum VS Reddy et al., "Genetic improvement of sorghum in the semi-arid tropics." http://bit.ly/2ydg2AQ.

1259 G. Qiao et al., "The enhancement of drought tolerance for pigeon pea inoculated by arbuscular mycorrhizae fungi." Plant Soil Environ., 57 (2011): 541-546.

1260 M. E. Emefiene et al., "The use of Pigeon pea (Cajanus cajan) for drought mitigation in Nigeria." International Letters of Natural Sciences. Volume 1, 6-16. doi:10.18052/www.scipress.com/ILNS.1.6.

1261 United States Department of Agriculture. Natural Resources Conservation Service. Warm Season Cover Crops and Planting Specifications Plant Materials Technical Notes. https://www.nrcs.usda.gov/Internet/FSE_PLANTMATERIALS/publications/etpmctn12917.pdf [Cowpea, page 5.].

1262 Peter J. Matlon, 1990, "Improving Productivity in Sorghum and Pearl Millet in Semi-Arid Africa." Food Research Institute Studies, Stanford University, Food Research Institute, issue 01.

1263 Belum VS Reddy et al., "Genetic improvement of sorghum in the semi-arid tropics." http://bit.ly/2ydg2AQ.

1264 United States Department of Agriculture. Natural Resources Conservation Service. Fact Sheet. Cover Crop (340) Tennessee. 2015. https://efotg.sc.egov.usda.gov/references/public/TN/CoverCrop_340_FactSheet_Final_July2015.pdf. [See Table 1 Comparison of Maturity and Cold Tolerance of Small Grains and Annual Ryegrass, page 2.].

1265 International Rice Research Institute. Flood-tolerant rice saves farmers livelihoods. http://irri.org/our-impact/increase-food-security/flood-tolerant-rice-saves-farmers-livelihoods.

1266 Suzanne K. Redfern et al., "Rice in Southeast Asia: facing risks and vulnerabilities to respond to climate change." http://bit.ly/2JN7AtE.

1267 M.E. Brown et al., 2015, "Climate Change, Global Food Security, and the U.S. Food System." http://www.usda.gov/oce/climate_change/FoodSecurity2015Assessment/FullAssessment.pdf.

1268 Tony Fischer et al., "Crop yields and global food security. Will yield increase continue to feed the world?" Australia Centre for International Agricultural Research. Grains Research and Development Corporation. https://www.aciar.gov.au/node/12101. [See page 17].

1269 H. Charles et al., "Food security and sustainable intensification." Phil. Trans. R. Soc. B 2014 369 20120273; DOI:10.1098/rstb.2012.0273. Published 17 February 2014.

1270 Christos Stefanis, "Global Food Security: An Agricultural Perspective." Journal of Agriculture and Sustainability. ISSN 2201-4357. Volume 6, Number 1, 2014, 69-87.

1271 Tony Fischer et al., "Crop yields and global food security. Will yield increase continue to feed the world?" Australia Centre for International Agricultural Research. Grains Research and Development Corporation. https://www.aciar.gov.au/node/12101. [See page 17].

1272 H.C.J. Godfray et al., "Food Security: The Challenge of Feeding 9 Billion People." Science12 Feb 2010 : 812-818. DOI: 10.1126/science.1185383.

1273 Sarah K. Lowder et al., "The Number, Size, and Distribution of Farms, Smallholder Farms, and Family Farms Worldwide." World Development, Volume 87, 2016, 16-29. https://doi.org/10.1016/j.worlddev.2015.10.041.

1274 M.A. Altieri et al., "Agroecologically efficient agricultural systems for smallholder farmers: contributions to food sovereignty." Agronomy for Sustainable Development (2012) 32: 1. https://doi.org/10.1007/s13593-011-0065-6.

1275 John F. Morton, "The impact of climate change on smallholder and subsistence agriculture." Proceedings of the National Academy of Sciences Dec 2007, 104 (50) 19680-19685; DOI: 10.1073/pnas.0701855104.

1276 M.A. Altieri et al. "Agroecologically efficient agricultural systems for smallholder farmers: contributions to food sovereignty." Agronomy for Sustainable Development (2012) 32: 1. https://doi.org/10.1007/s13593-011-0065-6.

1277 Bancy M. Mati, "100 Ways to Manage Water for Smallholder Agriculture in Eastern and Southern Africa." A Compendium of Technologies and Practices March 2007. SWMnet Working Paper 13. IMAWESA.

1278 J.N. Pretty et al., "Resource-conserving agriculture increases yields in developing countries." Environ. Sci. Technol. 40, 1114 (2006). doi:10.1021/es051670d pmid:16572763.

1279 Bancy M. Mati, "100 Ways to Manage Water for Smallholder Agriculture in Eastern and Southern Africa." A Compendium of Technologies and Practices March 2007. SWMnet Working Paper 13. IMAWESA.

1280   Crop Stakeholder Organizations: 1) International Maize and Wheat Improvement Center (CIMMYT), https://www.cimmyt.org/, 2) International Rice Research Institute (IRRI), http://irri.org/. 3) The Pan-African Bean Research Alliance (PABRA), http://www.pabra-africa.org/seeds-systems. 4) International Crops Research Institute for the Semi-Arid Tropics (ICRISAT), http://www.icrisat.org. 5) International Potato Centre, https://cipotato.org/.

1281 M.E. Brown et al., 2015, "Climate Change, Global Food Security, and the U.S. Food System." http://www.usda.gov/oce/climate_change/FoodSecurity2015Assessment/FullAssessment.pdf. [See page 10, third paragraph].

1282 A.A. Kader, 2005, "Increasing Food Availability By Reducing post-Harvest Losses of Fresh Produce." Acta Horticulturae 682, 2169-2176. DOI: 10.17660/ActaHortic.2005.682.296.

1283 Julian Parfitt et al., "Food waste within food supply chains: quantification and potential for change to 2050." Phil. Trans. R. Soc. B 2010 365 3065-3081; DOI: 10.1098/rstb.2010.0126.

1284 A.A. Kader, 2005, "Increasing Food Availability By Reducing post-Harvest Losses of Fresh Produce." Acta Hortic. 682, 2169-2176. DOI: 10.17660/ActaHortic.2005.682.296.

1285 S. Mark Howden et al., "Adapting agriculture to climate change." PNAS December 11, 2007. 104 (50) 19691-19696; https://doi.org/10.1073/pnas.0701890104.

1286 Christopher D. Golden et al., "Nutrition: Fall in fish catch threatens human health." Nature. Comment. 534, 317–320 (16 June 2016) doi:10.1038/534317a.

1287 Christophe Béné et al., "Feeding 9 billion by 2050 – Putting fish back on the menu." Food Sec. (2015) 7: 261. https://doi.org/10.1007/s12571-015-0427-z.

1288 Ling Cao et al., "China's aquaculture and the world's wild fisheries." Science 09 Jan 2015: Volume 347, Issue 6218, 133-135. DOI: 10.1126/science.1260149.

1289 Max Troell et al., "Does aquaculture add resilience to the global food system?" Proceedings of the National Academy of Sciences Sep 2014, 111 (37) 13257-13263; DOI: 10.1073/pnas.1404067111.

1290 H. Charles et al., "Food Security: The Challenge of Feeding 9 Billion People." Science12 Feb 2010 : 812-818. DOI: 10.1126/science.1185383.

1291 Max Troell et al., "Does aquaculture add resilience to the global food system?" Proceedings of the National Academy of Sciences Sep 2014, 111 (37) 13257-13263; DOI: 10.1073/pnas.1404067111.

1292 Rosamond L. Naylor et al., "Effect of aquaculture on world fish supplies." Nature Volume 405, 1017–1024 (29 June 2000).

1293 S.J. Hall et al., 2011, "Blue Frontiers: Managing the Environmental Costs of Aquaculture." The WorldFish Center, Penang, Malaysia.

1294 A.G.J. Tacon and M. Metian, 2008, "Global Overview on the Use of Fish Meal and Fish Oil in Industrially Compounded Aquafeeds." Trends and Future Prospects. Aquaculture, 285, 146-158. http://dx.doi.org/10.1016/j.aquaculture.2008.08.015.

1295 Rosamond L. Naylor et al., "Effect of aquaculture on world fish supplies." Nature Volume 405, 1017–1024 (29 June 2000).

1296 Max Troell et al., "Does aquaculture add resilience to the global food system?" Proceedings of the National Academy of Sciences Sep 2014, 111 (37) 13257-13263; DOI: 10.1073/pnas.1404067111.

1297 Christophe Béné et al., "Feeding 9 billion by 2050 – Putting fish back on the menu." Food Sec. (2015) 7: 261. https://doi.org/10.1007/s12571-015-0427-z.

1298 Rosamond L. Naylor et al., "Effect of aquaculture on world fish supplies." Nature Volume 405, 1017–1024 (29 June 2000).

1299 Rosamond L. Naylor et al., "Effect of aquaculture on world fish supplies." Nature Volume 405, 1017–1024 (29 June 2000).

1300 Rosamond L. Naylor et al., "Effect of aquaculture on world fish supplies." Nature Volume 405, 1017–1024 (29 June 2000).

1301 Max Troell et al., "Does aquaculture add resilience to the global food system?" Proceedings of the National Academy of Sciences Sep 2014, 111 (37) 13257-13263; DOI: 10.1073/pnas.1404067111.

1302 Christopher D. Golden et al., "Nutrition: Fall in fish catch threatens human health." Nature. Comment. 534, 317–320 (16 June 2016) doi:10.1038/534317a.

1303 Max Troell et al., "Does aquaculture add resilience to the global food system?" Proceedings of the National Academy of Sciences Sep 2014, 111 (37) 13257-13263; DOI: 10.1073/pnas.1404067111.

1304 Rosamond L. Naylor et al., "Effect of aquaculture on world fish supplies." Nature Volume 405, 1017–1024 (29 June 2000).

1305 Max Troell et al., "Does aquaculture add resilience to the global food system?" Proceedings of the National Academy of Sciences Sep 2014, 111 (37) 13257-13263; DOI: 10.1073/pnas.1404067111.

1306 R. E. Evenson and D. Gollin, "Assessing the Impact of the Green Revolution, 1960 to 2000." Science 02 May 2003: 758-762.

1307 Prabhu L. Pingali, "Green Revolution: Impacts, limits, and the path ahead." PNAS. July 31, 2012, Volume 109, no. 31. www.pnas.org/cgi/doi/10.1073/pnas.0912953109.

1308 Francesco Orsini et al., "Urban agriculture in the developing world: A review." October 2013Agronomy for Sustainable Development 33(4):695-720. DOI: 10.1007/s13593-013-0143-z.

1309 Francesco Orsini et al., "Urban agriculture in the developing world: A review." October 2013Agronomy for Sustainable Development 33(4):695-720. DOI: 10.1007/s13593-013-0143-z.

1310 Alberto Zezza, Luca Tasciotti, "Urban agriculture, poverty, and food security: Empirical evidence from a sample of developing countries." Food Policy 35 (2010) 265–273. doi:10.1016/j.foodpol.2010.04.007.

1311 Francesco Orsini et al., "Urban agriculture in the developing world: A review." October 2013Agronomy for Sustainable Development 33(4):695-720. DOI: 10.1007/s13593-013-0143-z.

1312 Francesco Orsini et al., "Urban agriculture in the developing world: A review." October 2013Agronomy for Sustainable Development 33(4):695-720. DOI: 10.1007/s13593-013-0143-z.

1313 Francesco Orsini et al., "Urban agriculture in the developing world: A review." October 2013Agronomy for Sustainable Development 33(4):695-720. DOI: 10.1007/s13593-013-0143-z.

1314 Philippe Tixier and Hubert de Bon, Urban Horticulture. CIRAD, France. First published as Chapter 11 of the RUAF publication "Cities Farming for the Future; Urban Agriculture for Green and Productive Cities" by René van Veenhuizen (ed.), RUAF Foundation, the Netherlands, IDRC, Canada and IIRR publishers, the Philippines, 2006 (460 pages).

1315 Kathrin Specht et al., "Urban agriculture of the future: An overview of sustainability aspects of food production in and on buildings." March 2014. Agriculture and Human Values 31(1). DOI: 10.1007/s10460-013-9448-4.

1316 S. Thomaier et al., 2015, "Farming in and on urban buildings: Present practice and specific novelties of Zero-Acreage Farming (ZFarming)." Renewable Agriculture and Food Systems, 30(1), 43-54. doi:10.1017/S1742170514000143.

1317 Ali AlShrouf, "Hydroponics, Aeroponic and Aquaponic as Compared with Conventional Farming." American Scientific Research Journal for Engineering, Technology, and Sciences (ASRJETS) (2017) Volume 27, No 1, 247-255.

1318 P. Moustier, 2007, "Urban Horticulture in Africa and Asia, An Efficient Corner Food Supplier." ISHS Acta Horticulturae, 762 :145-148.

1319 Francesco Orsini et al., "Urban agriculture in the developing world: A review." October 2013Agronomy for Sustainable Development 33(4):695-720. DOI: 10.1007/s13593-013-0143-z.

1320 Kathrin Specht et al., "Urban agriculture of the future: An overview of sustainability aspects of food production in and on buildings." March 2014. Agriculture and Human Values 31(1). DOI: 10.1007/s10460-013-9448-4.

1321 Alberto Zezza and Luca Tasciotti, "Urban agriculture, poverty, and food security: Empirical evidence from a sample of developing countries." Food Policy 35 (2010) 265–273. doi:10.1016/j.foodpol.2010.04.007.

1322 P. Moustier, 2007, "Urban Horticulture in Africa and Asia, An Efficient Corner Food Supplier." ISHS Acta Horticulturae, 762 :145-148.

1323 Philippe Tixier and Hubert de Bon, Urban Horticulture. CIRAD, France. First published as Chapter 11 of the RUAF publication "Cities Farming for the Future; Urban Agriculture for Green and Productive Cities" by René van Veenhuizen (ed.), RUAF Foundation, the Netherlands, IDRC, Canada and IIRR publishers, the Philippines, 2006 (460 pages).

1324 Francesco Orsini et al., "Urban agriculture in the developing world: A review." October 2013Agronomy for Sustainable Development 33(4):695-720. DOI: 10.1007/s13593-013-0143-z.

1325 Kathrin Specht et al., "Urban agriculture of the future: An overview of sustainability aspects of food production in and on buildings." March 2014. Agriculture and Human Values 31(1). DOI: 10.1007/s10460-013-9448-4.

1326 Devi Buehler and Ranka Junge, "A Review. Global Trends and Current Status of Commercial Urban Rooftop Farming." Sustainability 2016, 8(11), 1108; doi:10.3390/su8111108.

1327 Ali AlShrouf, "Hydroponics, Aeroponic and Aquaponic as Compared with Conventional Farming." American Scientific Research Journal for Engineering, Technology, and Sciences (ASRJETS) (2017) Volume 27, No 1, 247-255.

1328 Kathrin Specht et al., "Urban agriculture of the future: An overview of sustainability aspects of food production in and on buildings." March 2014. Agriculture and Human Values 31(1). DOI: 10.1007/s10460-013-9448-4.

1329 Ali AlShrouf, "Hydroponics, Aeroponic and Aquaponic as Compared with Conventional Farming." American Scientific Research Journal for Engineering, Technology, and Sciences (ASRJETS) (2017) Volume 27, No 1, 247-255.

1330 Devi Buehler and Ranka Junge., "A Review. Global Trends and Current Status of Commercial Urban Rooftop Farming." Sustainability 2016, 8(11), 1108; doi:10.3390/su8111108.

1331 David C. Love et al., "Energy and water use of a small-scale raft aquaponics system in Baltimore, Maryland, United States." Aquacultural Engineering, Volume 68, 2015. 19-27. https://doi.org/10.1016/j.aquaeng.2015.07.003.

1332 Lim Yinghui Astee and Nirmal T. Kishnani, 2010, "Building Integrated Agriculture: Utilizing Rooftops for Sustainable Food Crop Cultivation in Singapore." Journal of Green Building: Spring 2010, Volume 5, No. 2, 105-113.

1333 Devi Buehler and Ranka Junge, "A Review. Global Trends and Current Status of Commercial Urban Rooftop Farming." Sustainability 2016, 8(11), 1108; doi:10.3390/su8111108.

1334 S. Thomaier et al., 2015, "Farming in and on urban buildings: Present practice and specific novelties of Zero-Acreage Farming (ZFarming)." Renewable Agriculture and Food Systems, 30(1), 43-54. doi:10.1017/S1742170514000143.

1335 Ali AlShrouf, "Hydroponics, Aeroponic and Aquaponic as Compared with Conventional Farming." American Scientific Research Journal for Engineering, Technology, and Sciences (ASRJETS) (2017) Volume 27, No 1, 247-255.

1336 Ali AlShrouf, "Hydroponics, Aeroponic and Aquaponic as Compared with Conventional Farming." American Scientific Research Journal for Engineering, Technology, and Sciences (ASRJETS) (2017) Volume 27, No 1, 247-255.

1337 Dionysios Touliatos et al., "Vertical farming increases lettuce yield per unit area compared to conventional horizontal hydroponics." Food and Energy Security 2016; 5(3): 184–191. https://doi.org/10.1002/fes3.83.

1338 Devi Buehler and Ranka Junge., "A Review. Global Trends and Current Status of Commercial Urban Rooftop Farming." Sustainability 2016, 8(11), 1108; doi:10.3390/su8111108.

1339 R.R. Shamshiri et al., "Advances in greenhouse automation and controlled environment agriculture: A transition to plant factories and urban agriculture." International Journal of Agricultural and Biological Engineering, 2018; 11(1): 1–22.

1340 Erdem Cuce et al., "Renewable and sustainable energy saving strategies for greenhouse systems: A comprehensive review." Renewable and Sustainable Energy Reviews 64 (2016) 34–59. http://dx.doi.org/10.1016/j.rser.2016.05.077.

1341 Dries Waaijenberg, "Design, Construction and Maintenance of Greenhouse Structures." Proc. IS on Greenhouses, Environmental Controls & In-house Mechanization for Crop Production in the Tropics and Sub-tropics. Eds. Rezuwan Kamaruddin, Ibni Hajar Rukunuddin & Nor Raizan Abdul Hamid. Acta Hort. 710, ISHS 2006.

1342 Gene Giacomelli et al., "Innovation in greenhouse engineering." July 2007. Acta horticulturae. DOI:10.17660/ActaHortic.2008.801.3.

1343 Examples of Hi-Tech Indoor Farms: 1) PlantLab. http://www.plantlab.nl/ 2) Plenty Unlimited Inc. https://www.plenty.ag/. 3) Sundrop Farms. http://www.sundropfarms.com/. 4) AeroFarms. http://aerofarms.com/.

1344 R.R. Shamshiri et al., "Advances in greenhouse automation and controlled environment agriculture: A transition to plant factories and urban agriculture." International Journal of Agricultural and Biological Engineering, 2018; 11(1): 1–22. [See Table 1, page 15 for examples of vertical farming, adapted from and citing F. Kalantari et al., "A Review of Vertical Farming Technology: A Guide for Implementation of Building Integrated Agriculture in Cities." Advanced Engineering Forum, Volume 24, 76-91, 2017].

1345 F. Kalantari et al., "A Review of Vertical Farming Technology: A Guide for Implementation of Building Integrated Agriculture in Cities." Advanced Engineering Forum, Volume 24, 76-91, 2017.

1346 R.R. Shamshiri et al., "Advances in greenhouse automation and controlled environment agriculture: A transition to plant factories and urban agriculture." International Journal of Agricultural and Biological Engineering, 2018; 11(1): 1–22.

1347 F. Kalantari et al., ."A Review of Vertical Farming Technology: A Guide for Implementation of Building Integrated Agriculture in Cities." Advanced Engineering Forum, Volume 24, 76-91, 2017.

1348 A.T. Nasseri et al., "Single Cell Protein: Production and Process." February 2011, American Journal of Food Technology 6(2). DOI 10.3923/ajft.2011.103.116.

1349 Single-cell protein. https://www.sciencedirect.com/topics/biochemistry-genetics-and-molecular-biology/single-cell-protein.

1350 A.T. Nasseri et al., "Single Cell Protein: Production and Process." February 2011, American Journal of Food Technology 6(2). DOI 10.3923/ajft.2011.103.116.

1351 U.S. Energy Information Administration (EIA). Global energy by fuel type and delivered energy by sector (not including 25.5% of electricity and heat related losses during energy conversion). EIA, Annual Energy Outlook 2017, DOE/EIA-0383(2017) (Washington, DC: January 2017). Data used; https://www.eia.gov/outlooks/ieo/excel/appf_tables.xlsx.

1352 New Zealand Government. Ministry of Business, Innovation, and Employment. Smart Guides. Hikina Whakatutuki. Practical Advice on Smarter Home Essentials. https://www.smarterhomes.org.nz/smart-guides/. https://www.smarterhomes.org.nz/smart-guides/siting-and-location/house-orientation/. https://www.smarterhomes.org.nz/smart-guides/design/. https://www.smarterhomes.org.nz/smart-guides/power-lighting-and-energy-saving/, https://www.smarterhomes.org.nz/smart-guides/heating-cooling-and-insulation/.

1353 US Department of Energy. Office of Energy Efficiency and Renewable Energy. https://www.energy.gov/eere/efficiency/homes, https://www.energy.gov/energysaver/about-us

1354 New Zealand Government. Ministry of Business, Innovation, and Employment. Smart Guides. Hikina Whakatutuki. Practical Advice on Smarter Home Essentials. https://www.smarterhomes.org.nz/smart-guides/. https://www.smarterhomes.org.nz/smart-guides/siting-and-location/house-orientation/. https://www.smarterhomes.org.nz/smart-guides/design/. https://www.smarterhomes.org.nz/smart-guides/power-lighting-and-energy-saving/, https://www.smarterhomes.org.nz/smart-guides/heating-cooling-and-insulation/.

1355 US Department of Energy. Office of Energy Efficiency and Renewable Energy. https://www.energy.gov/eere/efficiency/homes, https://www.energy.gov/energysaver/about-us.

1356 US Department of Energy. Geothermal Heat Pumps. https://www.energy.gov/energysaver/heat-and-cool/heat-pump-systems/geothermal-heat-pumps.

1357 US Department of Energy. Air-Source Heat Pumps. https://www.energy.gov/energysaver/heat-pump-systems/air-source-heat-pumps.

1358 US Department of Energy. Small Wind Electric Systems. https://www.energy.gov/energysaver/buying-and-making-electricity/small-wind-electric-systems.

1359 US Department of Energy. Hybrid Wind and Solar Electric Systems. https://www.energy.gov/energysaver/buying-and-making-electricity/hybrid-wind-and-solar-electric-systems.

1360 US Department of Energy. Microhydropower Systems. https://www.energy.gov/energysaver/buying-and-making-electricity/microhydropower-systems.

1361 Mother Earth News Staff, Handbook of Homemade Power. Bantam; 11th edition (1980). ISBN-10: 0553143107. https://www.amazon.com/Mother-Earth-Handbook-Homemade-Power/dp/0553143107/.

1362 US Department of Energy. Solar Water Heaters. https://www.energy.gov/energysaver/water-heating/solar-water-heaters.

1363 US Department of Energy. Heat Pump Water Heaters. https://www.energy.gov/energysaver/water-heating/heat-pump-water-heaters.

1364 US Department of Energy. Drain-Water Heat Recovery. https://www.energy.gov/energysaver/water-heating/drain-water-heat-recovery.

1365 US Environmental Protection Agency. Energy and the Environment. Electricity Customers. Electricity Customers. https://www.epa.gov/energy/electricity-customers.

1366 U.S. Energy Information Administration (2018). Use of electricity in the USA. https://www.eia.gov/energyexplained/index.cfm?page=electricity_use.

1367 California Drought Contingency Plan. State of California, California Natural Resources Agency, Department of Water Resources. http://drought.unl.edu/archive/plans/drought/state/CA_2010.pdf.

1368 New Zealand Government. Ministry of Business, Innovation, and Employment. Smart Guides. Hikina Whakatutuki. Practical Advice on Smarter Home Essentials. https://www.smarterhomes.org.nz/smart-guides/water-and-waste/.

1369 New Zealand Government. Ministry of Business, Innovation, and Employment. Smart Guides. Hikina Whakatutuki. Practical Advice on Smarter Home Essentials. https://www.smarterhomes.org.nz/smart-guides/water-and-waste/efficient-use-of-water/.

1370 US Depertment of energy. Drain-Water Heat Recovery. https://www.energy.gov/energysaver/water-heating/drain-water-heat-recovery.

1371 New Zealand Government. Ministry of Business, Innovation, and Employment. Smart guides. Reusing greywater. Water and waste: Efficient use of water, hot water options, reusing greywater, using rainwater, managing storm water, on-site sewage systems. https://www.smarterhomes.org.nz/smart-guides/water-and-waste/re-using-greywater/.

1372 New Zealand Government. Ministry of Business, Innovation, and Employment. Smart Guides. Hikina Whakatutuki. Practical Advice on Smarter Home Essentials. https://www.smarterhomes.org.nz/smart-guides/water-and-waste/collecting-and-using-rainwater/

1373 New Zealand Government. Ministry of Business, Innovation, and Employment. Smart Guides. Hikina Whakatutuki. Practical Advice on Smarter Home Essentials. https://www.smarterhomes.org.nz/smart-guides/water-and-waste/efficient-use-of-water/.

1374 US Government Ready Website for all types of emergencies and natural disasters. Drought. https://www.ready.gov/drought.

1375 Hikmet Ş. Aybar, "A review of desalination by solar still." May 2007. NATO Security through Science Series C: Environmental Security. In book: Solar Desalination for the 21st Century. DOI: 10.1007/978-1-4020-5508-9_15.

1376 Manchanda and Kumar, "A comprehensive decade review and analysis on designs and performance parameters of passive solar still." Renewables (2015) 2:17. DOI 10.1186/s40807-015-0019-8.

1377 Rasika R Dahake et al., "A Review on Solar Still Water Purification." International Journal for Innovative Research in Science & Technology Volume 3 Issue 9 2017 59-63.

1378 A.Z.A. Saifullaha et al., "Solar pond and its application to desalination." Asian Transactions on Science & Technology (ATST ISSN: 2221-4283) Volume 02 Issue 03.

1379 Edward Bryant, Natural Hazards. Second Edition. https://www.amazon.com/Natural-Hazards-Professor-Edward-Bryant/dp/0521537436.

1380 Anthony J. McMichael, "Insights from past millennia into climatic impacts on human health and survival." Proceedings of the National Academy of Sciences Mar 2012, 109 (13) 4730-4737; DOI: 10.1073/pnas.1120177109.

1381 David D. Zhang et al., "Global climate change, war, and population decline in recent human history." Proceedings of the National Academy of Sciences Dec 2007, 104 (49) 19214-19219; DOI: 10.1073/pnas.0703073104.

1382 D. Collet and M. Schuh (eds.), "Famines During the 'Little Ice Age' (1300–1800) ." DOI 10.1007/978-3-319-54337-6_2.

1383 Geoffrey Parker, "Global Crisis. War, Climate Change and Catastrophe in the Seventeenth Century." Yale University Press. New Haven and London. British Library Catalogue Record, 10987654321.

1384 Liangcheng Tan et al., "Precipitation variations of Longxi, northeast margin of Tibetan Plateau since AD 960 and their relationship with solar activity." Climate of the Past, 4, 19–28, 2008, https://doi.org/10.5194/cp-4-19-2008, 2008.

1385 Jared Diamond, Collapse: How Societies Choose to Fail or Succeed. https://www.amazon.com/Collapse-Societies-Choose-Succeed-Revised/dp/0143117009/.

1386 Sustainable Agriculture Research & Education. https://www.sare.org/Learning-Center/Books/Building-Soils-for-Better-Crops-3rd-Edition/Text-Version/Cover-Crops.

1387 Pinterest, vegetable vertical garden. https://bit.ly/2Jg0FIO.

1388 Google Translate Web. http://itools.com/tool/google-translate-web-page-translator.

1389 New Zealand. Ministry of Civil Defense and Emergency Management. Household Emergency Plan. This link covers; Disaster Category Links to Earthquakes, Storms, Floods, Tsunamis, Volcanic explosions, Landslides, and Other. http://getthru.govt.nz/how-to-get-ready/household-emergency-plan/.

1390 New Zealand. Ministry of Civil Defense and Emergency Management. Household Emergency Plan. This link covers; Disaster Category Links to Earthquakes, Storms, Floods, Tsunamis, Volcanic explosions, Landslides, and Other. http://getthru.govt.nz/how-to-get-ready/household-emergency-plan/. http://getthru.govt.nz/assets/Uploads/GRG-Checklist.pdf to be completed with all members of your household.

1391 New Zealand. Ministry of Civil Defense and Emergency Management. Household Emergency Plan. This link covers; 1). Learn about disasters and how to keep safe. 2). Create and practice a household emergency plan. 3). Assemble and maintain emergency survival items. 4). Have a getaway kit in case you must leave in a hurry. http://getthru.govt.nz/how-to-get-ready.

1392 New Zealand. Ministry of Civil Defense and Emergency Management. http://getthru.govt.nz/how-to-get-ready/emergency-survival-items/.

1393 US Government. Be Informed, https://www.ready.gov/be-informed. Make a Plan, https://www.ready.gov/make-a-plan. Take Action-Get Involved, https://www.ready.gov/get-involved.

1394 Conduct a Google search, "WHO Public health preparedness."

1395 US Government. Pandemic Flu Outbreak. https://www.ready.gov/pandemic. https://www.cdc.gov/flu/pandemic-resources/index.htm.

1396 New Zealand Ministry of Health. https://www.health.govt.nz/our-work/emergency-management/pandemics. Pandemic planning and response: Prepare yourself for a pandemic, Health sector pandemic influenza guidance. New Zealand Influenza Pandemic Plan. Pandemic influenza legislation.

1397 US Government. Be Informed. https://www.ready.gov/winter-weather.

1398 Manchanda and Kumar, A comprehensive decade review and analysis on designs and performance parameters of passive solar still. Renewables (2015) 2:17. DOI 10.1186/s40807-015-0019-8.

1399 Rasika R Dahake et al., "A Review on Solar Still Water Purification." International Journal for Innovative Research in Science & Technology Volume 3 Issue 9 2017 59-63.

1400 Hikmet Ş. Aybar, "A review of desalination by solar still." May 2007. NATO Security through Science Series C: Environmental Security. In book: Solar Desalination for the 21st Century. DOI: 10.1007/978-1-4020-5508-9_15.

1401 A.Z.A. Saifullaha et al., "Solar pond and its application to desalination." Asian Transactions on Science & Technology (ATST ISSN: 2221-4283) Volume 02 Issue 03.

1402 Global mean surface temperature data, commonly referred to as HadCRUT4. https://www.metoffice.gov.uk/hadobs/hadcrut4/data/current/download.html. [Exposé: Look at the bottom left hand or first column for the current year-to-date temperature. Subtract that from the 2016 total to see the magnitude of the fall. Global Data: https://bit.ly/2nCgctz. Northern Hemisphere Data: https://bit.ly/2MRt75G, Southern Hemisphere Data: https://bit.ly/2nBfYTA. Tropics Data: https://bit.ly/2nFXJMM. [last downloaded 25/07/2018].

1403 S. Yamayoshi et al., "Enhanced Replication of Highly Pathogenic Influenza A(H7N9) Virus in Humans." Emerging Infectious Diseases Journal. 2018;24(4):746-750. https://dx.doi.org/10.3201/eid2404.171509.

1404 Qi Tang et al., "China is closely monitoring an increase in infection with avian influenza A (H7N9) virus." BioScience Trends. 2017; 11(1):122-124. DOI: 10.5582/bst.2017.01041.

1405 Artois J et al., "Changing Geographic Patterns and Risk Factors for Avian Influenza A(H7N9) Infections in Humans, China." Emerging Infectious Diseases Journal 2018;24(1):87-94. https://dx.doi.org/10.3201/eid2401.171393

1406 Centers for Disease Control and Prevention. "Highly Pathogenic Asian Avian Influenza A (H5N1) in People." https://www.cdc.gov/flu/avianflu/h5n1-people.htm

1407 L.O. Durand et al., "Timing of Influenza A(H5N1) in Poultry and Humans and Seasonal Influenza Activity Worldwide, 2004–2013." Emerging Infectious Diseases Journal. 2015;21(2):202-208. https://dx.doi.org/10.3201/eid2102.140877.

1408 Centers for Disease Control and Prevention. First Global Estimates of 2009 H1N1 Pandemic Mortality Released by CDC-Led Collaboration. 1918 Spanish flu mortality rate estimates. https://www.cdc.gov/flu/spotlights/pandemic-global-estimates.htm.

1409 Roy M Anderson and Robert M May, "Vaccination and Herd Immunity to Infectious Diseases." Nature, Volume 318, 28 November 1985. DOI: 10.1038/318323a0.

1410 T.J. John and R. Samuel, "Herd immunity and herd effect: new insights and definitions." European Journal of Epidemiology (2000) 16: 601. https://doi.org/10.1023/A:1007626510002.

1411 T.H. Kim et al., "Vaccine herd effect." Scandinavian Journal of Infectious Diseases. 2011;43(9):683-689. doi:10.3109/00365548.2011.582247.

1412 R.H. Borse et al., "Effects of Vaccine Program against Pandemic Influenza A(H1N1) Virus, United States, 2009–2010." Emerging Infectious Diseases Journal. 2013;19(3):439-448. https://dx.doi.org/10.3201/eid1903.120394. https://wwwnc.cdc.gov/eid/article/19/3/12-0394_article#r3.

1413 Xiao-Feng Liang et al., "Safety of Influenza A (H1N1) Vaccine in Postmarketing Surveillance in China." New England Journal of Medicine 364;7 nejm.org February 17, 2011.

1414 Harvey V. Fineberg, "Pandemic Preparedness and Response — Lessons from the H1N1 Influenza of 2009." April 3, 2014. New England Journal of Medicine2014; 370:1335-1342. DOI: 10.1056/NEJMra1208802.

1415 Rino Rappuoli and Philip R. Dormitzer, "Influenza: Options to Improve Pandemic Preparation." Science 22 Jun 2012: Volume 336, Issue 6088, 1531-1533. DOI: 10.1126/science.1221466.

1416 Ali H. Ellebedy et al., "H5N1 immunization broadens immunity to influenza." Proceedings of the National Academy of Sciences Sep 2014, 111 (36) 13133-13138; DOI: 10.1073/pnas.1414070111.

1417 Paul Gillard et al., "Long-term booster schedules with AS03A-adjuvanted heterologous H5N1 vaccines induces rapid and broad immune responses in Asian adults." BMC Infectious Diseases201414:142. https://doi.org/10.1186/1471-2334-14-142.

1418 Isabel Leroux-Roels et al., "Broad Clade 2 Cross-Reactive Immunity Induced by an Adjuvanted Clade 1 rH5N1 Pandemic Influenza Vaccine." PLOS. Published: February 27, 2008. https://doi.org/10.1371/journal.pone.0001665.

1419 Grazia Galli et al., "Fast rise of broadly cross-reactive antibodies after boosting long-lived human memory B cells primed by an MF59 adjuvanted prepandemic vaccine." Proceedings of the National Academy of Sciences May 2009, 106 (19) 7962-7967; DOI:10.1073/pnas.0903181106.

1420 Lopez P et al., "Combined Administration of MF59-Adjuvanted A/H5N1 Prepandemic and Seasonal Influenza Vaccines: Long-Term Antibody Persistence and Robust Booster Responses 1 Year after a One-Dose Priming Schedule." Clinical and Vaccine Immunology : CVI. 2013;20(5):753-758. doi:10.1128/CVI.00626-12.

1421 Banzhoff A et al., "MF59®-adjuvanted vaccines for seasonal and pandemic influenza prophylaxis. Influenza and Other Respiratory Viruses." 2008;2(6):243-249. doi:10.1111/j.1750-2659.2008.00059.x.

1422 P Gillard et al., "An assessment of prime-boost vaccination schedules with AS03A-adjuvanted prepandemic H5N1 vaccines: a randomized study in European adults." Influenza and Other Respiratory Viruses. 2013;7(1):55-65. doi:10.1111/j.1750-2659.2012.00349.x.

1423 Anuradha Madan et al., "Immunogenicity and Safety of an AS03-Adjuvanted H7N9 Pandemic Influenza Vaccine in a Randomized Trial in Healthy Adults." The Journal of Infectious Diseases, Volume 214, Issue 11, 1 December 2016, 1717–1727, https://doi.org/10.1093/infdis/jiw414.

1424 Stadlbauer D et al., 2017, "Vaccination with a recombinant H7 hemagglutinin-based influenza virus vaccine induces broadly reactive antibodies in humans." mSphere 2:e00502-17. https://doi.org/10.1128/mSphere.00502-17.

1425 Christopher Bren d'Amour et al., "Teleconnected food supply shocks." Environment Research Letter 11 (2016) 035007 doi:10.1088/1748-9326/11/3/035007. (see table 1, page 2 for the list of countries for each main crop).

1426 Martin Enserink, "The Challenge of Getting Swine Flu Vaccine to Poor Nations." Question and Answer session; discussing vaccine supply versus serving national interests first. Nov. 3, 2009 , 3:22 PM . Accessed 01/04/2018. http://www.sciencemag.org/news/2009/11/challenge-getting-swine-flu-vaccine-poor-nations.

1427 Mark Turner, 2015, "Vaccine procurement during an influenza pandemic and the role of Advance Purchase Agreements: Lessons from 2009-H1N1." Global Public Health, 11:3, 322-335, DOI: 10.1080/17441692.2015.1043743.

1428 D.P. Fidler, 2010, "Negotiating Equitable Access to Influenza Vaccines: Global Health Diplomacy and the Controversies Surrounding Avian Influenza H5N1 and Pandemic Influenza H1N1." PLoS Med 7(5): e1000247. https://doi.org/10.1371/journal.pmed.1000247.

1429 Mark A. Miller et al., "The Signature Features of Influenza Pandemics —Implications for Policy." June 18, 2009. New England Journal of Medicine 2009; 360:2595-2598. DOI: 10.1056/NEJMp0903906.

1430 Fatimah S Dawood et al., "Estimated global mortality associated with the first 12 months of 2009 pandemic influenza A H1N1 virus circulation: a modeling study." The Lancet Infectious Diseases , Volume 12, No. 9, p687–695, September 2012. DOI: https://doi.org/10.1016/S1473-3099(12)70121-4. [Data: Taking the study's estimated mortality range divided by the 2009 world population estimate of 6.8 billion people, obtained from the U.S. Energy Information Administration, International Data. World population statistics. https://bit.ly/2LF7Gom].

1431 Centers for Disease Control and Prevention. First Global Estimates of 2009 H1N1 Pandemic Mortality Released by CDC-Led Collaboration. 1918 Spanish flu mortality rate estimates. https://www.cdc.gov/flu/spotlights/pandemic-global-estimates.htm.

1432 C. Potter 2001, "A history of influenza." Journal of Applied Microbiology, 91: 572-579. doi:10.1046/j.1365-2672.2001.01492.x.

1433 J.K. Taubenberger and D.M. Morens, "Pandemic influenza – including a risk assessment of H5N1." Revue scientifique et technique (International Office of Epizootics). 2009;28(1):187-202.

1434 M. Gilbert et al., "Climate change and avian influenza." Revue scientifique et technique (International Office of Epizootics). 2008;27(2):459-466.

1435 Mark A. Miller et al., "The Signature Features of Influenza Pandemics —Implications for Policy." New England Journal of Medicine2009; 360:2595-2598. DOI: 10.1056/NEJMp0903906.

1436 Paul Gillard et al., "Long-term booster schedules with AS03Aadjuvanted heterologous H5N1 vaccines induces rapid and broad immune responses in Asian adults." BMC Infectious Diseases201414:142. https://doi.org/10.1186/1471-2334-14-142.

1437 Hai-Nv Gao et al., "Clinical Findings in 111 Cases of Influenza A (H7N9) Virus Infection." New England Journal of Medicine2013; 368:2277-2285. DOI: 10.1056/NEJMoa1305584.

1438 Qi Tang et al., "China is closely monitoring an increase in infection with avian influenza A (H7N9) virus." BioScience Trends. 2017; 11(1):122-124. DOI: 10.5582/bst.2017.01041.

1439 Yamayoshi S et al., "Enhanced Replication of Highly Pathogenic Influenza A(H7N9) Virus in Humans." Emerging Infectious Diseases Journal 2018;24(4):746-750. https://dx.doi.org/10.3201/eid2404.171509.

1440 European Centre for Disease Prevention and Control. "Human infection with avian influenza A(H7N9) virus–fifth update." 27 February 2017. Stockholm: ECDC; 2017.

1441 Artois J et al., "Changing Geographic Patterns and Risk Factors for Avian Influenza A(H7N9) Infections in Humans, China." Emerging Infectious Diseases Journal 2018;24(1):87-94. https://dx.doi.org/10.3201/eid2401.171393.

1442 N. Xiang et al., "Assessing Change in Avian Influenza A(H7N9) Virus Infections During the Fourth Epidemic — China." September 2015–August 2016. MMWR Morb Mortal Wkly Rep 2016;65:1390–1394. DOI: http://dx.doi.org/10.15585/mmwr.mm6549a2.

1443 U.S. Department of Health and Human Services. Pandemic Influenza Plan 2017 Update. https:// www.cdc.gov/flu/pandemic-resources/pdf/pan-flu-report-2017v2.pdf.

1444 Centers for Disease Control and Prevention. "Highly Pathogenic Asian Avian Influenza A (H5N1) in People." https://www.cdc.gov/flu/avianflu/h5n1-people.htm.

1445 Kumnuan Ungchusak et al., "Probable Person-to-Person Transmission of Avian Influenza A (H5N1). 2005." New England Journal of Medicine2005; 352:333-340. DOI: 10.1056/ NEJMoa044021.

1446 L.O. Durand et al., "Timing of Influenza A(H5N1) in Poultry and Humans and Seasonal Influenza Activity Worldwide, 2004–2013." Emerging Infectious Diseases Journal 2015;21(2):202-208. https://dx.doi.org/10.3201/eid2102.140877.

1447 B. Lina, 2008, "History of Influenza Pandemics." In: Raoult D., Drancourt M. (eds) Paleomicrobiology. Springer, Berlin, Heidelberg. https://doi.org/10.1007/978-3-540-75855-6_12.

1448 E. Tognotti, "Influenza pandemics: a historical retrospect." The Journal of Infection in Developing Countries 2009, 3: 331-334. DOI: https://doi.org/10.3855/jidc.239.

1449 C. Potter, 2001, "A history of influenza." Journal of Applied Microbiology, 91: 572-579. doi:10.1046/j.1365-2672.2001.01492.x.

1450 J.K. Taubenberger and D.M. Morens, "1918 Influenza: the Mother of All Pandemics." Emerging Infectious Diseases. 2006;12(1):15-22. doi:10.3201/eid1201.050979.

1451 Eugenia Tognotti, "Emerging Problems in Infectious Diseases Influenza pandemics: a historical retrospect." The Journal of Infection in Developing Countries2009; 3(5):331-334.

1452 Yu-Chia Hsieh et al., "Influenza Pandemics: Past, Present and Future." J. Formos Med Assoc. 2006 Jan;105(1):1-6. DOI:10.1016/S0929-6646(09)60102-9.

1453 Xiang N et al., "Assessing Change in Avian Influenza A(H7N9) Virus Infections During the Fourth Epidemic — China." September 2015–August 2016. MMWR Morb Mortal Wkly Rep 2016;65:1390–1394. DOI: http://dx.doi.org/10.15585/mmwr.mm6549a2.

1454 C. Potter, 2001, "A history of influenza." Journal of Applied Microbiology, 91: 572-579. doi:10.1046/j.1365-2672.2001.01492.x.

1455 Cécile Viboud et al., "Global Mortality Impact of the 1957–1959 Influenza Pandemic." The Journal of Infectious Diseases, Volume 213, Issue 5, 1 March 2016, 738–745, https://doi. org/10.1093/infdis/jiv534.

1456 A. Mark et al., "The Signature Features of Influenza Pandemics —Implications for Policy." New England Journal of Medicine 2009; 360:2595-2598. DOI: 10.1056/NEJMp0903906.

1457 B. Lina, 2008, "History of Influenza Pandemics." In: Raoult D., Drancourt M. (eds) Paleomicrobiology. Springer, Berlin, Heidelberg. https://doi.org/10.1007/978-3-540-75855-6_12.

1458 E. Tognotti, 2009, "Influenza pandemics: a historical retrospect." Journal of Infection in Developing Countries, 3:331-334. doi: https://doi.org/10.3855/jidc.239.

1459 C. Potter, 2001, "A history of influenza." Journal of Applied Microbiology, 91: 572-579. doi:10.1046/j.1365-2672.2001.01492.x.

1460 J.K. Taubenberger and D.M. Morens, "1918 Influenza: the Mother of All Pandemics." Emerging Infectious Diseases. 2006;12(1):15-22. doi:10.3201/eid1201.050979.

1461 Edwin D. Kilbourne, Influenza. Chapter 1; History of Influenza. Springer Science & Business Media, 6/12/2012 - Medical. ISBN 978-1-4684-5239-6.

1462 Svenn-Erik Mamelund, "Influenza, Historical." December 2008. International Encyclopedia of Public Health, First Edition (2008), vol. 3, pp. 597-609. DOI: 10.1016/B978-012373960-5.00372-5.

1463 Data: (1) Figure 14.1.A: Yearly mean total sunspot number (1700 – 2017). Sunspot data from the World Data Center SILSO, Royal Observatory of Belgium, Brussels. http://sidc.be/silso/ datafiles#total. Downloaded 05/05/2018. (2) Figure 14.1.B: T. Kobashi et al., 2013. Causes of Greenland temperature variability over the past 4000 year: implications for northern hemispheric temperature changes. Climate of the Past, 9(5), 2299-2317. doi: 10.5194/cp-9-2299-2013. National Centers for Environmental Information, NESDIS, NOAA, U.S. Department of Commerce. Northern Hemisphere 4000 Year Temperature Reconstructions. https://www. ncdc.noaa.gov/paleo/study/15535. Downloaded 05/05/2018. (3) TSI Reconstruction was based on NRLTSI2 (Coddington et al., BAMS, 2015 doi: 10.1175/BAMS-D-14-00265.1). http://spot. colorado.edu/~koppg/TSI/TIM_TSI_Reconstruction.txt. (4) Influenza pandemic and epidemic publications: (a) B. Lina, 2008, History of Influenza Pandemics. In: Raoult D., Drancourt M. (eds) Paleomicrobiology. Springer, Berlin, Heidelberg. https://doi.org/10.1007/978-3-540-75855-

6_12. (b) E. Tognotti, 2009, Influenza pandemics: a historical retrospect. Journal of Infection in Developing Countries, 3:331-334. doi: https://doi.org/10.3855/jidc.239. (c) C. Potter, 2001, A history of influenza. Journal of Applied Microbiology, 91: 572-579. doi:10.1046/j.1365-2672.2001.01492.x. (d) J.K. Taubenberger and D.M. Morens, 1918 Influenza: the Mother of All Pandemics. Emerging Infectious Diseases. 2006;12(1):15-22. doi:10.3201/eid1201.050979. (e) Edwin D. Kilbourne, Influenza. Chapter 1; History of Influenza. Springer Science & Business Media, 6/12/2012 - Medical. ISBN 978-1-4684-5239-6. (f) Svenn-Erik Mamelund, Influenza, Historical. December 2008. International Encyclopedia of Public Health, First Edition (2008), vol. 3, pp. 597-609. DOI: 10.1016/B978-012373960-5.00372-5.

1464 Data: (C) Beryllium-10 concentration anomaly data is based on; A.M. Berggren et al., 2009, A 600-year annual 10Be record from the NGRIP ice core, Greenland. Geophysical Research Letters, 36, L11801, doi:10.1029/2009GL038004. National Centers for Environmental Information, NESDIS, NOAA, U.S. Department of Commerce. North GRIP - 600 Year Annual 10Be Data. https://www.ncdc.noaa.gov/paleo-search/study/8618. Downloaded 05/05/2018. (D) TSI Reconstruction was based on NRLTSI2 (Coddington et al., BAMS, 2015 doi: 10.1175/BAMS-D-14-00265.1). http://spot.colorado.edu/~koppg/TSI/TIM_TSI_Reconstruction.txt. (E) Usoskin, I.G., et al. 2008. Cosmic Ray Intensity Reconstruction. IGBP PAGES/World Data Center for Paleoclimatology. Data Contribution Series # 2008-013. NOAA/NCDC Paleoclimatology Program, Boulder CO, USA. Original References: 1) I.G. Usoskin et al., 2002, A physical reconstruction of cosmic ray intensity since 1610. Journal of Geophysical Research, 107(A11), 1374. Downloaded May 2018. ftp://ftp.ncdc.noaa.gov/pub/data/paleo/climate_forcing/solar_variability/usoskin-cosmic-ray.txt. (F) Jochen Halfar et al., 2013, Arctic sea-ice decline archived by multicentury annual-resolution record from crustose coralline algal proxy. Proceedings of the National Academy of Sciences. doi: 10.1073/pnas.1313775110. National Centers for Environmental Information, NESDIS, NOAA, U.S. Department of Commerce. Arctic Northwest Atlantic 646 Year Coralline Algae Sea Ice Record. https://www.ncdc.noaa.gov/paleo/study/15454. (G) Influenza pandemic and epidemic publications: See the publications in the preceding citation; (4) a-f.

1465 Jiangwen Qu, "Is sunspot activity a factor in influenza pandemics?" Reviews in Medical Virology 2016 Sep;26(5):309-13. doi: 10.1002/rmv.1887. Epub 2016 May 2.

1466 J. Qu et al., 2016, "Sunspot Activity, Influenza and Ebola Outbreak Connection." Journal of Astrobiology and Outreach 4:154. doi: 10.4172/2332-2519.1000154.

1467 J.M. Vaquero and M.C. Gallego, "Sunspot numbers can detect pandemic influenza A: The use of different sunspot numbers." Medical Hypotheses, Volume 68, Issue 5, 1189 – 1190. DOI: 10.1016/j.mehy.2006.10.021.

1468 Guang Wu and Shaomin Yan, "Searching of Main Cause Leading to Severe Influenza A Virus Mutations and Consequently to Influenza Pandemics/ Epidemics." American Journal of Infectious Diseases 1 (2): 116-123, 2005. ISSN 1553-6203.

1469 J. Qu et al., 2016, "Sunspot Activity, Influenza and Ebola Outbreak Connection." Journal of Astrobiology and Outreach4:154. doi: 10.4172/2332-2519.1000154.

1470 Derek Gatherer, "The Little Ice Age and the emergence of influenza A." Medical Hypotheses, Volume 75, Issue 4, 359 – 362. https://doi.org/10.1016/j.mehy.2010.03.032.

1471 V. Zaporozhan and A. Ponomarenko, "Mechanisms of Geomagnetic Field Influence on Gene Expression Using Influenza as a Model System." Basics of Physical Epidemiology. International Journal of Environmental Research and Public Health. 2010;7(3):938-965. doi:10.3390/ijerph7030938.

1472 Jan Houseman and Alan Fehr, "Listening for Cosmic Rays!" The Inuvik Neutron Monitor. http://neutronm.bartol.udel.edu//listen/main.html.

1473 Roy M Anderson and Robert M May, "Vaccination and Herd Immunity to Infectious Diseases." Nature, Volume 318, 28 November 1985. DOI: 10.1038/318323a0.

1474 T.J. John and R. Samuel. "Herd immunity and herd effect: new insights and definitions." European Journal of Epidemiology (2000) 16: 601. https://doi.org/10.1023/A:1007626510002.

1475 T.H. Kim et al., "Vaccine herd effect." Scandinavian Journal of Infectious Diseases. 2011;43(9):683-689. doi:10.3109/00365548.2011.582247.

1476 R.H. Borse et al., "Effects of Vaccine Program against Pandemic Influenza A(H1N1) Virus."
United States, 2009–2010. Emerging Infectious Diseases Journal 2013;19(3):439-448. https://
dx.doi.org/10.3201/eid1903.120394. https://wwwnc.cdc.gov/eid/article/19/3/12-0394_article#r3.

1477 Xiao-Feng Liang et al., "Safety of Influenza A (H1N1) Vaccine in Postmarketing Surveillance
in China." New England Journal of Medicine 364;7 nejm.org February 17, 2011.

1478 Harvey V. Fineberg, "Pandemic Preparedness and Response — Lessons from the H1N1
Influenza of 2009." April 3, 2014. New England Journal of Medicine2014; 370:1335-1342. DOI:
10.1056/NEJMra1208802.

1479 R.H. Borse et al., "Effects of Vaccine Program against Pandemic Influenza A(H1N1) Virus."
United States, 2009–2010. Emerging Infectious Diseases Journal 2013;19(3):439-448. https://
dx.doi.org/10.3201/eid1903.120394.

1480 Rino Rappuoli and Philip R. Dormitzer, "Influenza: Options to Improve Pandemic
Preparation." Science 22 Jun 2012: Volume 336, Issue 6088, 1531-1533. DOI: 10.1126/
science.1221466.

1481 R.H. Borse et al., "Effects of Vaccine Program against Pandemic Influenza A(H1N1) Virus."
United States, 2009–2010. Emerging Infectious Diseases Journal 2013;19(3):439-448. https://
dx.doi.org/10.3201/eid1903.120394. https://wwwnc.cdc.gov/eid/article/19/3/12-0394_article#r3.

1482 Rino Rappuoli and Philip R. Dormitzer, "Influenza: Options to Improve Pandemic
Preparation." Science 22 Jun 2012: Volume 336, Issue 6088, 1531-1533. DOI: 10.1126/
science.1221466.

1483 R.H. Borse et al., "Effects of Vaccine Program against Pandemic Influenza A(H1N1) Virus."
United States, 2009–2010. Emerging Infectious Diseases Journal 2013;19(3):439-448. https://
dx.doi.org/10.3201/eid1903.120394. https://wwwnc.cdc.gov/eid/article/19/3/12-0394_article#r3.

1484 Mark Turner, 2015, "Vaccine procurement during an influenza pandemic and the role of
Advance Purchase Agreements." Lessons from 2009-H1N1, Global Public Health, 11:3, 322-
335, DOI: 10.1080/17441692.2015.1043743.

1485 D.P. Fidler, 2010, "Negotiating Equitable Access to Influenza Vaccines: Global Health
Diplomacy and the Controversies Surrounding Avian Influenza H5N1 and Pandemic Influenza
H1N1." PLoS Med 7(5): e1000247. https://doi.org/10.1371/journal.pmed.1000247.

1486 D.P. Fidler, 2010, "Negotiating Equitable Access to Influenza Vaccines: Global Health
Diplomacy and the Controversies Surrounding Avian Influenza H5N1 and Pandemic Influenza
H1N1." PLoS Med 7(5): e1000247. https://doi.org/10.1371/journal.pmed.1000247.

1487 Martin Enserink, "The Challenge of Getting Swine Flu Vaccine to Poor Nations. Question and
Answer session; discussing vaccine supply versus serving national interests first." Nov. 3, 2009
, 3:22 PM . Accessed 01/04/2018. http://www.sciencemag.org/news/2009/11/challenge-getting-
swine-flu-vaccine-poor-nations.

1488 Donald G. McNeil Junior, 2009, Nation Is Facing Vaccine Shortage for Seasonal Flu. The
Australian government pressured its national vaccine supplier (CSL) to serve national
needs ahead of fulfilling its swine flu vaccine contract with the USA. https://www.nytimes.
com/2009/11/05/health/05flu.html (accessed 01/04/2018).

1489 European Commission. Science for Environment Policy. DG Environment New Service Alert.
Causes of the 2007 - 2008 global food crisis identified. 20 January 2011. http://ec.europa.eu/
environment/integration/research/newsalert/pdf/225na1_en.pdf.

1490 Headey, D. and Fan, S. (2008), Anatomy of a crisis: the causes and consequences of surging
food prices. Agricultural Economics, 39: 375-391. doi:10.1111/j.1574-0862.2008.00345.x.

1491 Philip C. Abbott, Export Restrictions as Stabilization Responses to Food Crisis, American
Journal of Agricultural Economics, Volume 94, Issue 2, 1 January 2012, 428–434, https://doi.
org/10.1093/ajae/aar092.

1492 "How long would it take to develop a new pandemic vaccine?" Centers for Disease Control
and Prevention, National Center for Immunization and Respiratory Diseases (NCIRD). https://
www.cdc.gov/flu/pandemic-resources/basics/faq.html.

1493 Graham K. MacDonald et al., "Rethinking Agricultural Trade Relationships in an Era of
Globalization." BioScience, Volume 65, Issue 3, 1 March 2015, 275–289, https://doi.org/10.1093/
biosci/biu225.

1494 https://web.archive.org/web/20140309015656/http://www.novartisvaccines.com/downloads/
diseases-products/MF59-Adj-fact-sheet.pdf.

1495 Pandemrix - European Medicines Agency - Europa EU: See AS03's adjuvant detail from within GlaxoSmithKline's Pandemrix influenza vaccine (H1N1) (split virion, inactivated, adjuvanted) www.ema.europa.eu/docs/en_GB/document_library/.../WC500038121.pdf.

1496 D. Stadlbauer et al., 2017, "Vaccination with a recombinant H7 hemagglutinin-based influenza virus vaccine induces broadly reactive antibodies in humans." mSphere 2:e00502-17. https://doi.org/10.1128/mSphere.00502-17.

1497 Paul Gillard et al., "Long-term booster schedules with AS03A-adjuvanted heterologous H5N1 vaccines induces rapid and broad immune responses in Asian adults." BMC Infectious Diseases201414:142. https://doi.org/10.1186/1471-2334-14-142.

1498 Isabel Leroux-Roels et al., "Broad Clade 2 Cross-Reactive Immunity Induced by an Adjuvanted Clade 1 rH5N1 Pandemic Influenza Vaccine." PLOS. Published: February 27, 2008. https://doi.org/10.1371/journal.pone.0001665.

1499 Grazia Galli et al., "Fast rise of broadly cross-reactive antibodies after boosting long-lived human memory B cells primed by an MF59 adjuvanted prepandemic vaccine." Proceedings of the National Academy of Sciences May 2009, 106 (19) 7962-7967; DOI:10.1073/pnas.0903181106.

1500 P. Lopez et al., "Combined Administration of MF59-Adjuvanted A/H5N1 Prepandemic and Seasonal Influenza Vaccines: Long-Term Antibody Persistence and Robust Booster Responses 1 Year after a One-Dose Priming Schedule." Clinical and Vaccine Immunology : CVI. 2013;20(5):753-758. doi:10.1128/CVI.00626-12.

1501 A. Banzhoff et al., "MF59® adjuvanted vaccines for seasonal and pandemic influenza prophylaxis. Influenza and Other Respiratory Viruses." 2008;2(6):243-249. doi:10.1111/j.1750-2659.2008.00059.x.

1502 P. Gillard et al., An assessment of prime-boost vaccination schedules with AS03A-adjuvanted prepandemic H5N1 vaccines: a randomized study in European adults. Influenza and Other Respiratory Viruses. 2013;7(1):55-65. doi:10.1111/j.1750-2659.2012.00349.x.

1503 Anuradha Madan et al., "Immunogenicity and safety of an AS03-adjuvanted H7N1 vaccine in healthy adults: A phase I/II, observer-blind, randomized, controlled trial." Vaccine, Volume 35, Issue 10, 2017, 1431-1439, https://doi.org/10.1016/j.vaccine.2017.01.054.

1504 Anuradha Madan et al., "Immunogenicity and Safety of an AS03-Adjuvanted H7N9 Pandemic Influenza Vaccine in a Randomized Trial in Healthy Adults." The Journal of Infectious Diseases, Volume 214, Issue 11, 1 December 2016, 1717–1727, https://doi.org/10.1093/infdis/jiw414.

1505 D. Stadlbauer et al., 2017, "Vaccination with a recombinant H7 hemagglutinin-based influenza virus vaccine induces broadly reactive antibodies in humans." mSphere 2:e00502-17. https://doi.org/10.1128/mSphere.00502-17.

1506 Paul Gillard et al., "Long-term booster schedules with AS03A-adjuvanted heterologous H5N1 vaccines induces rapid and broad immune responses in Asian adults." BMC Infectious Diseases201414:142. https://doi.org/10.1186/1471-2334-14-142.

1507 Jesse L. Goodman, "Investing in Immunity: Prepandemic Immunization to Combat Future Influenza Pandemics." Clinical Infectious Diseases, Volume 62, Issue 4, 15 February 2016, 495–498, https://doi.org/10.1093/cid/civ957.

1508 Banzhoff A et al., "MF59®-adjuvanted vaccines for seasonal and pandemic influenza prophylaxis. Influenza and Other Respiratory Viruses." 2008;2(6):243-249. doi:10.1111/j.1750-2659.2008.00059.x.

1509 Isabel Leroux-Roels et al., "Broad Clade 2 Cross-Reactive Immunity Induced by an Adjuvanted Clade 1 rH5N1 Pandemic Influenza Vaccine." PLOS. Published: February 27, 2008. https://doi.org/10.1371/journal.pone.0001665.

1510 Rino Rappuoli and Philip R. Dormitzer, "Influenza: Options to Improve Pandemic Preparation." Science 22 Jun 2012: Volume 336, Issue 6088, 1531-1533. DOI: 10.1126/science.1221466.

1511 Grazia Gallia, "Fast rise of broadly cross-reactive antibodies after boosting long-lived human memory B cells primed by an MF59 adjuvanted prepandemic vaccine." 7962–7967 PNAS May 12, 2009 Volume 106 no. 19 www.pnas.org cgi doi 10.1073 pnas.0903181106.

1512 Grazia Gallia, "Fast rise of broadly cross-reactive antibodies after boosting long-lived human memory B cells primed by an MF59 adjuvanted prepandemic vaccine." 7962–7967 PNAS May 12, 2009 Volume 106 no. 19 www.pnas.org cgi doi 10.1073 pnas.0903181106.

1513 Rino Rappuoli and Philip R. Dormitzer, "Influenza: Options to Improve Pandemic Preparation." Science 22 Jun 2012: Volume 336, Issue 6088, 1531-1533. DOI: 10.1126/science.1221466.

1514 Pio Lopez et al., "Combined Administration of MF59-Adjuvanted A/H5N1 Prepandemic and Seasonal Influenza Vaccines: Long-Term Antibody Persistence and Robust Booster Responses 1 Year after a One-Dose Priming Schedule." May 2013 Volume 20 Number 5 Clinical and Vaccine Immunology p. 753–758 cvi.asm.org 753.

1515 Personal Comment: This perspective comes from my experience as a former biotech CEO and vaccine innovator (2002-2012), who coinnovated a synthetic universal pandemic flu vaccine able to immunologically target all potential influenza-A strains H1, 2, 3, 5, 7 and H9 with a single vaccine. I also raised £23 million of investment from pharmaceutical corporate and life science investors, and developed a vaccine company de novo and its vaccine programs from a concept through to the early clinical stage of development. Through this experience I heard the views and opinions firsthand of innumerable government vaccine regulatory professionals (FDA, EMA, MHRA, regulatory affairs professionals), pharmaceutical and vaccine company executives, vaccine manufacture and formulation development experts, key scientific key opinion leaders, business and corporate development executives, investors, government representatives, etc. My opinion sums up the attitudes, sentiments, and the opinions that I was privy to through this decade-long career experience.

1516 Vaccines involved in human safety issues, public inquiries, and safety investigations in the last decade. Prepandemrix, Gardasil, Measles Mumps and Rubella vaccines.

1517 Harvey V. Fineberg, "Pandemic Preparedness and Response — Lessons from the H1N1 Influenza of 2009." April 3, 2014. New England Journal of Medicine2014; 370:1335-1342. DOI: 10.1056/NEJMra1208802.

1518 K.A. McLean et al., "The 2015 global production capacity of seasonal and pandemic influenza vaccine." Vaccine. 2016;34(45):5410-5413. doi:10.1016/j.vaccine.2016.08.019.

1519 James T. Matthews, "Egg-Based Production of Influenza Vaccine: 30 Years of Commercial Experience." The Bridge. National Academy of Engineering. September 1, 2006 Volume 36 Issue 3.

1520 Rino Rappuoli and Philip R. Dormitzer, "Influenza: Options to Improve Pandemic Preparation." Science 22 Jun 2012: Volume 336, Issue 6088, 1531-1533. DOI: 10.1126/science.1221466.

1521 T. Jefferson et al., "Vaccines for preventing influenza in healthy adults." Cochrane Database of Systematic Reviews 2010, Issue 7. Art. No.: CD001269. DOI: 10.1002/14651858.CD001269.pub4.

1522 Kristin L. Nichol, "Efficacy and effectiveness of influenza vaccination." Vaccine. Volume 26, Supplement 4, 12 September 2008, D17-D22. https://doi.org/10.1016/j.vaccine.2008.07.048.

1523 Lamberto Manzoli et al., 2012, "Effectiveness and harms of seasonal and pandemic influenza vaccines in children, adults and elderly." Human Vaccines & Immunotherapeutics, 8:7, 851-862, DOI: 10.4161/hv.19917.

1524 V. Tisa et al., "Quadrivalent influenza vaccine: a new opportunity to reduce the influenza burden." J Prev. Med. Hyg. 2016, 57: E28-E33. https://www.ncbi.nlm.nih.gov/pmc/articles/PMC4910440/.

1525 European Medicines Agency Evaluation of Medicines for Human Use. Guideline on dossier structure and content for pandemic influenza vaccine marketing authorization application. EMEA/CPMP/VEG/4717/2003 ©EMEA 2008. 1.2. Legal framework Directive 2001/83/EC lays down in Article 8 the requirements for a marketing authorization application and Regulation (EC) No 726/2004 lays down the procedure for submission to the EMEA via the centralized route. http://www.ema.europa.eu/docs/en_GB/document_library/Scientific_guideline/2009/09/WC500003869.pdf.

1526 The European Agency for the Evaluation of Medicinal Products Evaluation of Medicines for Human Use. Guideline on submission of marketing authorization applications for pandemic influenza vaccines through the centralized procedure. April 2004, EMEA/

CPMP/VEG/4986/03. http://www.ema.europa.eu/docs/en_GB/document_library/Scientific_ guideline/2009/09/WC500003815.pdf.

1527 U.S. Department of Health and Human Services. Food and Drug Administration Center for Biologics Evaluation and Research. May 2007 FDA Guidance for Industry Clinical Data Needed to Support the Licensure of Pandemic Influenza Vaccines. Data supporting the safety of the adjuvanted formulation and its added benefit over the unadjuvanted formulation is submitted in the BLA (42 U.S.C. 262(a)(2)(C)(i); 21 CFR 601.2). https://www.fda.gov/ downloads/biologicsbloodvaccines/guidancecomplianceregulatoryinformation/guidances/ vaccines/ucm091985.pdf.

1528 Prepandemic influenza vaccine (H5N1) (surface antigen, inactivated, adjuvanted) Novartis Vaccines and Diagnostics prepandemic influenza vaccine (H5N1) (surface antigen, inactivated, adjuvanted). https://bit.ly/2GqDWwL.

1529 See FDA website Influenza Virus Vaccine Safety & Availability. https://www.fda.gov/ BiologicsBloodVaccines/SafetyAvailability/VaccineSafety/ucm110288.htm.

1530 S. Kommareddy et al., Immunopotentiators in Modern Vaccines (Second Edition), 2017, via Science Direct, Learn more about MF59. https://www.sciencedirect.com/topics/neuroscience/ mf59.

1531 Pandemrix - European Medicines Agency - Europa EU: Prepandrix prepandemic influenza vaccine (H5N1) (split virion, inactivated, adjuvanted). https://bit.ly/2InG1FV. www.ema.europa. eu/docs/en_GB/document_library/.../WC500038121.pdf.

1532 Assessment report for Aflunov. Common name: prepandemic influenza vaccine (H5N1) (surface antigen, inactivated, adjuvanted). Procedure No.: EMEA/H/C/002094/II/0007/G http:// www.ema.europa.eu/docs/en_GB/document_library/EPAR_-_Assessment_Report_-_Variation/ human/002094/WC500144788.pdf.

1533 ID Biomedical Corporation. Request to supplement your biologics license application (BLA) for Influenza A (H5N1) Virus Monovalent Vaccine, Adjuvanted. https://www.fda.gov/downloads/ BiologicsBloodVaccines/Vaccines/ApprovedProducts/UCM520303.pdf.

1534 Vittoria Cioce, "Pandemic Influenza Vaccine Stockpile." BARDA Industry Day 2015. October 14, 2015 https://www.medicalcountermeasures.gov/media/36830/07_cioce_-pandemic- influenza-vaccine-stockpile.pdf.

1535 Vittoria Cioce, Section Chief, "Vaccine Stockpile. Influenza Vaccine and Adjuvant Stockpiles." Biomedical Advanced Research and Development Authority (BARDA). HHS/ASPR/BARDA/ ID. https://www.medicalcountermeasures.gov/BARDA/documents/Day1_VCioce-InfluenzaVac cineAndAdjuvantStockpiles-508.pdf.

1536 V. Tisa et al., "Quadrivalent influenza vaccine: a new opportunity to reduce the influenza burden." J PREV MED HYG 2016; 57: E28-E33. https://www.ncbi.nlm.nih.gov/pmc/articles/ PMC4910440/.

1537 Rino Rappuoli and Philip R. Dormitzer, "Influenza: Options to Improve Pandemic Preparation." Science 22 Jun 2012: Volume 336, Issue 6088, 1531-1533. DOI: 10.1126/ science.1221466.

1538 Pio Lopez et al., "Combined Administration of MF59-Adjuvanted A/H5N1 Prepandemic and Seasonal Influenza Vaccines: Long-Term Antibody Persistence and Robust Booster Responses 1 Year after a One-Dose Priming Schedule." May 2013 Volume 20 Number 5 Clinical and Vaccine Immunology p.753–758 cvi.asm.org 753.

1539 FDA Vaccines, Blood & Biologics. Prevnar13 (Pfizer Inc.,) is a Pneumococcal 13-valent Conjugate Vaccine (Diphtheria CRM197 Protein) for the vaccine prevention of invasive disease caused by Streptococcus pneumoniae serotypes 1, 3, 4, 5, 6A, 6B, 7F, 9V, 14, 18C, 19A, 19F and 23F in children 6 weeks through 5 years of age. https://www.fda.gov/biologicsbloodvaccines/ vaccines/approvedproducts/ucm201667.htm.

1540 FDA Vaccines, Blood & Biologics. Gardasil 9 (Merck & CO., Inc.) is a human papillomavirus (HPV) vaccine composed of purified virus-like particles containing the major capsid (L1) protein from 9 different subtypes of the HPV virus, 6, 11, 16, 18, 31, 33, 45, 52, and 58. Gardasil also contains an aluminum-based adjuvant to boost immune responses. https://www.fda.gov/ biologicsbloodvaccines/vaccines/approvedproducts/ucm426445.htm.

1541 The Brundtland Report "Our Common Future." To view the report, click the link. http://www. sustainabledevelopment2015.org/AdvocacyToolkit/index.php/earth-summit-history/historical-documents/92-our-common-future. [Exposé: See Our Common Future, Chapter 2: Towards Sustainable Development, sub-Section 4: "Ensuring a sustainable level of population."].

1542 Global mean surface temperature data, commonly referred to as HadCRUT4. https://www. metoffice.gov.uk/hadobs/hadcrut4/data/current/download.html. [Exposé: Look at the bottom left hand or first column for the current year-to-date temperature. Subtract that from the 2016 total to see the magnitude of the fall. Global Data: https://bit.ly/2nCgctz. Northern Hemisphere Data: https://bit.ly/2MRt75G, Southern Hemisphere Data: https://bit.ly/2nBfYTA. Tropics Data: https://bit.ly/2nFXJMM. [last downloaded 25/07/2018].

www.ingramcontent.com/pod-product-compliance
Lightning Source LLC
Chambersburg PA
CBHW071856090426
42811CB00004B/625